Global Climate Justice

Theory and Practice

EDITED BY FAUSTO CORVINO AND TIZIANA ANDINA

E-INTERNATIONAL
RELATIONS
PUBLISHING

E-International Relations
Bristol, England
2023

ISBN: 978-1-910814-68-0

Production: Michael Tang
Cover Image: Michal Bednarek/Shutterstock

A catalogue record for this book is available from the British Library.

E-International Relations

Editor-in-Chief and Publisher: Stephen McGlinchey
Books Editor: Bill Kakenmaster
Editorial assistance: Goda Skiotytė, Milena Marjanovic

E-International Relations is the world's leading International Relations website. Our daily publications feature expert articles, reviews and interviews – as well as student learning resources. The website is run by a non-profit organisation based in Bristol, England and staffed by an all-volunteer team of students and academics. In addition to our website content, E-International Relations publishes a range of books.

As E-International Relations is committed to open access in the fullest sense, free versions of our books are available on our website https://www.e-ir.info/

Abstract

This book offers philosophical and interdisciplinary insights into global climate justice with a view to climate neutrality by the middle of the twenty-first century. The first section brings together a series of introductory contributions on the state of the climate crisis, covering scientific, historical, diplomatic and philosophical dimensions. The second section focuses on the challenges of justice and responsibility to which the climate crisis exposes and will expose the global community in the coming years: on the one hand, aiming for the ambitious mitigation target of 1.5°C and, on the other hand, securing resources for adaptation and for climate-damage compensation to the most vulnerable. The third section investigates normative aspects of the transition towards a fossil-fuel free society, from the responsibility of oil companies to the gender-differentiated effects of climate change, passing through what is owed to transition losers and the legal protection of future generations.

Acknowledgments

We are grateful to the editors and editorial assistants of E-International Relations for their comments and advice. In particular, we are infinitely grateful to Bill Kakenmaster for reading and commenting on the manuscript several times, and for his useful recommendations, as well as for an incredible job of fact checking and proof reading. We would also like to thank several anonymous reviewers, who we believe should be kept anonymous, for their contribution to the drafting of all chapters.

About the editors

Fausto Corvino is Postdoctoral Researcher in Financial Ethics in the Department of Philosophy, Linguistics and Theory of Science at the University of Gothenburg. Prior to this, he was Postdoctoral Researcher in Moral Philosophy in the DIRPOLIS Institute (Law, Politics and Development) at Sant'Anna School of Advanced Studies, Pisa (2018–2020; 2021–2022), and Postdoctoral Researcher in Theoretical Philosophy in the Department of Philosophy and Educational Sciences at the University of Turin (2020–2021). He holds a PhD in Politics, Human Rights and Sustainability (2017) from Sant'Anna School of Advanced Studies. His main research interests lie in theories of justice (including global and intergenerational justice), philosophy of economics and applied ethics (in particular, the ethics of climate change and of the energy transition).

Tiziana Andina is Full Professor of Philosophy at the University of Turin (Italy) and the director of Labont – Center for Ontology at the University of Turin. Previously, she has been a fellow at Columbia University (2008–2009) and Käte Hamburger Kolleg, University of Bonn (2015) – as well as Visiting Professor at ITMO University, Russia (2014), University of Nanjing and Wuhan, China (2018, 2019). She has published articles on social philosophy and the philosophy of art and her recent works concerns the definition of art and social ontology. Her publications include: *The Philosophy of Art: The Question of Definition. From Hegel to Post-Dantian Theories* (Bloomsbury Academy 2013), *An Ontology for Social Reality* (Palgrave Macmillan 2016), *What is Art? The Question of Definition Reloaded* (Brill 2017), *A Philosophy for Future Generations* (Bloomsbury Academic 2022) and the edited volumes *Post-Truth: Philosophy and Law* (Routledge 2019) and *Institutions in Actions: The Nature and Rule of the Institutions in the Real World* (Springer Nature 2020).

Contents

Section Three: Normative Perspectives on the Climate Neutrality Ambition

Contributors

Silvia Bacchetta holds a PhD in Political Studies (2021) from NASP (Network for the Advancement of Social and Political studies) at the University of Milan (Italy) with a dissertation on the problem of individuals' compliance with climate change-related provisions. She graduated from the University of Eastern Piedmont (Italy) with a thesis on global justice. Her research interests intersect between political philosophy and climate justice.

Roberto Buizza serves as Scientific Attaché at the Italian Embassy in London where he supports the work of the Italian diplomacy and leads the embassy's scientific activities. He is Full Professor in Physics at Scuola Superiore Sant'Anna in Pisa and Honorary Research Fellow at Imperial College London's Grantham Institute – Climate Change and the Environment. He has a degree in Physics, a PhD in Mathematics and a Master's in Business Administration. From 1991–2018, he worked at the European Centre for Medium-Range Weather Forecasts, where he served as Head of the Predictability Division and Lead Scientist. Since 2018, he has been working at Scuola Superiore Sant'Anna of Pisa, where he led a research centre on climate change. An expert in numerical weather prediction, Earth-system modelling and ensemble methods and predictability, he has more than 230 publications.

Daniel Burkett is a Visiting Fellow at the University of New South Wales, where he previously completed a Postdoctoral Research Fellowship with the Climate Justice stream of the Practical Justice Initiative. He completed a law degree and a Master's in Philosophy at Victoria University of Wellington in New Zealand, followed by a PhD in Philosophy from Rice University. His current research interests are in applied ethics and social and political philosophy – particularly where issues surrounding punishment and climate change intersect with considerations of state policy.

Elena Casetta is an Assistant Professor at the Department of Philosophy and Education at the University of Turin (Italy), where she teaches Philosophy of Nature and Philosophy of Biology. Trained in theoretical philosophy, she then specialized in philosophy of biology at the IHPST (Institut d'Histoire et de Philosophie des Sciences et des Techniques), CNRS/Paris 1/ENS Paris, and at the Centre for Philosophy of Sciences of the University of Lisbon. She works mainly on environmental philosophy and the philosophy of biodiversity with a focus on conservation sciences. Her latest books are (with A. Borghini) *Brill's Companion to the Philosophy of Biology* (Brill 2019), and *From Assessing to Conserving Biodiversity* (Springer 2019), edited with Davide Vecchi and Jorge Marques da Silva.

Marco Grasso is Professor of Political Geography at the University of Milan-Bicocca. His interdisciplinary research contributed to the investigation of the governance, politics, and ethics of climate change at the national and global levels with regards to non-state actors, and to the theorization and empirical scrutiny of climate policy and politics in order to understand how to favour collective action towards the carbon transition. He embeds political, ethical, geographic, and economic analysis into the socio-political aspects of climate change with the objective of reframing climate issues in ways that make action more feasible. He is the author of *From Big Oil to Big Green. Holding the Oil Industry to Account for the Climate Crisis* (MIT Press 2022), alongside a number of journal articles.

Tahseen Jafry is Professor of Climate Justice and Director of the Mary Robinson Centre for Climate Justice at Glasgow Caledonian University. She has extensive research and development experience on the social justice and equity aspects of climate change: climate migration and conflict, gender and poverty targeting, the management of natural resources, the geo-political nature of climate justice discourse and the psycho-social impacts of climate change. She has published widely in multidisciplinary journals, is Editor of the *Routledge Climate Justice Handbook* (Routledge 2018) and Chair of the World Forum on Climate Justice.

Rutger Lazou is a PhD candidate in the Department of Philosophy at the University of Graz where he investigates how the transition towards a low-carbon society and economy can be realized in an ethical way. Prior to this, he obtained a Master's in Moral Sciences at the University of Ghent in which he specialized in animal ethics. He is the author of *De toekomst van de kat* [The Future of Cats] (Uitgeverij Houtekiet 2019).

Kirk Lougheed is Assistant Professor of Philosophy and Director of the Center for Faith and Human Flourishing at LCC International University. He is also a Research Associate at the University of Pretoria. He has published over 30 peer-reviewed articles or book chapters, appearing in such places as *Philosophia, Ratio, Res Philosophica* and *Synthese*. He is author of *The Epistemic Benefits of Disagreement* (Springer 2020), *The Axiological Status of Theism and Other Worldviews* (Palgrave 2020), *Ubuntu and Western Monotheism* (Routledge 2022) and is editor of *Four Views on the Axiology of Theism: What Difference Does God Make?* (Bloomsbury 2020) and co-editor (with Jonathan Matheson) of *Epistemic Autonomy* (Routledge 2022).

Elias Moser is a Postdoctoral Assistant at the Section Moral and Political Philosophy at the Karl-Franzens University of Graz. He acquired his PhD at the University of Berne in 2017 with an inquiry into the concept of inalienable rights. In his early career as postdoctoral scholar, he broadened his scope towards the fields of ethics of technology and economic philosophy. In 2018–2019, he worked as Researcher at the Institute for Technology Assessment (ITA) at the Austrian Science Funds (OEAW) and for the Research Platform Nano-Norms-Nature at the University of Vienna. In 2021, he was a Research Fellow at the Centre for Philosophy of Natural and Social Science (CPNSS) at London School of Economics and Political Science (LSE).

Samantha Noll is an Assistant Professor in The School of Politics, Philosophy, and Public Affairs (PPPA) at Washington State University. She is also the bioethicist affiliated with the Functional Genomics Initiative and the Center for Reproductive Biology. She is the co-author or co-editor of two books: *A Field Guide to Formal Logic* (Great River Learning 2020) and the *Routledge Handbook of Philosophy of the City* (Routledge 2019). She is also author or co-author of more than 30 other publications in journals ranging from the *Journal of Agricultural and Environmental Ethics* to *Environmental Ethics* and *Pragmatism Today*. She is currently working on projects that engage with ethical considerations at the intersection of philosophy of food, environmental ethics and emerging technologies.

Gianluca Ronca holds a PhD in Philosophy (Universidad de Castilla-La Mancha, Toledo, Spain - University of Rome 'Tor Vergata', Italy) and is Adjunct Professor of Political Philosophy (Collegio Ghislieri, Pavia, Italy). An an alumnus of the Collegio Ghislieri, his work spans across philosophy of human rights, critical theory and philosophy of international law. He has dealt with transitional justice and published several essays in international volumes and scientific journals. In 2017, he resided as a researcher and embedded reporter at the UNIFIL mission in South Lebanon. His next project is a book manuscript on the legal and philosophical foundations of political forgiveness.

Nishtha Singh works as an Environment Specialist for the Delhi Government. She is also a social entrepreneur who has worked at grassroots with Indian artisans for almost 6 years. She has previously worked for Deloitte Consulting for three years. She studied for a Master's in Carbon Management at the University of Edinburgh and then worked with the Edinburgh City Council in its strategy to make Edinburgh net zero by 2030.

Sue Spaid is Associate Editor of *Aesthetic Investigations* since 2014. She is the author of *The Philosophy of Curatorial Practice: Between Work and World* (Bloomsbury 2020), and the author of five books on art and ecology, Her environmental philosophy papers that connect wellbeing and values to biodiversity, climate justice, degraded lands, hydrological justice, stinky foods' superpowers and urban farming have appeared in *Ethics & the Environment*,

Journal of Somaesthetics, Rivista di Estetica, Philosophica, Popular Inquiry and *Art Inquiry: Recherche sur les arts* and *Rethinking Art and Creativity in an Era of Ecocide*.

Olle Torpman is a Postdoctoral Researcher in climate justice at the Institute for Futures Studies (Sweden). Between 2018 and 2021, he occupied an adjunct position in animal ethics at the Swedish University of Agricultural Sciences. He has published on questions related to principles of just allocation of climate responsibility, as well as on issues of environmental and animal ethics and is currently investigating which climate measures can be recommended on the basis different ethical principles with a special interest in the ethical principles of conservative ideology.

Vera Tripodi is Assistant Professor at Politecnico di Torino (Italy). She received her PhD in Logic and Epistemology from Sapienza University of Rome. Previously, she worked as Assistant Professor at University of Milan, was a Postdoctoral Researcher at University of Barcelona and at University of Oslo and a Visiting Postdoctoral Research Fellow at Columbia University. She specializes in feminist philosophy and ethics, bioethics, ethics of technology and social ontology. She is a Founding Member and Vice President of the SWIP ITALIA (The Society for Women in Philosophy – Italy). She is co-editor (with Enrico Terrone) of *Being and Value in Technology* (Palgrave 2022).

Foreword

TAHSEEN JAFRY

The scientific data is clear. Over the last decade, we have witnessed the hottest years on record, with 2020 being one of the five hottest years since records began. The World Meteorological Organisation has warned that the world is not on track to keep global warming to within 1.5°C above pre-industrial levels. The current rate of greenhouse gas (GHG) emissions trends suggests the world is heading towards a global warming between 3°C and 5°C by 2100. The authors of the landmark Intergovernmental Panel on Climate Change Special Report on the Impacts of Global Warming of 1.5°C (2018) say urgent and unprecedented changes are needed to hold the increase in global average temperature between 1.5°C and 2°C. The Paris Agreement (2015) has set out bold ambitions to limit GHG emissions so as to prevent the worst impacts of climate change. Through their Nationally Determined Contributions (NDCs), many countries have committed to reduce their carbon emissions by 2050; notably the United Kingdom has committed to get carbon neutral by 2050, and Scotland even adopted a 5-year lower target, 2045. This global ambition was vital in the run up to the 26th United Nations Climate Change Conference (COP26) and was even more pertinent as we approached COP27. However, climate ambition comes with pressing questions. Who bears the burden of climate action as societies around the world transit to a low-carbon regime and to sustainable ways of life? Relatedly, who bears the burden of climate adaptation as societies adjust to with new climate scenarios? Moreover, with reference to climate justice, we understand that the impacts of climate change will affect some communities more severely than others. Low-income, marginalised and other disadvantaged groups will be hit the most. The inherent climate injustice is so profound that these communities will bear the economic, social and environmental burdens of climate change and will also be denied the right to a decent quality of life, despite not being significantly at fault for climate change.

The title of this book, *Global Climate Justice: Theory and Practice*, is pertinent as we head from COP26 and COP27 (which was also coined as the Africa COP) to COP28, and as the world grapples with identifying workable solutions to the climate crisis whilst protecting the most vulnerable.

Understanding the theory of climate justice is fundamental for arriving at practical ways forward that meet the needs, wants and aspirations of communities on the front line of the climate crisis. Thus, the chapters in this book provide rich and deep insights on the nexus between theory and practice of climate justice. In saying that, I would like to turn my attention here to two critical issues that lie at the heart of finding participatory and practical ways to advance climate justice by putting theory into practice.

The first is *gender and just climate financing*. Finance is rapidly becoming the defining issue in the fight against climate change. Many countries are developing Long Term Strategies and NDCs for which finance is needed to implement, whilst at the same time, donor countries are under extreme pressure to meet the pledge of delivering $100 billion per annum by 2023 (3 years later than promised at COP15 in Copenhagen in 2009) and to agree on the New Collective Quantified Goal on Climate Finance from 2025 onwards. Fair and equitable access to finance will be vitally important for ensuring that financial resources reach all sectors of society and address both mitigation and adaptation needs. Currently, access to climate finance comes through large multilateral institutions, such as Global Climate Facility and Global Environment Facility, as well as smaller scale bilateral funds. Most of this climate finance goes to mitigation projects (often for renewable energy development) and to supporting infrastructure. Often the main recipients of funds are governments and private sector actors that manage to seize large grants, loans and guarantees from development organisations such as the World Bank and United States Agency for International Development to the tune of between $10–$50 million. Comparatively little funding flows to adaptation projects, especially when they do not provide significant financial return on investment. Consequently, access to climate finance is neither fair nor equitable, resulting in women, youth and vulnerable people missing out despite often bearing a disproportionate share of the impacts of climate change. What is required is a call for a significant overhaul of the global, bilateral and national climate finance flow architecture and of climate change financing policies, so as to integrate climate justice considerations in the design and implementation of climate adaptation programmes. Raising awareness and building in processes for fair and equitable access to finance are vital for ensuring that climate funds are allocated to those who have benefited least from them so far, which can help them develop skills and build resilience in the fight against climate change. This would be a significant step towards climate justice.

The second point is *Non-Economic Loss and Damage (NEL&D)*. According to the United Nations Development Programme (UNDP) report, "The human cost of disasters: an overview of the last 20 years (2000–2019)", since 1980 the world has seen an 80% increase in climate related disasters. These disasters disproportionately impact the poorest countries, i.e., the people who have contributed the least to climate change. Although it is often hard to pin down any specific disaster to climate change, there is evidence of an increase in the frequency of extreme weather events. Predictions that heatwaves, heat stress and heavy rainfall will become more intense and frequent in Sub-Saharan Africa have implications for the safety and survival of the poorest people in the world. Due to the loss of shelter, livestock, food, income and basic provisions, including access to water, many people, especially women, have been left distraught and broken. At the same time, recent evidence shows that extreme poverty is expected to rise for the first time since 1998 in Sub-Saharan Africa. In part due to the economic decline of the Covid-19 pandemic, over 500 million people are living in extreme poverty, in an intertwined climate and health crisis. There is also growing evidence that women are the largest group of people to be affected by Post Traumatic Stress Disorder (PTSD) caused by extreme weather, and this is also linked to sexual and gender-based violence. Women tend to face a higher risk of depression, anxiety, distress, suicide and grief following extreme weather. This correlates with research being conducted at the Mary Robinson Centre for Climate Justice at Glasgow Caledonian University and has led to the development and rollout of a programme of work to provide vital safety nets and support to women in Malawi.

NEL&D cannot be reckoned easily, in either monetary or other ways; hence the issue is often neglected, hidden and difficult to deal with. However, this does not mean that we should shy away from addressing NEL&D. In the interest of achieving climate justice, it is imperative that we 'understand better' NEL&D, and this will pose a novel challenge. Looking at this through a gendered lens, very often women and girls face increasing pressures due to climate change and the onslaught of extreme weather events, due to their roles and responsibilities: they are left vulnerable, they have to walk longer and farther to collect water and fuel wood, with the risk trafficking, sexual exploitation and resultant child pregnancy. Climate change exacerbates gender-based violence, and we know that in many developing countries women are on the brink of a mental health crisis. Accordingly, NEL&D is about going beyond repairing infrastructure, bridges and roads that have been destroyed to start 'repairing' people whose lives have been left damaged and broken. NEL&D is about the 'safety' not just the survival of the poorest and the most vulnerable. Although none of this is easy to quantify and qualify, NEL&D is an area of climate justice research which warrants much attention.

Looking ahead, I do not know what will be achieved at COP28 and beyond. But we must look towards ensuring a healthy and sustainable environment for all of humanity and delivering on our commitments to achieve climate justice.

The protection of human rights and especially of the most vulnerable – socially, legally, financially and environmentally – is vital. This is not simply a moral duty of care or an act of compassion, but rather an obligation for the world's richest economic actors to put humanity in the conditions to live healthy lives with dignity.

Introduction

The Global Philosophy of the (Global) Carbon Budget

FAUSTO CORVINO AND TIZIANA ANDINA

The global management of a limited carbon budget calls for a reassessment of climate justice and responsibility. Sticking to the 1.5°C mitigation target requires the global community to nearly halve carbon dioxide (CO_2) emissions by 2030 and achieve carbon neutrality by 2050 at the latest. This in turn raises three sets of normative issues. First, in a scenario of limited future emissions (consistent with the 1.5°C target), the notion of differentiated responsibilities needs to be reframed in terms of duties of economic and technological assistance, aimed both at financing mitigation and adaptation projects in the countries most exposed to climate damage and at compensating for this damage where it can no longer be avoided. Second, the climate crisis is now so severe that the welfare of future generations will depend on how present generations manage the global carbon budget (GCB). This in turn calls for the notion of intergenerational responsibility to be fleshed out in clear decarbonisation targets. Third, unequal emission levels within countries mean that future global emissions will be unevenly appropriated across social classes. Global climate justice thus intersects with the problem of carbon inequality and individual responsibility for climate change.

1. The 1.5°C Mitigation Target and Net Zero Emissions

Scientists unequivocally agree that greenhouse gas (GHG) emissions in general and CO_2 emissions in particular are causing global warming and consequently climate change. The Industrial Revolution kick-started anthropogenic climate change, leading to a progressive increase in the concentration of CO_2 in the atmosphere, from 278 ppm in pre-industrial times to 417 ppm in 2021 (Betts 2021). The consequences of climate change are there for all to see, ranging from rising sea levels to seawater intrusion threatening coastlines, hotter summer days and bigger and more frequent wildfires, droughts causing famines, an intensification of the number and severity of abnormal weather events and so on. The economic costs of global warming are substantial. According to a recent stress test conducted by the European Central Bank (2021, 19) on 4,000 companies and 1,600 banks in the euro area, in the absence of climate mitigation policies, global warming would lead to a 10% reduction in European gross domestic product (GDP) by 2100. And the Swiss Re Institute (2021, 10) estimates that in a no-mitigation scenario, where emissions follow the current trajectory, global GDP would fall by 18% by 2050, and of course most of these losses would occur in developing countries.

A practical and communicatively powerful way to measure climate change is through the so-called (global) carbon budget – the amount of CO_2 that humanity can still emit to maintain compatibility with a given mitigation target. The carbon budget is mainly a function of two factors: the mitigation target we choose and the probability (to meet the target) we aim for (e.g., either a 50% or 66% probability that mitigation policies will achieve the desired mitigation target). For many years, the climate target endorsed by the international community has been to limit global warming to no more than 2°C above pre-industrial levels. The latter target was adopted as indicating the maximum risk that could be tolerated, which in turn was associated, at least in early scientific climate studies, with a doubling of the concentration of atmospheric CO_2 compared to pre-industrial levels (see Tschakert 2015). The first to put the spotlight on the more ambitious goal of keeping global warming below 1.5°C above pre-industrial levels was the Alliance of Small Island States (AOSIS), an intergovernmental organization founded in 1990 and representing 44 (mostly developing) countries, both islands and low-lying coastal states, which were (and unfortunately still are) at the risk of being literally submerged by rising sea levels as a consequence of global warming (see Ourbak and Magnan 2018; Bolon 2018).

On the eve of COP15 (Copenhagen 2009), AOSIS called upon the Potsdam Institute for Climate Impact Research to carry out a study on the consequences of a global warming of 2°C. Armed with the expectation that global warming of more than 1.5°C could result in substantial threats to the very existence of its members, AOSIS sought to draw the world's attention to the need for more ambitious mitigation efforts by the international community (see CICERO, n.d.). Despite these efforts, both developing and developed countries largely turned a deaf ear to the demand for more robust mitigation coming from AOSIS, and thus the Copenhagen Accord stuck to the 2°C target. The parties, however, opened the door to a revision of the collective mitigation target by committing to decide by 2015 whether the 2°C target should be reduced to 1.5°C based on empirical evidence to be gathered in the meantime (UNFCCC 2009, art.12) – the same concept of periodic assessment and eventual reinforcement of the long-term mitigation target is reiterated in the Cancun Agreements (UNFCCC 2010, art.4, art.138). COP21 (which produced the 2015 Paris Agreement) is therefore the moment when the 1.5°C target enters climate negotiations alongside the 2°C target. In the famous and groundbreaking wording of the Paris Agreement, the parties committed to 'holding the increase in the global average temperature to well below 2°C above pre-industrial levels and pursuing efforts to limit the temperature increase to 1.5°C above pre-industrial levels' (UNFCCC 2015, art. 2). This was followed in 2018 by the Intergovernmental Panel on Climate Change (IPCC) publishing a special report on the impacts of 1.5 °C of global warming. The report's key message is that containing global warming to below 1.5°C, rather than 2°C, would substantially reduce the negative impacts of climate change on ecosystems and human well-being. The report also sets out a series of mitigation pathways that are compatible with the 1.5°C target, thus providing policymakers with practical information on how to effectively reduce greenhouse gas (GHG) emissions (IPCC 2018).

Climate science is obviously an extremely complex subject about which it is often impossible to make clear-cut statements. Even after a specific mitigation target has been identified in a given global warming threshold, it is not always possible to say with certainty how many tons of CO_2 can be emitted before the earth's thermometer exceeds this threshold; however, it is possible to define this in probabilistic terms. Accordingly, the IPCC (2021, 29) tells us that to have a 50% chance of meeting the 1.5°C target, from 2020 onwards, humanity can only emit 500 gigatons (Gt) of CO_2 – decreasing to 460 Gt beginning in 2021 if we account for global emissions in 2020 (Hausfather 2021b). If we consider that global GHG emissions were about 50 $GtCO_{2e}$ in 2021 (Climate Action Tracker 2021, 1), it is evident that under a business-as-usual scenario, humanity is going to burn literally the 1.5°C carbon budget in less than 10 years, starting from 2022 (CO_{2e} stands for 'carbon dioxide equivalent' and it is used to measure the climate effects of different greenhouse gases, by converting emissions of any other greenhouse gases into the equivalent of CO_2 emissions with the same warming potential, see Eurostat, n.d.).

This does not mean that the 1.5°C target is out of reach, but the inertia of the past certainly makes it difficult to achieve. Essentially, humanity would have to halve global CO_2 emissions every decade until emissions reach net zero in 2050 – and then bring emissions of other GHGs to net zero by 2070 at the latest (Levin et al. 2019). Net zero CO_2 emissions do not mean zero CO_2 emissions, but rather no more anthropogenic CO_2 emissions than cannot be offset by anthropogenic CO_2 removals – i.e., 'through deliberate human activities' (IPCC 2018, 555) – before they build up in the atmosphere and increase the existing CO_2 stock. Anthropogenic CO_2 removals can take place either through nature-based solutions or in a technological way (see EASAC 2018). Nature-based solutions consist of expanding so-called natural carbon sinks, such as plants and forests, which can sequester CO_2 through the process of photosynthesis – this can be done either by planting new trees in addition to existing ones or by replanting felled forests. Technological solutions range from the capture of CO_2 at the point of emission in power plants with subsequent transport and underground storage (CCS, or carbon capture and storage) to the production of energy with the combustion of biomass and subsequent storage of the CO_2 produced (BECCS, or bioenergy with carbon capture and storage). In addition, there are also geo-engineering solutions, i.e., the more or less invasive modification of the earth's natural systems by technological means to mitigate climate change (see Gardiner and McKinnon 2020). Examples range from the direct capture of CO_2 from the atmosphere (DAC, or direct air capture) to the fertilisation of the oceans (to increase their CO_2 absorption capacity), as well as artificially increasing the reflectivity of clouds so that they can reflect solar energy back into space (SRM, or solar radiation management).

Once carbon neutrality is achieved, by balancing anthropogenic CO_2 (and more generally GHG) emissions with negative emissions, global warming will stop, and the earth's temperature will gradually stabilize (Hausfather 2021a). This is also because, we must not forget, part of the CO_2 emitted by human beings is absorbed naturally by the biosphere through the carbon cycle. Therefore, once anthropogenic emissions become net zero, carbon sinks such as land and ocean will not have to compensate for the anthropogenic flux of emissions (which will have been reduced to zero) and they will be able to gradually reduce the stock of CO_2 in the atmosphere, which is the result of past emissions (IPCC 2018, 159). Obviously, the greater the margin of negative emissions, not least through the development and refinement of technologies that have been pioneering to date, the greater the chances of meeting the 1.5°C mitigation target. This is for two main reasons. One is that some GHG emissions cannot be abated, either because – at least for now – there is no valid green substitute, such as in the case of aircraft engines, or because they are linked to biological activities that are difficult if not impossible to modify, such as methane emitted by grazing animals. The other reason is that we are not sure that we will achieve net zero emissions within 30 years, and if we miss the target negative emissions will give us a chance, in a sense, to go back and correct the mistake by eliminating the part of the CO_2 stock in the atmosphere that exceeds the 1.5°C global carbon budget.

2. The Three Pillars of Global Climate Justice

While effective and equitable mitigation would initially have been sufficient to ensure climate justice, this is no longer the case. Climate change is happening, has already caused widespread harm and will continue to do so in the years to come. Accordingly, on the one hand, there is a need to invest in adaptation, i.e., in modifying or, if you like, strengthening ecological and socio-economic systems in order to increase their resilience to climate threats (see Baatz 2018; Boran and Heat 2016). Examples of adaptation are changes in agricultural practices or even the adoption of climate-resistant crops, constructing buildings that are insulated against excessive heat and/or able to withstand flooding and/or can capture rainwater and make it available in times of drought, strengthening urban infrastructure against abnormal weather phenomena, but also industrial policies such as economic diversification and so on (see Acevedo et al. 2020; Sloat et al. 2020; UNEP 2021a; Steuteville 2021). Where neither mitigation nor adaptation succeed, however, environmental harms become unavoidable, and then the complex issue of compensation for loss and damage arises, especially where the economically most fragile individuals are hit the hardest (see Mechler et al. 2020; Pidcock and Yeo 2017; Page and Heyward 2017).

Moreover, the normative research on climate mitigation has progressively changed its focus. Initially, scholars applied both backward-looking claims, focused on the responsibility for past actions, and forward-looking claims, based on distributive principles (e.g., equality), directly to the distribution of the carbon budget among countries (see Gajevic Sayegh 2017, 344-350). That is, the idea was that the main normative challenge raised by climate change consisted in the allocation of future permissions to emit. The responsibility for the past mattered as long as you believed that someone's future permits to emit were to be assigned by what they and their compatriots had emitted in the past. And even those who considered global wealth inequality as sufficient justification for a differentiated allocation of climate burdens focused mainly on the question of 'differentiation in mitigation' (Gajevic Sayegh 2017, 345; see Caney 2021). This was because there was, at least in the 2000s, relative optimism about the success of climate mitigation, and it was believed that once a mitigation target had been set, there was still room for those who had polluted less in the past to pollute more in the future.

Today, however, the mitigation challenge is so pressing that the distributive issue concerning the remaining share of the carbon budget has become secondary (see also Robinson and Shine 2018; Prattico 2019). Mitigation requires an immediate and rapid reduction in the levels of global CO_2 emissions, which in turn takes the form of a series of intermediate decarbonisation targets – the ultimate goal being to add no more CO_2 to the atmosphere after 2050. In short, the global emissions gap, i.e., the difference between the current emission trend and the one we should follow to successfully mitigate climate change, is so large that every country is called upon to swiftly reduce GHG emissions (UNEP 2021b, 3). Furthermore, until the last decade green technologies were not yet at an advanced stage of application (think of recent developments in the electricity and renewable energy markets), so it was difficult to imagine that meeting a number of basic needs, especially in poorer countries, could be done without GHG emissions. Now, however, it is clear that energy demand can be met without polluting – provided, of course, that the economic resources are available to do so (see Shue 2014, 6–9; Argyriou 2017).

The global justice discourse regarding mitigation has therefore shifted from the allocation of future emission permits to the global redistribution of wealth and technology, with the aim of supporting those with fewer economic resources and less technology to undertake climate mitigation (Shankar 2021; Delgado 2021). This is obviously accompanied by a similar concern about supporting developing countries in adapting to climate change and about offsetting loss and damage (see also Light and Taraska 2014; Baatz 2018; Gajevic Sayegh 2017, 350–361). As a result, much of global climate justice is now played out, within international negotiations, in terms of climate finance. At COP15 (Copenhagen 2009), developed countries made the historic pledge to provide $100 billion per year to developing countries to finance mitigation and adaptation activities beginning in 2020. Climate finance is to be understood as 'new and additional resources' coming 'from a wide variety of sources, public and private, bilateral and multilateral, including alternative sources of finance' (UNFCCC 2009, art. 8). To date, the $100 billion pledge has been breached, which is especially shameful considering the extra resources developed countries were able to inject in their own economies (about $20 trillion) in response to the Covid-19 pandemic (see Sachs 2021). At COP26 (in Glasgow, UK), the parties acknowledged their regret for this failure and went so far as to promise almost $96 billion per year by 2022, with the idea of relaunching more ambitious targets after 2025 (see Mountfor et al. 2021). This is obviously not enough. On the one hand, the costs of adaptation in developing countries could be much higher than previously thought, at about $300 billion in 2030 and $500 billion in 2050 (UNCTAD 2021). On the other hand, more than half of climate finance continues to be directed to mitigation projects, which are more profitable for lenders, rather than to adaptation, and less than a third of public climate finance is given as grants, while almost all the rest is given as loans (OECD 2021, 7-8; see also Sheridan and Jafry 2019; Timperley 2021).

3. The Normative Implications of the Global Carbon Budget

The positive emissions (i.e., those emissions exceeding the counterbalancing capacity of anthropogenic negative emissions) that humanity can still afford compatibly with the 1.5°C mitigation target, and which are exemplified in the global carbon budget (GCB), are a set of scarce, rival, not easily excludable, global and intergenerational goods. Why the GCB is a set of scarce and rival goods is obvious. Each emission erodes a share of the GCB that cannot be appropriated by another person. The non-excludability, or rather the difficult excludability, derives from the fact that every person continuously emits GHGs, even when breathing, and it is very complicated if not impossible to regulate all access to the GCB through bans, quotas or sanctions. The GCB is global because every emission, regardless of where it occurs, consumes a part of it. The GCB is intergenerational both because its current size is the result of emission-generating activities that took place in the past, including by people who are no longer with us, and because the fate of future generations depends on its management. All of this poses a number of normative issues, which we hope to address at least in part in this volume.

The first range of normative issues interweaves global and backward-looking intergenerational concerns. Developing countries are subject to double climate disadvantage. On the one hand, they are usually more exposed to the negative effects of climate change, both for geographical and economic reasons. On the other hand, they are being asked to forego industrialisation driven by fossil fuels – which are currently cheaper and often available in large quantities in developing countries. Both disadvantages are the direct consequence of past emissions, which have caused the climate change we are experiencing today and determined the magnitude of the current 1.5°C GCB. It needs to be explained whether and to what extent current duties of climate justice are a function of past emissions, even in the face of two non-trivial theoretical problems: people who emitted in the past, at least until the end of the twentieth century, were generally unaware of the effects that pollution was having on global warming; moreover, many past polluters are no longer around (see Moss and Robyn 2019; García-Portela 2019; Bell 2011; Gosseries 2004). Obviously, some of the benefits of past emissions have accrued to people in the present, i.e., emissions have helped to reinforce global inequalities, so many believe that it is unjust enrichment (where unjust can mean either that it has caused intergenerational harm or that it has been achieved by appropriating an excessive share of the GCB) that drives current duties of climate justice (see Page 2012). Still others believe that it is the ability to contribute, regardless of past responsibility, that should give us the measure of how much developed countries owe to developing countries in terms of climate justice (see Caney 2010).

There are, furthermore, two technical issues relating to emission accounting methods that have significant ethical implications, at least at the global level. First, when we say that a given country is responsible for X quantity of emissions in the period $t - t_1$, we are normally referring to data that account for the total emissions that occurred in that country in that time segment.

However, not all emissions in a country go to meet domestic demand. Many emissions are used to make products or services that are then exported. Thus, on the one hand, a country may give the impression that it is reducing its emissions, simply because it is not producing as much as before, while on the other hand, its emission consumption is increasing through imports (see Helm 2020, 58-81). And of course, where it is the emissions produced that count, there is a strong incentive for companies to move production abroad and then import (so-called 'carbon leakage'). This is, for example, one of the reasons why the European Commission (2021) proposed the introduction of a carbon border adjustment mechanism (i.e., a carbon tax at the EU border) as part of the "Fit for 55" climate-policy package and the European Parliament (2022) recently voted to bring it into force from 2027 for fuller implementation in 2032. On the other hand, production-based accounting relieves the moral responsibility of net CO_2-importing countries and inflates the responsibility of citizens in exporting countries (see Duus-Otterström and Hjorthen 2019; Grasso 2016). The same applies to the second technical issue, i.e., calculation of total vs. per capita emissions. If you only look at the total emissions of a country, you do not take into account how many people have received the economic benefits of these emissions.

To simplify, we could say that combining production-based accounting with total emissions favours western countries at the expense of eastern ones. Countries such as Russia, Kazakhstan, Vietnam, China, India and Mongolia are net exporters of CO_2, i.e., they consume less CO_2 than they produce. China, for example, exports about 10% of its domestic emissions, while India exports more than 8% and Russia almost 15%. The United States, on the other hand, imports 7% of the emissions they produce, while countries like Italy, Spain and Austria import more than 30% (see Ritchie 2020). At the same time, China's total emissions are almost double those of the United States (US), but the US's per capita emissions are almost double those of China.

The second range of normative issues are intergenerational in a forward-looking respect. Global warming projections for the end of the twenty-first century depend on climate policy choices in the present. In other words, depending on how effectively people living now mitigate climate change, people born between the end of the twenty-first and the beginning of the twenty-second century will find themselves living on a warmer or cooler planet, with all the consequences that will follow. At the moment, we are already experiencing global warming of 1.2°C above pre-industrial levels – which in turn is a direct consequence of the increase in the concentration of CO_2 in the atmosphere. All the climate policies currently in place in the world will lead to a warming of around 2.7°C by the end of the twenty-first century. If all countries were to meet their mid-term (2030) mitigation targets, taking updated nationally determined contributions (NDCs) at COP26 (Glasgow 2021) as a reference point, global warming at the end of the century would be 2.4°C. Only in the optimistic, and therefore unlikely scenario in which all countries not only implemented submitted and binding long-term targets but also announced ones would global warming be limited to 1.8°C. A message that encapsulates the urgency of raising the bar on climate commitments is this: the climate pledges for 2030 collected at COP26 would lead us to emit more than twice as much in 2030 as would be needed to stay in line with a 1.5°C consistent mitigation pathway (Climate Action Tracker 2021).

To understand the extent of the economic damage that sub-optimal mitigation would cause to future generations, it is useful to consider a recent stress-test analysis by the Swiss Re Institute (2021). It analyses the effects of global warming of 2.6°C by mid-century – which would occur in the absence of any mitigation policy, and which comes close to the global warming we would have by the end of the century as a result of the mitigation policies that are currently implemented in the world. Under a 2.6°C global warming scenario in 2050, global GDP would fall by about 14% (during the peak of the Covid-19 pandemic, global GDP contracted by 3.5% – see Levy Yeyati and Filippini 2021), and for some countries such as China (18%), Malaysia (36%), Indonesia (30%) and Singapore (35%) the loss would even be higher (Swiss Re Institute 2021, 11).

It is evident that radical mitigation consistent with the 1.5°C target is optimal in a diachronic perspective because the future costs of global warming above 2°C are greater than the costs we would face, starting now, to avoid it. But it is also true that radical mitigation implies that people living now make investments and in some cases economic sacrifices to yield net benefits that will be enjoyed mainly by future people (see Burke et al. 2018). This requires a redefinition of the concept of intergenerational responsibility to match the climate challenge we face. This also means clarifying whether and what the present generation owes to future, not necessarily overlapping, generations in terms of climate justice and why intergenerational duties of climate justice are of no less normative relevance than intra-generational duties of socio-economic justice (see Mulgan 2018; Caney 2014; Gardiner 2011).

Finally, the third range of normative issues concern individuals, as consumers, investors and/or entrepreneurs. For climate mitigation to be effective, everyone needs to do their bit by reducing individual emissions, at least the superfluous ones – i.e., those related to activities that can be carried out without polluting or polluting less. However, two major problems arise. One is related to carbon inequality. A small minority of people have emitted and continue to emit more than everyone else. Of all the CO_2 emitted from 1990 to 2015, more than half is attributable to the activities of the richest 10%

of the world's population, while more than 15% of CO_2 was emitted by the richest 1%. Given that the largest emissions in history occurred from 1990 onwards, this implies that the richest 10% of the world's population has appropriated 31% of the 1.5°C GCB in just 25 years, while the richest 1% has appropriated 9% of it (Oxfam 2020, 2). A figure that epitomises the relevance of carbon inequality not only in terms of moral responsibility for past emissions but also for the effectiveness of mitigation is the following: if the richest 10% of the world's population continued to emit at current levels, while the emissions of the remaining 90% of the world's population magically became zero from now, the 1.5°C GCB would be consumed shortly after 2030 (Oxfam 2020, 2). Furthermore, since 1990 the emissions of the super-rich, the richest 0.01% of the world's population, have increased by 110%, while those of the middle 40% of the global class (people who are neither part of the poorest 50% nor the richest 10%) have only increased by 4%, and those of the poorest 50% have increased by 32%, mainly due to their being lifted out of poverty (Chancel 2021, 17; see also Kenner 2019). Add to this the responsibility of companies, especially oil producers. It is estimated, for example, that 35% of methane and CO_2 emissions over the last 50 years are attributable to the production and consumption of fossil fuel products coming from just 20 companies (Taylor and Watts 2019; see also Grasso 2022).

Carbon inequality leads to two major concerns. Firstly, since climate change is common knowledge, the rich have been emitting far more than the middle and lower classes, and they have obviously also been making large profits from emission-generating activities – one only has to look at data on growing intra-country inequality to see that carbon inequality and income inequality go hand in hand. This raises the question of whether and to what extent the rich have a greater duty to compensate the victims of climate change than the middle and lower classes. Of course, the concept of class responsibility does not replace that of country responsibility, but in a sense contributes to specifying it. When we say, for example, that the United States has appropriated a certain share of the GBC in a certain historical period, we may state that the US has a duty, as an international actor, to transfer resources to those who are damaged by climate change and/or forced to give up an equivalent share of the GCB. The concept of class responsibility, on the other hand, can indicate how the global climate duties of the US should be translated into domestic tax burdens, to be distributed among the different social classes, not least in relation to their respective climate responsibility. Secondly, radical carbon inequality, such as we now have at a global level, means that for climate mitigation to succeed, decarbonisation efforts by the richest worldwide must be disproportionately greater than that of the rest of the world's population. And this clearly implies that we need to think about public policies that target the richest and prompt them to change their lifestyle and investment choices (see Otto et al. 2019; Benoit 2020).

The second normative problem related to the individual dimension of climate change concerns the moral link between individual emissions and climate harms. Stephen Gardiner (2011, 45–48) famously defined 'moral corruption' as the tendency for individuals to hide somewhat behind the empirical complexities of climate change in order to distort the way one should approach the phenomenon and shrug off moral responsibility for one's emissions, thus not taking the issue seriously, neither privately nor politically. Climate change has, in a certain sense, posed a novel challenge to the philosophical concept of moral responsibility, where the relationship between cause and effect extends both across a global geographical scale and across a timeframe extending into the future as far as linking generations that are not overlapping (see also Jamieson 2007; Persson 2016). Climate mitigation, as we have seen, requires everyone to reduce their carbon footprint, but for this to happen people need to be motivated to do so (see Peeters et al. 2019). On the one hand, it is necessary to defend a genuinely cosmopolitan and intergenerational ethos, which inspires individuals to take on part of the climate-transition burden in the interests of individuals they will never meet, whether because they live in distant places or have not yet been born (see Meyer 2021). On the other hand, reasons must be given why everyone is called upon to play their part, even in the face of such a complex and wide-ranging problem that no one person, company, city or state can solve alone, and in respect to which every individual contribution seems to be irrelevant if taken on its own (see Broome 2019; Banks 2013; Nolt 2011).

4. The Book's Objective and Why Now

The book *Global Climate Justice: Theory and Practice* has two main purposes, dictated by the historical (and climatic) moment we are living. On the one hand, the book aims to propose a series of reflections on the distributive aspects of global climate justice, in its three components mentioned above, for the post-COP26 period. The hope that has been pinned on COP26 (Glasgow 2021) was to keep the 1.5°C mitigation target alive. To understand whether this has been achieved, it will be necessary to see whether the decarbonisation commitments and announcements are translated into concrete action. What is important to emphasise, however, is that we are going through a crucial phase in the history of humankind, in which we must decide whether we will take the earth beyond irreversible climate risk thresholds or stop short of the cliff. Global CO_2 emissions have been on a steeply rising curve since 1950, increasing from just over 5 Gt per year to over 36 Gt in the immediate pre-Covid-19 period (Ritchie and Roser 2020). Effective climate mitigation requires this curve to take the shape of a bell between now and the next 30 years. We are now at the top of this hypothetical bell. We need ethical and political theories to guide us on the way down.

The second objective of the book is to trigger a series of parallel and complementary normative reflections on the climate neutrality ambition. Decarbonising the world's economy, and doing so in a just way, requires a series of significant social and political interventions. Perhaps the most obvious case concerns food and agriculture. Current styles of food consumption are simply incompatible with medium- and long-term climate mitigation goals. It is therefore essential to rethink both the production and consumption of many foods, red meat being one example. Obviously, this could have major social consequences, as well as encountering the obvious implementation hurdles that arise when people are called upon to change deep-rooted habits (see Murdock and Noll 215). Another key issue is gender. Women are often the most vulnerable to climate damage, especially in developing countries, both because they are usually in charge of providing food and water for their families, which climate change puts at risk, and because they have fewer socio-economic means to adapt to climate shocks. But they are also potential climate mitigation actors, especially in those social contexts where they play a key decision-making role in household energy choices. It is therefore clear that climate policies, be they aimed at mitigation, adaptation or compensation, need to be made gender-sensitive for them to be fair and effective (see Jafry 2016; Perkins 2019).

Another major challenge is the social responsibility of companies with respect to the goal of net zero emissions by 2050, especially companies that have contributed and still contribute the most to GHG emissions. On the one hand, companies are called upon to rethink their productive activities in order to limit their environmental footprint without negatively impacting their stakeholders (primarily their workers), and on the other hand, they have to account for the climate damage they have caused so far and from which they have gained considerable profits (Benjamin 2021, 20–45; Grasso 2022). Alongside the moral responsibility of polluters there arises the normative problem of stranded assets. To achieve climate neutrality by 2050, humanity will need to keep about 60% of gas and oil and 90% of coal underground (Welsby et al. 2021). This will mean, for those who control access to these energy assets, giving up conspicuous profits and liquidating the relevant infrastructure earlier than planned when it was built. The losers will not only be the companies but also the savers who have (more or less wittingly) invested in these companies and, of course, the workers. This raises both the issue of protecting the most vulnerable and whether and under what circumstances the winners of the climate transition owe something to the losers (see Caldecott et al. 2021). Finally, national and international legal systems are faced with a pressing challenge, i.e., to account for the rights of future individuals, who are the main victims of climate change, and also to reconceptualise the rights of non-human living beings vis-à-vis an existential threat such as climate change (see González-Ricoy and Gosseries 2016; Di Paola and Jamieson 2018).

We consider it useful to try to pursue both objectives in the same volume, despite the limited number of pages available, because the distributive challenge of climate justice inevitably intersects with a series of ethical, political and legal issues concerning the role of consumers, companies, policymakers and technology in the global undertaking towards long-term sustainability and carbon neutrality. At the same time, part of the upshot of the transition to a fossil-fuel free society depends on how the different components of climate distributive justice, in particular the aspects of effectiveness and equity, are combined.

References

Acevedo, M., K. Pixley, N. Zinyengere, *et al.* 2020. "A scoping review of adoption of climate-resilient crops by small-scale producers in low- and middle-income countries". *Nature Plants* 6: 1231–1241

Argyriou, M. 2017. "Developing countries can prosper without increasing emissions". *The Conversation*, September 21. https://theconversation.com/developing-countries-can-prosper-without-increasing-emissions-84044

Baatz, C. 2018. "Climate Adaptation Finance and Justice. A Criteria-Based Assessment of Policy Instruments". *Analyse & Kritik* 40 (1): 73–106.

Banks, M. 2013. "Individual Responsibility for Climate Change". *Southern Journal of Philosophy* 51 (1): 42–66.

Bell, D. 2011. "Global Climate Justice, Historic Emissions, and Excusable Ignorance". *The Monist* 94 (3): 391–411.

Benjamin, L. 2021. *Companies and Climate Change: Theory and Law in the United Kingdom*. Cambridge: Cambridge University Press.

Benoit, P. 2020. "A Luxury Carbon Tax to Address Climate Change and Inequality: Not All Carbon Is Created Equal". *Ethics & International Affairs*. https://www.ethicsandinternationalaffairs.org/2020/a-luxury-carbon-tax-to-address-climate-change-and-inequality-not-all-carbon-is-created-equal/

Betts, R. 2021. "Met Office: Atmospheric CO2 now hitting 50% higher than pre-industrial levels". *Carbon Brief*, March 16. https://www.carbonbrief.org/met-office-atmospheric-co2-now-hitting-50-higher-than-pre-industrial-levels

Bolon, C. 2018. "1.5 to Stay Alive: The Influence of AOSIS in International Climate Negotiations". *E-International Relations*, November 17. https://www.e-ir.info/2018/11/17/1-5-to-stay-alive-the-influence-of-aosis-in-international-climate-negotiations

Boran, I. and J. Heath. 2016. "Attributing Weather Extremes to Climate Change and the Future of Adaptation Policy". *Ethics, Policy and Environment* 19 (3): 239–255.

Broome, J. 2019. "Against Denialism". *The Monist* 102 (1): 110–129.

Burke, M., W. M. Davis, and N. S. Diffenbaugh. 2018. "Large potential reduction in economic damages under UN mitigation targets". *Nature* 557: 549–553.

Caldecott, B., A. Clark, K. Koskelo, E. Mulholland, C. Hickey (2021). "Stranded Assets: Environmental Drivers, Societal Challenges, and Supervisory Responses". *Annual Review of Environment and Resources* 46 (1): 417–447.

Climate Action Tracker 2021. "Warming Projections Global Update. November 2021". https://climateactiontracker.org/documents/997/CAT_2021-11-09_Briefing_Global-Update_Glasgow2030CredibilityGap.pdf

Caney, S. 2010. "Climate Change and the Duties of the Advantaged". *Critical Review of International Social and Political Philosophy* 13 (1): 203–228.

Caney, S. 2014. "Climate change, intergenerational equity and the social discount rate". *Politics, Philosophy and Economics* 13 (4): 320–342.

Caney, S. 2021. "Climate Justice", In *The Stanford Encyclopedia of Philosophy* (Winter 2021 Edition), edited by Edward N. Zalta. https://plato.stanford.edu/entries/justice-climate/

Chancel, L. 2021. "Climate change & the global inequality of carbon emissions, 1990-2020". *World Inequality Lab | Paris School of Economics*, October 18. https://wid.world/document/climate-change-the-global-inequality-of-carbon-emissions-1990-2020-world-inequality-lab-working-paper-2021-21/

CICERO. n.d. "The story of 1.5°C". https://cicero.oslo.no/en/understanding-one-point-five/the-story-of-15

Delgado, C. 2021. "How Developing Countries Can Reduce Emissions Without Compromising Growth". *Earth.org*, July 26. https://earth.org/how-developing-countries-can-reduce-emissions-without-compromising-growth/

Di Paola, M. and D. Jamieson. 2018. "Climate Change and the Challenges to Democracy", *University of Miami Law Review* 72 (2): 369–424.

Duus-Otterström, G. & F. D. Hjorthen. 2019. "Consumption-based emissions accounting: the normative debate". *Environmental Politics* 28 (5): 866–885.

EASAC. 2018. *Negative emission technologies: What role in meeting Paris Agreement targets?*. https://unfccc.int/sites/default/files/resource/28_EASAC%20Report%20on%20Negative%20Emission%20Technologies.pdf

European Central Bank. 2021. "ECB economy-wide climate stress test". Occasional Paper Series, N. 281, September 2021. https://www.ecb.europa.eu/pub/pdf/scpops/ecb.op281~05a7735b1c.en.pdf

European Commission. 2021. "Carbon Border Adjustment Mechanism: Questions and Answers". July 14. https://ec.europa.eu/commission/presscorner/detail/en/qanda_21_3661

European Parliament. 2022. "Revision of the EU Emissions Trading System", P9_TA(2022)0246. https://www.europarl.europa.eu/doceo/document/TA-9-2022-0246_EN.pdf

Eurostat. n.d. "Glossary: Carbon dioxide equivalent". https://ec.europa.eu/eurostat/statistics-explained/index.php?title=Glossary:Carbon_dioxide_equivalent

Gajevic Sayegh, A. 2017. "Climate justice after Paris: a normative framework". *Journal of Global Ethics* 13 (3): 344–365.

García-Portela, L. (2019). Individual Compensatory Duties for Historical Emissions and the Dead-Polluters Objection. *Journal of Agricultural and Environmental Ethics* 32 (4):591-609.

Gardiner, S. and C. McKinnon. 2020. "The Justice and Legitimacy of Geoengineering". *Critical Review of International Social and Political Philosophy* 23 (5): 557–563.

Gardiner, S. M. 2011. *A Perfect Moral Storm: The Ethical Tragedy of Climate Change*. Oxford/New York: Oxford University Press.

González-Ricoy, I. and Axel Gosseries (eds.). 2016. *Institutions For Future Generations*. Oxford: Oxford University Press.

Gosseries, A. 2004. "Historical Emissions and Free-Riding". *Ethical Perspectives* 11 (1): 36-60. http://dx.doi.org/10.2143/EP.11.1.504779

Grasso, M. 2016. "The Political Feasibility of Consumption-Based Carbon Accounting". *New Political Economy* 21 (4): 401–413.

Grasso, M. 2022. *From Big Oil to Big Green: Holding the Oil Industry to Account for the Climate Crisis*. Cambridge (MA): MIT Press.

Hausfather, Z. 2021a. "Will global warming 'stop' as soon as net-zero emissions are reached?". *Carbon Brief*, April 29. https://www.carbonbrief.org/explainer-will-global-warming-stop-as-soon-as-net-zero-emissions-are-reached

Hausfather, Z. 2021b. "Analysis: What the new IPCC report says about when world may pass 1.5C and 2C". *Carbon Brief*, August 10. https://www.carbonbrief.org/analysis-what-the-new-ipcc-report-says-about-when-world-may-pass-1-5c-and-2c

Helm, D. 2020. *Net Zero: How We Stop Causing Climate Change*. New York: Harper Collins

IEA 2021. *Net Zero by 2050: A Roadmap for the Global Energy Sector*. https://iea.blob.core.windows.net/assets/beceb956-0dcf-4d73-89fe-1310e3046d68/NetZeroby2050-ARoadmapfortheGlobalEnergySector_CORR.pdf

IPCC. 2018. "Global Warming of 1.5°C. An IPCC Special Report on the impacts of global warming of 1.5°C above pre-industrial levels and related global greenhouse gas emission pathways, in the context of strengthening the global response to the threat of climate change, sustainable development, and efforts to eradicate poverty". https://www.ipcc.ch/sr15/

IPCC. 2021. Summary for Policymakers. In: *Climate Change 2021: The Physical Science Basis. Contribution of Working Group I to the Sixth Assessment Report of the Intergovernmental Panel on Climate Change*. Cambridge University Press. https://www.ipcc.ch/report/ar6/wg1/downloads/report/IPCC_AR6_WGI_Full_Report.pdf

Jafry, T (ed.). 2020. *Routledge Handbook of Climate Justice*. London: Routledge.

Jafry, T. 2016. Making the case for gender sensitive climate policy – lessons from South Asia/IGP. *International Journal of Climate Change Strategies and Management*, 8 (4): 559–577.

Kenner, D. 2019. *Carbon Inequality: The Role of the Richest in Climate Change*. London & New York: Routledge.

Levin, K., T. Fransen, C. Schumer and C. Davis. 2019. "What Does "Net-Zero Emissions" Mean? 8 Common Questions, Answered". *World Resources Institute*, September 17. https://www.wri.org/insights/net-zero-ghg-emissions-questions-answered

Levy Yeyati, E. and F. Filippini. 2021. "Social and economic impact of COVID-19". *Brookings Institution*, June 8. https://www.brookings.edu/research/social-and-economic-impact-of-covid-19/

Light, A. and G. Taraska. 2014. "Climate Change, Adaptation, and Climate-Ready Development Assistance". *Environmental Values* 23 (2): 129–147.

Mechler, R., C. Singh, K. Ebi, R. Djalante, A. Thomas, R. James, P. Tschakert, M. Wewerinke-Singh, T. Schinko, D. Ley, J. Nalau, L. M. Bouwer, C. Huggel, S. Huq, J. Linnerooth-Bayer, S. Surminski, P. Pinho, R. Jones, E. Boyd, A. Revi. 2020. "Loss and Damage and limits to adaptation: recent IPCC insights and implications for climate science and policy". *Sustainability Science* 15: 1245–1251.

Meyer, L. 2021. "Intergenerational Justice". In *The Stanford Encyclopedia of Philosophy* (Summer 2021 Edition), edited by Edward N. Zalta. https://plato.stanford.edu/entries/justice-intergenerational/

Moss, Jeremy & Kath, Robyn (2019). Historical Emissions and the Carbon Budget. *Journal of Applied Philosophy* 36 (2): 268–289.

Mountford, H., D. Waskow, L. Gonzalez, C. Gajjar, N. Cogswell, M. Holt, T. Fransen, M. Bergen and R. Gerholdt. 2021. "COP26: Key Outcomes From the UN Climate Talks in Glasgow". *World Resources Institute.* https://www.wri.org/insights/cop26-key-outcomes-un-climate-talks-glasgow

Mulgan, T. 2018. "Answering to Future People: Responsibility for Climate Change in a Breaking World". *Journal of Applied Philosophy* 35 (3): 532–548.

Murdock, E., and S. Noll. 2015. "Beyond Access: Integrating Food Security and Food Sovereignty Models for Justice." In *Know Your Food: Food Ethics and* Innovation, edited by Helena Rocklinsberg and Per Sandin. Wageningen (NL): Wageningen Academic Publishers.

Nolt, J. 2011. "How Harmful Are the Average American's Greenhouse Gas Emissions?". *Ethics, Policy and Environment* 14 (1): 3–10.

OECD. 2021. Climate Finance Provided and Mobilised by Developed Countries: Aggregate trends updated with 2019 data. https://www.oecd-ilibrary.org/docserver/03590fb7-en.pdf?expires=1645198703&id=id&accname=guest&checksum=E762DD5AF1986A308112011292165E6A

Otto, I. M., K. Mi Kim, N. Dubrovsky, and W. Lucht 2019. "Shift the focus from the super-poor to the super-rich". *Nature Climate Change* 9: 82–87.

Ourbak, T. and A.K. Magnan (2018). "The Paris Agreement and climate change negotiations: Small Islands, big players". *Regional Environmental Change* 18: 2201–2207.

Oxfam. 2020. *Confronting Carbon Inequality.* https://oxfamilibrary.openrepository.com/bitstream/handle/10546/621052/mb-confronting-carbon-inequality-210920-en.pdf

Page, E. A. and C. Heyward. 2017. "Compensating for Climate Change Loss and Damage". *Political Studies* 65 (2): 356–372.

Page, E.A. 2012. "Give It up for Climate Change: A Defence of the Beneficiary Pays Principle". *International Theory* 4 (2): 300–330.

Peeters, W. L. Diependaele, and S. Sterckx 2019. "Moral Disengagement and the Motivational Gap in Climate Change". *Ethical Theory and Moral Practice* 22 (2): 425–447.

Perkins, P.E. 2019. "Climate justice, gender and intersectionality". In *Routledge Handbook of Climate Justice*, edited by Tahseen Jafry, 349–358. London: Routledge.

Persson, I. 2016. "Climate Change- The Hardest Moral Challenge?". *Public Reason* 8 (1-2): 3–13.

Pidcock R, S. Yeo. 2017. "Explainer: Dealing with the 'loss and damage' caused by climate change". *Carbon Brief*, May 9. https://www.carbonbrief.org/explainer-dealing-with-the-loss-and-damage-caused-by-climate-change

Prattico, E. 2019. "Sharing the burden of climate change via climate finance and business models". In *Routledge Handbook of Climate Justice*, edited by Tahseen Jafry (pp. 195-207). London: Routledge.

Ritchie, H. 2020. "How do CO2 emissions compare when we adjust for trade?". *Our World in Data*, April 30. https://ourworldindata.org/consumption-based-co2

Robinson, M. and T. Shine. 2018. "Achieving a climate justice pathway to 1.5 °C". *Nature Climate Change* 8: 564–569

Sachs, J. 2021. "Fixing Climate Finance. Project Syndicate". November 15. https://www.project-syndicate.org/commentary/fixing-climate-finance-requires-global-rules-by-jeffrey-d-sachs-2021-11

Shankar, V. 2021. "What Developing Countries Need to Reach Net Zero". *Project Syndicate*, October 11. https://www.project-syndicate.org/commentary/emerging-economies-net-zero-by-viswanathan-shankar-2021-10

Sheridan, T. and T. Jafry. 2019. "The inter-relationship between climate finance and climate justice in the UNFCCC". In *Routledge Handbook of Climate Justice*, edited by Tahseen Jafry (pp. 165-183). London: Routledge.

Shue, H. 2014. *Climate Justice: Vulnerability and Protection*. Oxford: Oxford University Press.

Sloat, L.L., S.J. Davis, J.S. Gerber, *et al*. 2020. "Climate adaptation by crop migration". *Nature Communications* 11 (1243): 1-9.

Steuteville, R. 20201. "Street trees: A wonder of climate adaptation". *Public Square: A CNU Journal*, October 19. https://www.cnu.org/publicsquare/2021/10/19/street-trees-wonder-climate-adaptation

Swiss Re Institute 2021. *The economics of climate change: no action not an option*. https://www.swissre.com/institute/research/topics-and-risk-dialogues/climate-and-natural-catastrophe-risk/expertise-publication-economics-of-climate-change.html

Taylor, M. and J. Watts. 2019. "Revealed: the 20 firms behind a third of all carbon emissions". *The Guardian*, October 9. https://www.theguardian.com/environment/2019/oct/09/revealed-20-firms-third-carbon-emissions

Timperley, J. 2021. "How To Fix the Broken Promises of Climate Finance". *Nature* 598: 400-402.

Tschakert, P. 2015. "1.5°C or 2°C: a conduit's view from the science-policy interface at COP20 in Lima, Peru". *Climate Change Responses* 2 (3): 1-11.

UNCTAD. 2021. "Scaling up climate adaptation finance must be on the table at UN COP26". October 28. https://unctad.org/news/scaling-climate-adaptation-finance-must-be-table-un-cop26

UNEP 2021a. "5 ways to make buildings climate change resilient". July, 7. https://www.unep.org/news-and-stories/story/5-ways-make-buildings-climate-change-resilient

UNEP 2021b. "Addendum to the Emissions Gap Report 2021". https://wedocs.unep.org/bitstream/handle/20.500.11822/37350/AddEGR21.pdf

UNFCCC. 2009. Copenhagen Accord. https://unfccc.int/resource/docs/2009/cop15/eng/11a01.pdf#page=4

UNFCCC. 2010. Cancun Agreements. https://unfccc.int/resource/docs/2010/cop16/eng/07a01.pdfUNFCCC. 2015. Paris Agreement. https://unfccc.int/sites/default/files/english_paris_agreement.pdf

Welsby, D., J. Price, S. Pye. et al. (2021). "Unextractable fossil fuels in a 1.5°C world". *Nature* 597: 230–234.

SECTION ONE

HISTORY, DIPLOMACY AND SCIENTIFIC EVIDENCE OF CLIMATE CHANGE

1

Climate Change: Scientific Evidence and Projected Warming

ROBERTO BUIZZA

Climate change is happening, and its impacts are increasingly felt by all communities, especially the vulnerable ones that do not have enough resources to adapt to the changes. Global warming, sea-ice melting and sea-level rise have all been accelerating, and with them the frequency and intensity of extreme weather events. Human activities bear most of the responsibility for what we have been observing since the Industrial Revolution. Greenhouse gas emissions must be cut dramatically by about 5-8% every year starting now if we want to achieve net-zero emissions by 2040–2050, and limit global average warming to below 2°C above the pre-industrial level as agreed in Paris in 2015. Failure to act swiftly and effectively will cause even more severe impacts, as indicated by the climate projections generated using state-of-the-art Earth-system models. This chapter, firstly, presents the evidence for climate change and explains the role of human beings in causing it. Secondly, it explains the concept of an Earth-system model and discusses its role as a tool for understanding the past climate evolution and generating projections of the future climate. Thirdly, it discusses the per capita greenhouse gas emissions of different countries and economic sectors. Finally, it presents climate projections to illustrate how Earth's climate could evolve under different emissions scenarios.

1. Introduction

The aim of this chapter is to provide a solid, objective and quantitative background to climate change. In Section 2 we will review the greenhouse effect, a key phenomenon that has made Earth a planet where complex lifeforms could evolve, and what are the main greenhouse gases. Understanding the greenhouse effect, the main gases that cause it and its role in the development of life on Earth is essential to be able to recognize what has been happening to Earth's climate in the 100 years following the Industrial Revolution. Sections 3 and 4 analyse the observed trends in greenhouse gas emissions over time and discuss the observed climate variations and their impacts. This first part of the chapter will provide a sound understanding of what has been happening to the Earth climate. Section 5 explains what an Earth-system model is and how it can be used to estimate future climate scenarios and their probabilities of occurrence. It discusses how these scenarios are generated and the role that emission assumptions play in determining the future climate. Section 6 presents an analysis of which countries have contributed mostly to the stock of greenhouse gases, and whether there is a clear link between greenhouse gas accumulation in the atmosphere and global warming. Section 7 discusses the quasi-linear relationship between accumulated emissions and average global warming. It also discusses how we must constrain the emissions if we want to limit the global warming below 1.5°C or 2°C degrees. Finally, Section 8 summarizes the key points discussed in this chapter, and discusses the global warming that we could experience in 2050 under four possible emissions scenarios: one with a continuous future increase of greenhouse gas emissions of 1% per year, and three with a continuous future decrease of greenhouse gas emissions by 1%, 3% or 5%.

2. The Greenhouse Effect

The Earth's temperature would be about 30°C colder if not for the atmosphere's greenhouse effect. The problem we face today is that too strong a greenhouse effect has been warming the planet in an unprecedented way. And with the warming, we have been experiencing an increased frequency and intensity of extreme weather events and sea-level rise, which have had negative impacts on tens of millions of people who do not have always have sufficient resources to adapt to climate change.

Before discussing climate change and the actions we can take to control it, let us briefly review the greenhouse effect by considering the Earth as a very simple system in equilibrium.

The Earth is a system that receives heat in the form of solar radiation and emits heat as a black body. If we apply the first law of thermodynamics (which says that energy must be conserved) to this simple model of the Earth, then there must be an equilibrium between the absorbed and the emitted energy.

The Earth, as any physical 'body' that has a temperature above absolute zero (−273.15°C), emits radiation with a certain characteristic (frequency, wavelength, spectrum). The sun, a much warmer body than the Earth, also emits radiation. We can apply the Stefan-Boltzmann law to compute the energy emitted per square meter by a black-body with a temperature *T*. Here, the term 'black body' means that we assume that the Earth and the sun behave as idealized, opaque and non-reflective bodies. The Stefan-Boltzmann law relates the emitted radiation per square meter to the black body's temperature:

(1) $E_{BB} = \sigma T^4$

where $\sigma = 5.67 \cdot 10^{-8} Wm^{-2}K^{-4}$.

For the sun, a star whose photosphere has a temperature of about 5,796°K, the amount of energy that each square meter of the photosphere (Figure 1) emits is:

(2) $E_{photo} = 5.67 \cdot 10^{-8} \cdot (5,796)^4 = 6.4 \cdot 10^7 Wm^{-2}$

From this value, we can calculate the energy emitted by the sun, in other words its luminosity L_0, which is given by the energy emitted per square meter E_{photo} multiplied by the surface the photosphere. Given that the photosphere radius is about 69,000 km (i.e., $r_{photo} = 6.96 \cdot 10^8 m$), we have:

(3)
$$L_0 = 4\pi r_{photo}^2 E_{photo} = 4 \cdot 3.14 \cdot (6.96 \cdot 10^8)^2 \cdot 6.4 \cdot 10^7 = 3.9 \cdot 10^{26} W$$

Equation (2) gives us the amount of energy emitted by a square meter of the photosphere, of a sphere with a radius of 69,000 km from the centre of the sun. We can now apply equations (2) and (3) to compute the amount of energy that crosses a square meter at the average distance of the Earth from the sun – about 150 million km ($r_d = 1.496 \cdot 10^{11} m$):

(4) $E_d = \dfrac{L_0}{4\pi r_d^2} = \dfrac{3.9 \cdot 10^{26}}{4 \cdot 3.14 \cdot (1.496 \cdot 10^{11})^2} = 1385\ Wm^{-2}$

The Earth's surface receiving the radiation from the sun has an area equal to a circle with the Earth radius, $\forall = 4.5_s^6 A = \pi r_e^2$ (Figure 2). If we multiply this area by the amount of radiation that hits a square meter at the average distance of the Earth from the sun, E_d, we can compute the amount of radiation that is absorbed by the Earth:

(5)   *absorbed solar radiation* $= \pi r_e^2 E_d$

To be more precise, we should also consider the fact that some solar radiation is reflected by the Earth's surface. This reflection depends on the surface characteristics, and, e.g., is stronger over surfaces covered by snow or sea-ice. On average, about 30% of all incoming solar radiation is reflected back into space. We can take this into account by multiplying the incoming solar radiation by $(1 - a_e)$, where a_e=0.3 is called the Earth albedo.

We can now compute the temperature of the Earth, assuming that it behaves as a black body with a temperature T_e, and assuming that the amount of energy emitted equals the amount of absorbed solar energy. In other words, we assume that it is in equilibrium and thus satisfies the following equation:

(6)   *absorbed solar radiation = emitted black body radiation*

where the emitted radiation is equal to the black-body radiation emitted per square meter times the surface the Earth $4\pi r_e^2$.

If we solve eq. (6), written as:

(7) $\pi r_e^2 (1 - a_e) E_d = \sigma T_e^4 4\pi r_e^2$

we can compute the equilibrium temperature of the Earth:

(8) $T_e = \sqrt[4]{\dfrac{(1-a_e)E_d}{4\sigma}} = \sqrt[4]{\dfrac{0.7 \cdot 1385}{4 \cdot 5.67 \cdot 10^{-8}}} = 255.7°K$

This result indicates that 255.7°K would be the temperature if the Earth behaved as a black body and if there was no atmosphere around it.

Let us now introduce a simple atmosphere around the Earth, characterized by a temperature T_A that allows solar radiation to cross it without being absorbed, but which absorbs the radiation emitted by the Earth (Figure 1.3). An atmosphere that acts as the real atmosphere allows most solar radiation, which has a short wavelength, to reach the Earth's surface and absorbs most of the long-wavelength radiation emitted by Earth. In other words, we assume the atmosphere can exert a greenhouse effect. This simple atmospheric layer also acts as a black body, and it emits (black body) radiation both towards the free space and towards the Earth surface's, depending on its temperature T_A.

We can write two energy balance equations, one for the atmosphere and one for the Earth's surface, that impose an equilibrium between the absorbed and the emitted radiation:

(9a)   *atmosphere:* $2\sigma T_A^4 = \sigma T_e^4$

(9b)   *surface*   : $\frac{1}{4}(1 - a_e)E_d + \sigma T_A^4 = \sigma T_e^4$

If we solve these two equations, we find that:

(10a) $T_A = \sqrt[4]{\frac{(1-a_e)E_d}{4\sigma}} = 255.7°K$

(10b) $T_e = \sqrt[4]{2}T_A = 304.1°K$

Thus, thanks to the presence of the atmosphere, which absorbs most of the long-wave radiation emitted by Earth's surface, and re-emits radiation as a black body, the Earth has an average surface temperature of 304°K, which is about 31°C. This temperature (31°C) is about 50°C higher than the temperature the Earth would have without a greenhouse effect.

We should point out that 31°C is higher than the observed average surface temperature, which is about 15°C. This difference is due to the fact that we have assumed that the Earth behaves as a simple system with a very simple atmosphere. In other words, we have over-simplified the Earth and did not consider the complexity of the Earth system. Thus, we should not be surprised if the completed and observed values are different.

Despite this, this simple model highlights the important role that an atmosphere has on the Earth's surface temperature. To make the model slightly more complex, we could assume that the atmosphere is compounded by few different layers instead of only one, with each layer characterized by a different temperature and find a solution for the Earth surface's temperature that is closer to the observed one.

The atmosphere allows most short-wave solar radiation to reach Earth's surface, while the amount of the long-wave radiation emitted by the Earth that it absorbs depends on its chemical composition. The principal greenhouse gases are water vapour, carbon dioxide, methane, nitrous oxide and fluorinated gases (Table 1). Note that their concentrations are very small, ranging from about 0.3% (in mass fraction with respect to dry air) for water vapour to much smaller values for carbon dioxide and methane.

There are two key differences between water vapour and the other gases. First, for any air mass, the water vapour concentration strongly depends on the pressure and the temperature of the air. If a mass of air that contains water vapour is cooled and/or compressed, water vapour would condense and precipitate; water continues to cycle, and as a consequence, the total concentration of water vapour in the atmosphere changes very little. Indeed, the time that a molecule of H_2O spends in the atmosphere is about 10 days. In the last 100 years, human beings have increased the concentration of carbon dioxide and methane in the atmosphere by more than 50%, from about 270 parts per million (ppm) at the end of the nineteenth century, to about 415 ppm today.

A further, important point to make is that the time that a molecule of CO_2 spends in the atmosphere is between 300 and 1,000 years, and the time that a molecule of CH_4 spends in the atmosphere is about 10 years. This means that greenhouse gases continue to exercise their effect long after they have been injected in the atmosphere. Because of these two differences, the continuous injection of large quantities of greenhouse gases in the atmosphere has and will for many tens and hundreds of years influence the Earth's climate.

3. Observed Trends in Greenhouse Gases

The concentration of the greenhouse gases has been increasing since the start of the Industrial Revolution. Figure 1.4 shows the annual mean concentrations of CO_2 and CH_4, computed using data collected at the Mauna Loa Observatory (Hawaii, United States).

Let us start by looking at CO_2. The atmospheric concentration of CO_2 has exceeded 400 ppm in 2015. It is worth stressing that the Earth has not seen such a high concentration in the past 2.5 million years (see, e.g., Wallace and Hobbs 2006; Hartmann 2016; IPCC 2021). Note also that the increase follows a roughly exponential curve, and that the annual percent increase has been growing nearly continuously. In fact, as shown in Figure 1.5, while in the 1960s and 1970s the annual percent increase in CO_2 was between 0.2-0.4%, in the last two decades it has been above 0.5%. Figure 1.5 also shows that the last two decades have seen a clear positive trend, with the growth rate increasing by about 0.01 every year, from about 0.47% in 2000 to about 0.65% in 2020.

CH_4 concentrations have also been increasing from about 1,645 ppb in 1985 (data before 1985 were not available) to 1,879 ppb in 2020 (Figure 1.4). Figure 1.5 shows that the annual growth rate decreased between 1985 and 2000, but since then values have been growing at a fast rate. The last two decades have seen a clear positive trend, with the growth rate increasing by about 0.03 every year, from about 0% in 2000 to 0.60% in 2020.

4. Observed Climate Change

As the concentration of the greenhouse gases increased, the atmosphere has been absorbing more long-wave radiation emitted by the Earth, and thus has been warming. A warmer atmosphere has been emitting more long-wave radiation towards the Earth's surface, which thus has also been warming. This cause-and-effect mechanism between increasing emissions of greenhouse gases and the global average temperature is a direct consequence of the greenhouse effect that was discussed in Section 1.

Figure 1.6 shows the anomaly of the global annual mean temperature with respect to the pre-industrial value. In other words, for each year, the figure shows the difference between the global annual mean temperature of that year and the global annual-mean temperature of the period 1850-1900. The solid line shows the annual values, while the straight line (which has a slope of ~0.02°C/year) shows the long-term warming trend. Note that superimposed over this linear warming trend of ~0.2°C per decade are natural oscillations of about 0.1-0.2°C. These natural oscillations are due to natural variability, for example linked to large volcanic eruptions or heat exchanged between the Earth's oceans and atmosphere. In this latter case, the occurrence of large-scale episodes in the tropical Pacific Ocean that causes the ocean temperature to warm (during the El Niño event) or to cool (during La Niña events) produce are natural oscillations in temperature.

Figure 1.6 is based on a new dataset produced by the European Union Copernicus Climate Change Service, the ERA-5 reanalysis (Hersbach et al. 2020), constructed by assimilating all available observations of the Earth system in a state-of-the-art model at the European Centre for Medium-Range Weather Forecasts (ECMWF). It covers the satellite era – the period from 1980 through the present, during which satellite data has allowed scientists to monitor the Earth's temperature accurately.

Figure 1.6 shows that in 2020 the global average temperature was about 1.2°C warmer than the pre-Industrial level, and that the six years from 2015 to 2020 have been the 6 warmest years since 1980.

This warming is not the only evidence that the climate is changing, as was very clearly summarized by the report from the Intergovernmental Panel on Climate Change (IPCC) published in August 2021. The Summary for Policy Makers (SPM) written by IPCC Working Group I confirms that many of the changes observed in the climate are unprecedented in thousands if not hundreds of thousands of years and that some of the changes already set in motion will continue to affect the climate over hundreds to thousands of years in the future (IPCC 2021). It also states, however, that strong and sustained reductions in emissions of CO_2 and other greenhouse gases could limit climate change.

The SPM provides updated estimates of the chances of crossing a global warming level of 1.5°C and 2°C (with respect to the pre-Industrial level) in the next decades. These two levels, 1.5°C and 2°C, were agreed by the 196 countries that signed the Paris Agreement in 2015 (UNFCCC 2015) as warming levels that should not be overcome. The SPM confirms that unless there are immediate and large-scale reductions in greenhouse gas emissions, limiting warming to 2°C will be beyond reach. It also confirms that the emissions of greenhouse gases from human activities are responsible for approximately 1.1°C of warming since 1850-1900, and reports that global temperature could reach a warming of between 2°C and 5°C unless immediate and concrete measures are taken to reduce emissions.

These conclusions are not new. They confirm the results published in the IPCC Fifth Assessment Report (AR5) and mentioned in earlier IPCC reports. What is new is that the SPM report of August 2021 is that it includes seven additional years of evidence (AR5 was published in 2014), it reports climate projections based on more realistic and higher resolution models, and it discusses results obtained using new techniques that merge different observations and allow scientists to assess the realism of Earth system models in a more sophisticated way. Furthermore, the report includes results from analyses that combine multi-model projections and thus provide more reliable uncertainty estimates.

It is worth summarizing some of the key impacts of climate change reported in the SPM:

- Climate change is intensifying the water cycle: this brings more intense rainfall and associated flooding, as well as more intense drought in many regions (examples are the floods that affected Central Europe in July 2021, Bangladesh and Australia in June/July 2022, and the forest fires linked to heat waves and droughts that have been affecting many countries during the past two summers of 2021 and 2022).
- Climate change is affecting rainfall patterns: at high latitudes, precipitation is likely to increase, while it is projected to decrease in the Tropics and the Mediterranean region.
- Climate change-induced sea-level rise has accelerated in the past decade, with sea-level rise rates reaching ~3.4mm/year: extreme events related to sea-level rise that previously occurred once every 100 years could happen every year by the end of this century.
- Climate change-induced warming will amplify permafrost thawing, the loss of seasonal snow cover, melting of glaciers (see, e.g., what has been happening in the Alps, including the disaster of the Marmolada glacier in July 2022) and ice sheets and the loss of Arctic Sea ice, which is projected to be ice-free in summer before the end of the century.

- Climate change led to ocean warming, ocean acidification and reduced oxygen levels that have affected ocean ecosystems.
- For cities, some aspects of climate change may be amplified, including heat waves (since urban areas are usually warmer than their surroundings), flooding from heavy precipitation events and sea-level rise in coastal cities.

The SPM also states that without immediate, drastic actions aimed at reducing greenhouse gas emissions, average global warming could reach more than 2.5°C by the end of the century, thus overtaking both warming levels that the 196 countries that signed the Paris Agreement in 2015 (UNFCCC 2015) agreed must not be overcome.

Let us now move to discuss, in the next sections, three key questions that will allow us to understand how we got here and by how much we have to reduce greenhouse gas emissions to limit warming to below 2°C.

- How can we estimate the warming levels we could reach in 10, 20, and 50 years?
- If we look back at the past, which countries have contributed most to greenhouse gas emissions?
- Is there a straightforward link between greenhouse gas emissions and average global warming?

5. Earth-System Models for Climate Studies

Climate projections, such as the ones reported in the IPCC assessment reports, are generated using state-of-the-art models of the Earth system. These models are designed to simulate as accurately as possible all the key processes that determine Earth's climate. In particular, they aim to simulate the interactions between the atmosphere, the oceans, the cryosphere and land processes. They are based on the laws of physics, such as the conservation of mass and energy and the fact that if a force is applied to a mass of air, it will change its velocity and position. They simulate how clouds are formed, move and cause precipitation. They simulate the interaction between the short-wave solar radiation and the long-wave radiation emitted by the Earth and masses of air in the atmosphere to determine how much energy reaches the Earth surface, how much is absorbed or reflected by the atmosphere. They simulate how air flows pass mountains and how ocean waves affect the presence and/or shape of sea ice.

These models see the Earth as a three-dimensional mesh of points with one point every few tens of kilometres on two horizontal axes and every few hundred metres on a vertical axis (see, e.g., Flato 2011, Gettelmann and Rodd 2016). In total, a model with one grid point every 10 km and with vertical levels every 500 m will have about 10^9 grid points. At each grid point, the state of the Earth is represented by a set of variables, the so-called state variables. For the atmosphere, for example, these variables are temperature, pressure, wind, specific humidity and cloud types. At the land surface, these variables include information about the type of vegetation and soil, soil temperature and humidity and snow cover. In general, the state variables are chosen so that they are the smallest number of variables that, if they are known, are sufficient to describe in a realistic way the state of the Earth system (e.g., whether it will rain or not, whether intense convective systems will develop or not and whether tropical storms could develop into hurricanes or not). The model aims to represent the state of the Earth system at each grid point at fixed time intervals, say about every 10 minutes.

These Earth system models can be used to predict how the state of the Earth system changes over time, from a few days ahead to years and even decades. The same models are used daily to predict the weather over the next days, weeks and months. To be able to predict the future states of the Earth system, the first step is to know as accurately as possible the state of the system now. This can be estimated by collecting as many observations as possible and by merging them in an objective way that considers their relative accuracy.

To predict the future states of the Earth system, these models, or more precisely the equations that describe how the different model variables interact and change over time, need to be solved. Since the equations are complex and the number of grid points onto which they need to be solved is huge (we said it is about one billion, i.e., 10^9), they can be solved only numerically, using very fast super-computers. The faster the super-computers, the finer the mesh of the model can be, and the more detailed the description of the physical processes is.

In the case of weather prediction, for example, every day more than 600 million observations are collected (95% using instruments on satellites) and about 100 million are assimilated to estimate the current state of the system. It takes between 2–4 hours for all observations to be received; then, the assimilation process (which uses an Earth system model) takes about one hour if one dedicates to it about half of the existing fastest super-computers.

Once the state of the atmosphere has been estimated, the model is integrated numerically on the super-computer and, in about one more hour (again, if one dedicates to this computation about half one of the fastest super-computers), one can generate a forecast valid for the next 10 days.

Thus, if one has access to one of the fastest super-computers, in about two hours since all observations have been received (or about six hours since the observations have been collected), one could know the state of the Earth system over the whole globe and know how it will change in the next 10 days.

One of the main differences between making a weather forecast and predicting the climate for the next 10, 20 or 50 years, is the fact that the model has to be integrated numerically for a much longer time. For example, if one wants to generate a climate prediction for the next 50 years with the same model used to make a weather forecast (with a mesh with one grid point every 10 km on the horizontal and every 500 metres on the vertical), one will need about 3 months if they could dedicate to this task half of one of the fastest super-computers. The problem is that although one could dedicate half a super-computer for one hour to the generation of a weather forecast, it is practically impossible to dedicate half of a super-computer for one climate prediction. Thus, more realistically, one should multiply this estimate of 3 months by a factor between 2 and 5, which means that generating a climate prediction for the next 50 years with a mesh with one grid point every 10 km on the horizontal and every 500 metres on the vertical, would take between 0.5 and 2.5 years.

To reduce the time required to complete one climate prediction, one could reduce the resolution from 10 to 50 km on the horizontal, and from 500 to 1,000 metres on the vertical. This reduces the number of grid points by a factor of 50 (5x5x2) and thus makes the whole computation faster by at least that factor, which means that one could generate a climate prediction for the next 50 years in a few weeks. The negative impact of using a reduced resolution is that a climate prediction, compared to a weather prediction, contains less details. Losing these details can have an impact on the capability of a climate prediction to inform us on whether extreme and small-scale events, such as windstorms and flash floods, could become more intense or more frequent in the future.

Despite the fact that Earth system models, when used to make climate predictions, are today characterized by a resolution of about 50 km on the horizontal and 1,000 metres on the vertical, they are still the best tools we have to understand how the climate could change, and what the probabilities that different climate scenarios will occur in the future are. If the models are realistic and simulate future scenarios reliably, they can provide better information than statistics based on the past events, i.e., on the climate of the past decades. This is why many efforts are made by the scientific community to make them increasingly more realistic by including more relevant processes and by improving the simulation of the processes they already simulate.

A second key difference between weather and climate predictions is due to the fact that the climate is very sensitivity to changes in the concentration of the greenhouse gases. Thus, to generate a prediction, one has to assume how the concentration of greenhouse gases will change in the future. In other words, the climate predictions are 'driven' by emissions scenarios: high emissions will induce an acceleration of the greenhouse effect and thus an increased warming, while low emissions will slow down the warming rate.

The SPM (IPCC 2021) discusses five possible future climate scenarios, defined by considering a range of possible future developments in anthropogenic drivers of climate change found in the scientific literature. Emissions vary between scenarios depending on socioeconomic assumptions, population growth, levels of climate change mitigation and, for aerosols and non-methane ozone precursors, air pollution controls. Climate projections indicate that, depending on the emissions, the global mean temperature anomaly in 2050 could stay below 1.5°C degrees or increase up to 2.5°C degrees (see IPCC 2021, Figure SPM.8).

The result of these efforts is better knowledge of how the system could behave in the future. For example, the SPM report talks explicitly about the probability that tipping points could be reached and irreversible changes to the Earth system could occur in the future. Tipping points, such as the ones linked to the melting of the Greenland ice sheets or to the deforestation of the Amazon Rainforest, can trigger an acceleration of global warming and are now mentioned more commonly because the scientific community has confidence that the Earth system models used today are capable of simulating how these key Earth system components could realistically evolve in the future.

Let us consider the first question we posed in Section 3:

• How do we know what level of warming we will reach in 10, 20, and 50 years?

The answer is: by integrating state-of-the-art Earth system models and using them to predict possible future scenarios that could occur as a function of assumed greenhouse gas emissions based on an estimated probability that each scenario could occur.

6. Greenhouse Gas Emissions Per Capita Between 1990–2016

Let us now address the second question that was posed in Section 3:

- If we look back at the past, which countries have contributed mostly to greenhouse gas emissions?

We will answer this question by looking at the CO_2 emissions data available from the World Bank between 1990 (a year that is often chosen as the reference with respect to which emission reduction targets are set) and 2016 (the last data point available in the World Bank data archive).

We often read that China is the major emitter of CO_2, and that Europe contributes to only about 9% of the total emissions. When we read these sentences, often it is not clear whether they refer to the total emissions of a country or a group of countries or to per capita emissions, and it is also not obvious whether they refer to a specific year or to emissions accumulated over a long time period.

Figure 1.7 shows both the total and per capita CO_2 emissions the top seven emitters accumulated between 1990–2016 (1990 was chosen since emission reduction targets are often based on a 1990 baseline; 2016 has been chosen since it is the last data point available in the World Bank data). The per capita values have been computed by dividing total emissions by total mid-year population.

In terms of total emissions (Figure 7, left panel), these seven countries (or groups of countries) emitted 68% of the world CO_2 emissions during the 1990–2016 period. China and the US contributed the most: 21% and 20% of the CO_2 emissions injected into the atmosphere during those 26 years came from these two countries. They are followed by the European Union, which injected about 12% of the world's total CO_2 emissions. The other major emitters followed, with contributions ranging from about 1% (Australia) to about 7% (Russian Federation).

By comparing these values, it should be evident that it is difficult to compare total emissions of countries with different populations, and that one should also consider per capita emissions. It should be noted that emissions per capita are a more 'just' measure of emissions since they reflect better whether each person has been given access to the same amount of energy.

The right panel of Figure 1.7 shows the accumulated per capita emissions, for the same period, 1990–2016. Note that now there is a wider difference between the countries. With respect to this measure, one person living in the US has emitted 500 tonnes of CO_2 over those 26 years, compared to about 120 tonnes for a person living in China. Australia, Canada and Russia now rank second, third and fourth among the major emitters. Note that an average person living in India has emitted only 30 tonnes of CO_2 in 26 years, about the same amount that persons living in the top four polluters injected in the atmosphere in just two years.

It is interesting to compare these numbers with the world average CO_2 emissions per capita – 115 tonnes: note that a person living in the top four polluters emitted about four times more than the world average, a person living in the European Union about two times more, a person living in China the same as the average and a person living in India four times less.

As economies developed and transformed throughout the years, the emission ranking changed. Figure 1.8 gives a snapshot of the most recent year, 2016, for which the data are available in the World Bank database. It shows that, in 2016 and in terms of total emissions, China is the top contributor with a contribution of about 29% of the world emissions, followed by the US with about 15% and the European Union with about 8.5%. But if we look at emissions per capita, the US, Canada and Australia remains the top polluters, with emissions per capita of about 15 tonnes in 2016, compared to about seven tonnes for a person living in China, 6.5 tonnes for a person living in the European Union and 1.8 tonnes for a person living in India. In 2016, the world average CO_2 emissions per capita was about 4.5 tonnes. Thus, it is evident that there are still countries that emit about four times more than the average and countries that emit about four times less.

Let us consider the second question posed in Section 3:

- If we look back at the past, which countries have contributed most to greenhouse gas emissions?

The answer depends on whether we use total or per capita emissions. China, the US and the European Union have contributed the most if we look at total emissions accumulated between 1990–2016. By contrast, if we look at per capita emissions accumulated between 1990–2016, the top emitters are the US, Australia, Canada and Russia, which emitted twice as much as an average person living in the European Union and four times as much as a person living in China.

7. Link between Greenhouse Gases and Global Warming

Let us know address the third of the three questions posed in Section 3:

- Is there a straightforward link between greenhouse gas emissions and average global warming?

The latest IPCC report (IPCC 2021) talks about a quasi-linear relationship between the amount of greenhouse gases that are released into the atmosphere and global warming. Indeed, if we contrast global warming, measured by the global annual average surface temperature anomaly from 1980 until the present (Figure 1.6) against the greenhouse gases accumulated since 1980 (Figure 1.9), we can detect a quasi-linear trend.

The fitted straight line shown in Figure 1.9 has a slope of 0.5°C/1,000 Gt, which means that during this period, each additional 1,000 Gt of greenhouse gases in the atmosphere has led to 0.5°C of warming on average. Note that the coefficient of determination of the linear fit, $R^2 = 0.82$, confirms the existence of a robust linear relationship. Note that there are variations around this linear relationship, reflecting the fact that each year's climate is influenced not only by greenhouse gas concentrations, but also by atmosphere and ocean dynamics (e.g., whether the year was characterized by a strong El Niño or La Niña event and whether other changes in the large-scale circulation caused heat waves over large areas of the globe).

The SPM talks about a slope of about 0.45°C/1,000 Gt, which is close to the relationship we found. We can use this quasi-liner relationship to estimate how much we could emit to limit warming.

For example, given that in 2018 the average global warming was 1.1°C above pre-Industrial levels, we can only emit about 2,000 Gt more of greenhouse gases if we want to keep warming below 2°C. In 2019, we emitted about 45 Gt of greenhouse gases (of which ~30 Gt was CO_2 and ~15 Gt was CH_4 and other greenhouse gases). If we continue to emit, on average, about 45 Gt per year, in about 45 years we would reach that amount.

If instead we want to keep global warming below 1.5°C, we have to limit emissions to less than 900 Gt. If we continue to emit as we did in 2019, i.e., about 45 Gt/year, we will surpass that value by 2038.

Let us consider the third of the three questions posed in Section 3:

- Is there a straightforward link between greenhouse gas emissions and average global warming?

The answer is yes: since there is a quasi-linear relationship between accumulated emissions and global average warming, we can estimate in a straightforward way how the climate will look like in the future.

8. Concluding Remarks

We started our discussion in Section 1 by examining the greenhouse effect and the key role it has played in the development of complex lifeforms on planet Earth.

Then, in Section 2, we discussed the observed, continued acceleration of increasing concentrations of greenhouse gases in the atmosphere. This acceleration is mainly due to human activities, as stated in the latest SPM report (IPCC 2021, see also Fig. SPM.1): 'It is unequivocal that human influence has warmed the atmosphere, ocean and land. Widespread and rapid changes in the atmosphere, ocean, cryosphere and biosphere have occurred'.

In Section 3, we documented the impact of the continued increase in greenhouse gas concentrations on the average global surface temperature between 1980 and 2020. We also briefly discussed some uncontroversial impacts of climate change, e.g., sea-level rise, sea-ice extension (especially in the Arctic) and increasingly intense and frequent extreme events.

In Section 4, we summarized the state of the climate, and discussed what has been happening to the global surface temperature since the Industrial revolution, and the impact of global warming on sea-level rise, the melting of sea-ice and glaciers and extreme weather events.

We then posed three key questions:

- How can we estimate the warming levels we could reach in 10, 20, and 50 years?
- If we look back at the past, which countries have contributed most to greenhouse gas emissions?
- Is there a straightforward link between greenhouse gas emissions and average global warming?

And, we addressed them in the forthcoming sections.

In Section 5, we addressed the first question and found that Earth system models are the best tools we have to predict future scenarios and their probabilities of occurrence. We have illustrated what an Earth system model is, the fact that they are based on the laws of physics and how they can be integrated numerically on super-computers to provide us with reliable information about the future climate.

In Section 6, we addressed the second question and found that the answer depends on how we measure emissions and whether emissions are accumulated over a long time period (in our case, we accumulated the emission over the 26 years, between 1990–2016) or over a short time period, such as the year 2016. We also said that the answer depends on whether one compares countries' and groups-of-countries' total emissions or their per capita emissions. We showed that if one looks at per capita emissions between 1990–2016, the ranking sees the US as the top contributor to climate change, followed by Australia, Canada, the Russian Federation, the European Union and then China.

In Section 7, we discussed the third question and showed that there is a quasi-linear relationship between greenhouse gas emissions and global warming. This relationship was also pointed out explicitly in the SPM report of 2021. Looking at data from 1980 to 2020, we showed that each additional 1,000 Gt of greenhouse gas emissions led 0.5°C of warming on average. This relationship can be used to make a simple, zero-dimensional prediction of the future climate. If we apply it, considering that the world injected about 45 Gt of greenhouse gases into the atmosphere in 2019, and if we assume that emissions will continue at this level on average for the next few decades, we can predict that in about 13 years the warming will be about 1.5°C warmer on average than it was before the Industrial Revolution. We can also estimate that in about 40 years the rate of global warming will be about 2°C.

It should now be clear what has been happening to the climate, and why it is necessary to start immediately reducing emissions of greenhouse gases. If we want to keep warming below 2°C, we have to reduce the emissions substantially – by 5-8% per year on average, until we reach net-zero emissions in 2050.

We can apply this simple quasi-linear relationship between greenhouse gas emissions and average global warming to look ahead into the future. If we look at the last 10 years, annual emissions grew from 39.8 Gt in 2009 to 45.8 Gt in 2018 (data from the World Bank).

Figure 1.10 shows, with the black diamonds, total greenhouse gas emissions and projected warming in 2050 computed applying the linear relationship discussed in Section 6, under four different emission scenarios:

- A continued average increase of 1% a year, as was the case between 2009–2018: this would cause a further 1,738 Gt of greenhouse gases to be injected into the atmosphere, which would reach 3,173 Gt and a warming of 2°C in 2050;

- A continued average decrease of 1% a year, starting from 2019: this would cause a further 1,249 Gt of greenhouse gases to be injected into the atmosphere, which would reach 2,684 Gt and a warming of 1.75°C in 2050;
- A continued average decrease of 3% a year, starting from 2019: this would cause a further 9994 Gt of greenhouse gases to be injected into the atmosphere, which would reach 2,429 Gt and a warming of 1.61°C in 2050;
- A continued average decrease of 5% a year, starting from 2019: this would cause a further 703 Gt of greenhouse gases to be injected into the atmosphere, which would reach 2,138 Gt and a warming of 1.47°C in 2050.

Thus, only by decreasing the emissions on average by ≥5% every year can we limit average global warming to below 1.5°C above pre-Industrial levels. But it is worth mentioning that even in this scenario, we will still be injecting about 9 Gt of greenhouse gases into the atmosphere every year by 2050. If we want to achieve 'almost' net-zero emissions, we must reduce average emissions by ≥8% every year: an average annual reduction of 8% a year would lower the emissions from the 45 Gt emitted in 2019, to about 3 Gt in 2050.

Tables and Figures

Table 1.1. Atmospheric concentration of the principal greenhouse gases: water vapour, carbon dioxide and methane.

	Total mass (in 10^{21} g)	Mass fraction (with respect to dry air)
Total atmosphere	5.136000	
Dry air	5.119000	
Nitrogen (N_2)	3.870000	75.6007%
Oxygen (O_2)	1.185000	23.1491%
Argon (Ar)	0.065900	1.2874%
Water vapour (H_2O)	0.017000	0.3321%
Carbon dioxide (CO_2)	0.002760	0.0539%
Methane (CH_4)	0.000005	0.0001%

Figure 1.1. Schematic of the calculation of the Sun radiation that passes through a square meter at the photosphere, which has a radius r_{photo} and at the average distance r_d of the Earth from the Sun.

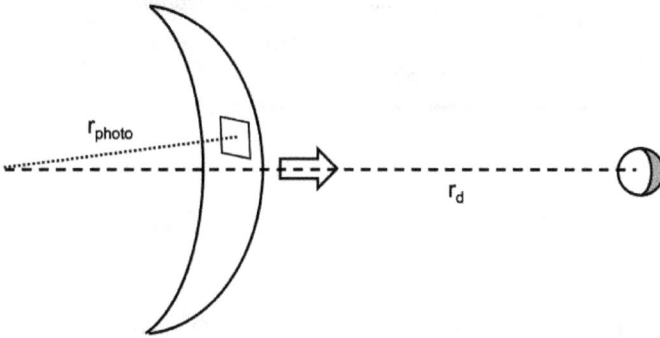

Figure 1.2. Schematic of the calculation of the sun radiation absorbed by the Earth.

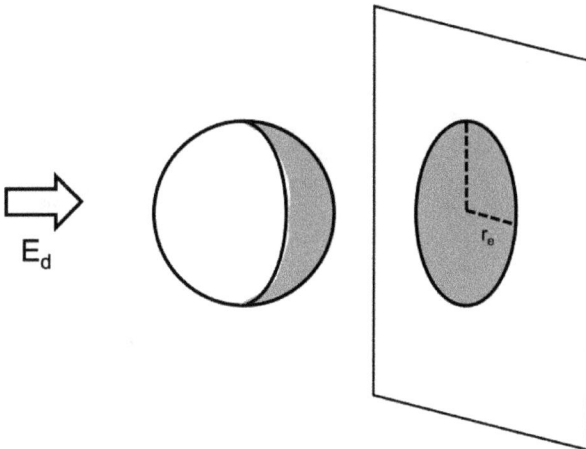

Figure 1.3. Schematic of the Earth atmosphere greenhouse effect.

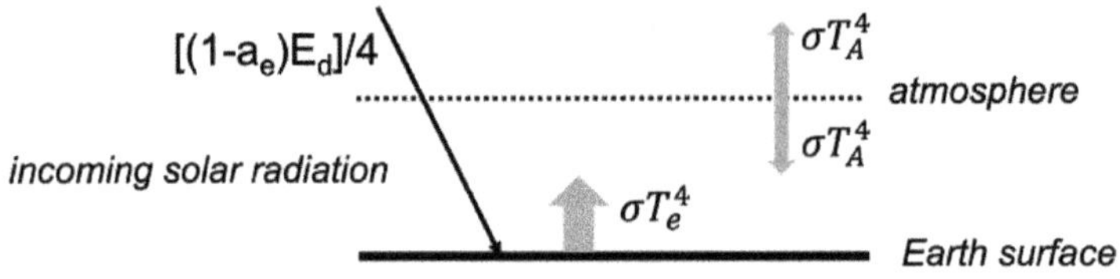

Figure 1.4. Annual mean concentration of carbon dioxide (CO_2, in parts per million, ppm; left panel) and of methane (CH_4, in parts per billion, ppb; right panel), measured at the Mauna Loa Observatory. CO_2 data from: Dr. Pieter Tans, NOAA/GML – gml.noaa.gov/ccgg/trends, and Dr. Ralph Keeling, Scripps Institution of Oceanography –scrippsco2.ucsd.edu/. CH_4 data from: Ed Dlugokencky, NOAA/GML – gml.noaa.gov/ccgg/trends_ch4/.

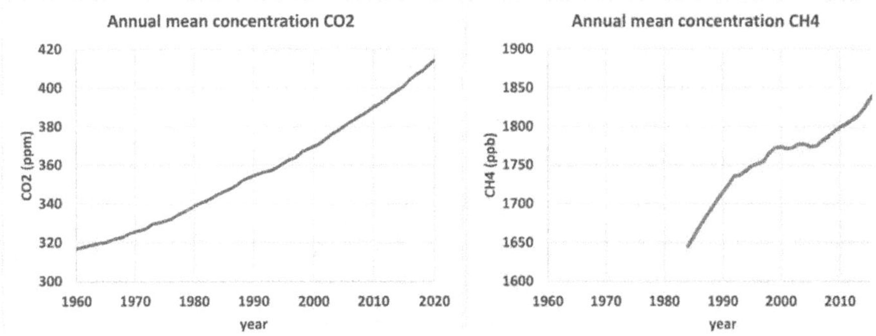

Figure 1.5. Annual percent increase in the concentration of carbon dioxide (CO_2, in parts per million, ppm; blue lines) and of methane (CH_4, in parts per billion, ppb; red lines), computed from the Mauna Loa Observatory data shown in Fig. 1.4.

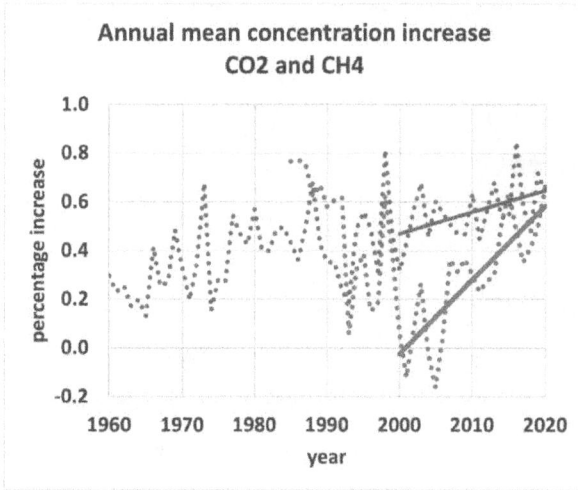

Figure 1.6. Global warming with respect to the pre-Industrial level. The solid line shows the anomaly of the global annual-mean surface-temperature with respect to the pre-industrial level. The dotted straight line shows the linear fit, which indicates a global warming trend of about 0.2°C every 10 years. Temperature data from the European Union Copernicus Climate Change Service – https://climate.copernicus.eu.

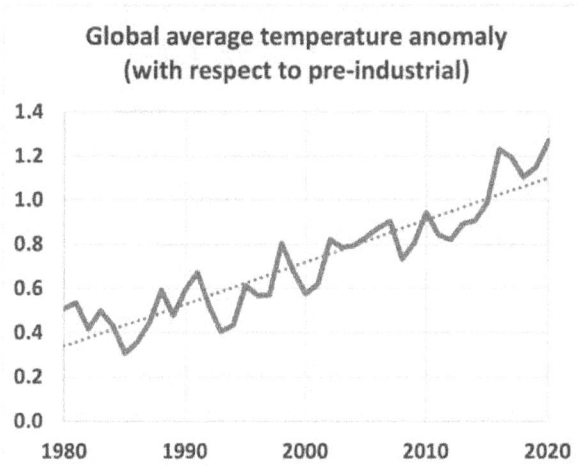

Figure 1.7. Left panel: total CO_2 emissions accumulated between 1990–2016 by the seven major emitters: United States, Russian Federation, India, European Union, China, Canada and Australia (values are expressed in gigatonnes). Right panel: per capita CO_2 emissions accumulated between 1990–2016 by the seven major emitters (in tonnes). Source: World Bank.

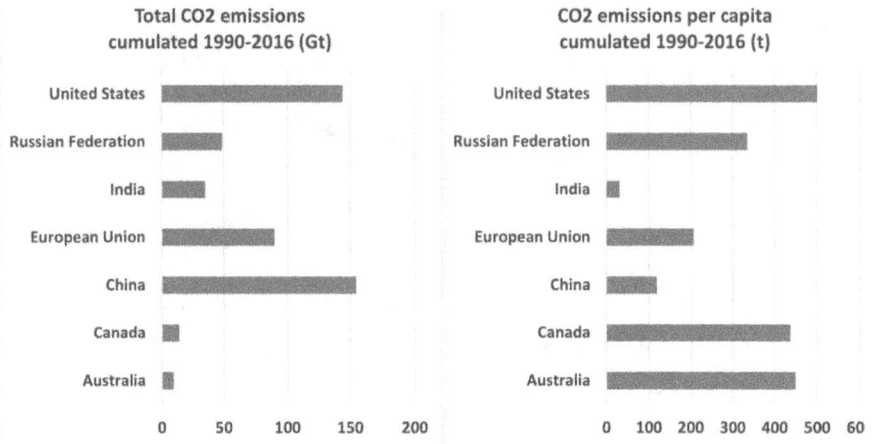

Figure 1.8. Left panel: total CO_2 emissions in 2016 by the seven major emitters: United States, Russia, India, European Union, China, Canada and Australia (values are expressed in gigatonnes). Right panel: per capita CO_2 emissions in 2016 by the seven major emitters (in tonnes). Source: World Bank.

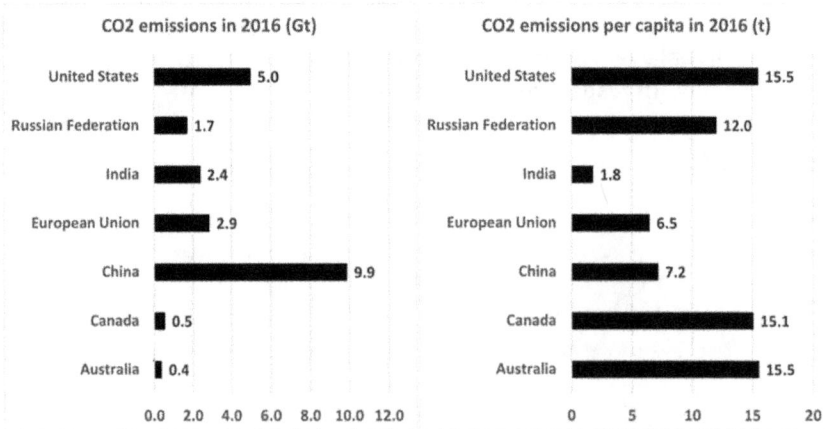

Figure 1.9. Total accumulated greenhouse gases emitted since 1980 (x-axis: data from the World Bank database) versus global annual average surface temperature anomaly with respect to the pre-Industrial level (y-axis: data from the European Union Copernicus Climate Change Service).

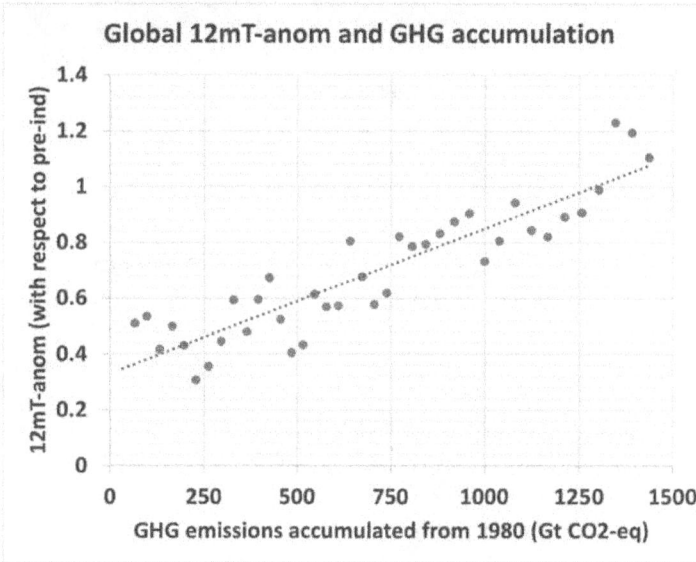

Figure 1.10. As Fig. 9 (up to accumulated emissions of about 1,500 Gt (red dots), with also projections of the state of the climate in 2050 (see text for details).

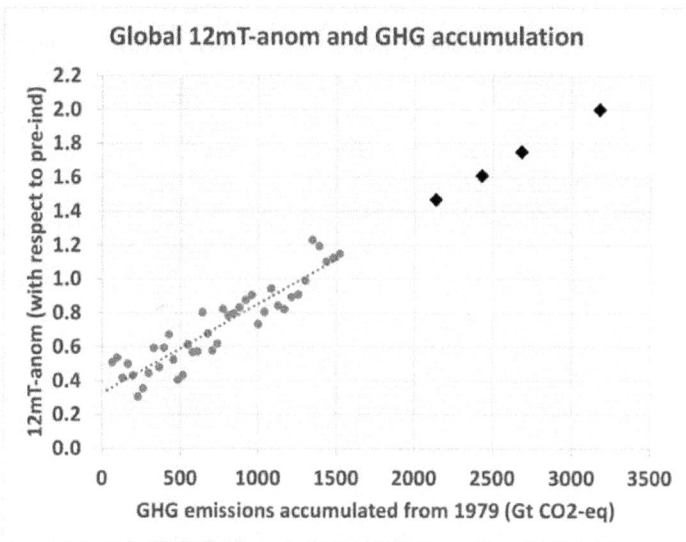

References

Flato, G. M. 2011: Earth system models: an overview. *WIRE's Climate Change*, 2: 6. pp. 783–800 https://www.doi.org/10.1002/wcc.148

Gettelmann, A. and Rodd, R.B. 2016. *Demystifying climate models. A User Guide to Earth System Models*. Berlin: Springer.

Hartmann, D. L. 2016: *Global Physical Climatology*. Second edition. Amsterdam: Elsevier.

Hersbach, H., Bell, B., Berrisford, P., Hirahara, S., Horanyi, A., Munoz-Sabater, J., Nicolas, J., Peubey, C., Radu, R., Schepers, D., Simmons, A., Soci, C., Abdalla, S., Abellan, X., Balsamo, G., Bechtond, P., Biavati, G., Bonavita, M., De Chiara, G., Dahlgren, P., Dee, D., Diamantakis, M., Dragani, R., Flemming, J., Forbes, R., Fuentes, M., Geer, A., Haimberger, L., Healy, S., Hogan, R. J., Holm, E., Janiskove, M., Keeley, S., Laloyaux, P., Lopez, P., Lupu, C., Radnoti, G., de Rosnay, P., Rozum, I., Vamborg, F., Villaume, S., Thepaut, J.-N., 2020: The ERA5 global reanalysis. *Q. J. Roy. Meteorol. Soc.*, 146, 1999–2049. https://www.doi.org/10.1002/qj.3830

IPCC, 2021: Summary for Policymakers. In: Climate Change 2021: The Physical Science Basis. Contribution of Working Group I to the Sixth Assessment Report of the Intergovernmental Panel on Climate Change [Masson- Delmotte, V., P. Zhai, A. Pirani, S.L. Connors, C. Péan, S. Berger, N. Caud, Y. Chen, L. Goldfarb, M.I. Gomis, M. Huang, K. Leitzell, E. Lonnoy, J.B.R. Matthews, T.K. Maycock, T. Waterfield, O. Yelekçi, R. Yu, and B. Zhou (eds.)]. Cambridge University Press.

United Nations Framework Convention on Climate Change (UNFCCC) 2015; The Paris Agreement. Available from the UNFCCC web site: https://unfccc.int/sites/default/files/english_paris_agreement.pdf

Wallace, J. M. and Hobbs, P. V. 2006: *Atmospheric Science – An introductory survey*. Second Edition. Amsterdam: Academic Press.

2

Climate Justice in the Global South: Understanding the Environmental Legacy of Colonialism

NISHTHA SINGH

The negative effects of climate change in developing countries cannot be examined in isolation. The present-day 'developed' global north and the 'reeling-under-its-problems' global south have their roots in history. The recent past carries significant weight in terms of climate change: on the one hand, it makes developing countries more vulnerable to climate threats and, on the other hand, it reduces the possibilities of these countries to implement both mitigation and adaptation strategies. Years of colonial exploitation have left the vast majority of the population in developing countries without basic health, education and food infrastructures. They have also led to the loss of cultures and essential techniques that people used to live in harmony with nature. Thousands of indigenous people were removed from their land and pushed into poverty. As the global north developed, more liberal societies came into being for the colonisers, but at the same time the seeds of regressive social practices and corrupt, authoritarian governments were sown for the colonised. This now creates often insurmountable obstacles for developing countries to address the climate challenge effectively and equitably. The global north has imposed on the global south a development model based on the unconditional exploitation of nature and human beings. Subsequently, rich global investors saw a great opportunity for profit in the vulnerability of the working class and of marginalized communities in developing countries. We cannot escape the misery that climate change will cause, and the most efficient way to fight it is through global collaboration. Colonial powers of yesterday and capitalists of today have benefitted, more or less directly, from unsustainable practices and exploitation, and therefore, the burden of mitigation and adaptation needs to be shared among the countries and communities responsible for inflicting nature's fury on the blameless victims who now have limited resources to address climate threats.

Introduction

Climate change is the most daunting challenge for the present generation, and yet humanity is still reckoning its devastating impacts (Palmer and Stevens 2019). The most obvious consequence of a warmer planet will be the rise in extreme weather events, evident over the past few years. The word apocalypse, which is often associated with an out-of-control climate change, originates in the Greek word 'apocalypsis', which means revelation (Diaz 2011). Disasters tend to unfold in a way that they peel the glossy surface of the society and reveal it, its institutions, its values and cracks in society (Diaz 2011). As a former secretary-general of the International Red Cross and Red Crescent Movement once noted: '[I] n many cases, nature's contribution to "natural" disasters is simply to expose the effects of deeper, structural causes' (Jackson 2005).

The Covid-19 pandemic gave a clear example of how severe social crises can have very asymmetrical effects, both at the national and the global level, i.e., the Covid-19 pandemic can work as a template to understand how unprepared humanity is to deal with the climate crisis. According to Oxfam's (2021, 9–10) inequality report, inequality is set to deepen in almost every country, the first time such a universal increase in inequality has happened since record-keeping began. Covid-19 will also likely increase gender inequality, as the pandemic hit women harder than men in economic terms. Given current economic contingencies, it will take the world's poorest almost a decade to recover from the damages brought by the Covid-19 pandemic. In contrast, the world's richest 1,000 billionaires took just nine months to recover. Moreover, in Brazil and in the United States (US), Latinx and Black people were more likely to die due to Covid-19 than white people.

Covid-19 provides a clear demonstration, moreover, of the failure of one-size-fits-all solutions and of the challenges faced by the global south in mitigating and adapting to global crises. For instance, as many countries did, India imposed a nationwide lockdown, resulting in a huge exodus of migrant workers from tier-one cities to villages (Paliath 2021). So bright was India's shining story that even the government was blind towards the potential issues that key engineers of this economic system could face. It was arguably the biggest humanitarian crisis that independent India had ever seen. Similarly, the pandemic led to school closures in 180 countries. While children in the poorest countries were deprived of schooling for four months, children in the developed countries, where there is better access to the internet and to online learning, were only kept home for six weeks, on average (Oxfam 2021,15).

In sum, the Covid-19 pandemic has acted as a litmus test for the resilience of different societies to global and asymmetrical crises. Climate change, moreover, risks unleashing even more epidemics, in addition to extreme weather events, further exacerbating existing inequalities related to gender, race, ethnicity and nationality (T.H. Chan School of Public Health 2021). Covid-19 exposed how grossly unequal our world is. The global fault lines have been systematically built over centuries. Economic might translated into a political system that favoured the rich and incentivized further human and natural exploitation (Oxfam 2021, 8).

Social development is the key to climate change mitigation and adaptation (World Bank 2021c, 3). The Covid-19 pandemic shows that the developed world is better prepared and can adapt better in the face of a pandemic or natural crisis. However, 'climate change' is not a very popular expression in developing countries like India (Soni 2020). One among several reasons for this unpopularity is that there are already so many pressing development challenges on the table. Yet both climate mitigation and adaptation processes are veiled under the same issues that the countries of the global south have set out to resolve (World Bank 2021c, 3). For instance, when scientists work together to figure out the most effective climate solutions, financially and environmentally, increased education for young girls often features in the top ten solutions along with energy transition. Similarly, guaranteeing indigenous peoples' legal rights to their land is a potent climate mitigation solution (Drawdown 2021). Key adaptation measures against climate change include building up basic health and school infrastructures. The Covid-19 pandemic also shed light on the necessity of scientific temperament in fighting against natural odds (Matta 2020). Essential ecosystems and economic systems need to be built so that the climate crisis does not worsen existing inequality.

Developing countries are dealing with entrenched and severe social problems like patriarchy, superstition, religious extremism, authoritarian and corrupt governments and exploitative capitalism. All of these issues distract from the climate crisis and a just energy transition. Amidst the above-mentioned social challenges, developing countries are grossly underprepared to fight climate change. While the template of development, responsible for ecological destruction, has been used and provided by the global north, the global south was exploited according to the same template in the pursuit of economic development until only recently. For instance, when the British left India in 1947, the literacy rate in India was 16%, with the literacy rate amongst the Indian female population being 8% (Tharoor 2016, 227). Moreover, the average life expectancy amongst the Indian population was 27 years (Tharoor 2016, 264). India had no domestic industry, and 90% of the Indian population lived below the poverty line. It took India 70 years to bring the literacy rate to 72% and pull 280 million people out of poverty (Tharoor 2016, 264).

The legacy of colonialism not only continues to exacerbate the social and economic impacts of climate change and of global threats more broadly, but it also affects the adaptation and mitigation strategies that are available in the global south. From a climate change perspective, we need to work on two fronts: reducing social injustice and abating greenhouse gas emissions (Voskoboynik 2018, 173). Both fronts ask for a reexamination of current social and economic practices and for compensation for historical wrongs. As Audre Lorde wrote: '[T]he master's tools will never dismantle the master's house'.

This chapter explores the history of colonialism and how it was responsible for establishing current structures of global power and expediting the process of climate change. Moreover, it analyses how, in most cases, colonialism introduced a capitalist organisation of society that benefitted only a few while robbing the largest part of the population's access to basic goods. This now creates significant obstacles for developing countries in adapting to climate change.

1. The History of Development and Underdevelopment

While the western part of the world was moving towards modernity, it prevented the global south from living in harmony with nature. Colonialism was a form of subjugation: it destroyed local knowledge and inflicted violence through cultural denial, exploitation of natural resources and political oppression (Apter 1991).

Nations were not turned into colonies all at once. Colonialism was a gradual project that started with empires and in some cases companies seeking territories from which they could extract natural resources, raw materials and labour, all out of a desire for wealth (Voskoboynik 2018, 37). Humans tend to be more empathetic towards the people who look like them. Similarly, we care more about our land and our environment than the land or environments of others (Voskoboynik 2018, 37). Such moral dissociation with people and places of different origins played a central role in colonialism insofar as it allowed colonial powers to give little importance to the people, cultures and lands of their colonies. Wealth meant power. Everything started to be looked at in terms of profits. This led to large-scale exploitation, which in turn led to cash crops replacing food crops, monoculture replacing multi-culture and slaves replacing free labour (Voskoboynik 2018, 38).

It is crucial to go through the colonial era to understand how we reached the current levels of atmospheric CO_2 concentration and thus also the current global warming of 1.1°C. Weizman and Sheikh (2015) once wrote: '[T]he current acceleration of climate change is not only an unintentional consequence of industrialization. The climate has always been a project for colonial powers, which have continually acted to engineer it'. The colonisers' main goal was to acquire wealth and power, and they intended to acquire them by acquiring crucial commodities like minerals, metals, crops and labourers (Crate 2009, 190). Colonisers sought to make their economy grow and strengthen their political power through industrialisation, and they asked their colonies to provide energy supplies, food, raw materials, labour and even to contribute to increasing the consumer demand of western products. Gold and silver became the most sought-after commodities for the colonisers (Brown 2012). Around 100 million kilogrammes (kg) of silver were mined in the world between the sixteenth and nineteenth centuries. Gradually, attention shifted from gold and silver to plant-based commodities like cotton, sugar, spices and coffee (Voskoboynik 2018, 38).

As colonies grew around the world, so did the agricultural system based on monoculture (Montgomery 2012). In India, the British completely reorganised the agricultural system to suit the international export market. Low-scale food crop systems were replaced by organised cash crops, such as tea and cotton. Other empires followed the same trend in their colonies (Voskoboynik 2018, 38). Cotton regimes were installed by the Portuguese empire across Mozambican, Brazilian and Angolan colonies. With changes in the pattern of agriculture, water management techniques also changed. Water started being managed by the state, which used advanced engineering techniques, and communal management of water was stalled (Adams 2003, 35). Traditional practices that communities had learnt over the years were now considered ineffective, outdated and even damaging (Voskoboynik 2018, 39). Extensive farming of cash crops, coupled with deforestation, exhausted the soil and made most of the land infertile. Barren patches of land also became breeding grounds for mosquitoes carrying yellow fever and malaria. Consequently, in the Caribbean and Brazil, epidemics killed vast sections of the population (Voskoboynik 2018, 40).

As pointed out by the historian Corey Ross (2014, 49),

> One of the recurring themes in the history of plantations is the perennial cycle of boom and bust. Whether the crop is sugar, tobacco, or cotton, the basic pattern is often the same: an initial frenzy of clearing and planting is followed by either a precipitous collapse of production or a gradual process of

creeping decline before eventually ending in soil exhaustion, abandonment, and relocation elsewhere.

Colonialism led to mass destruction that had never been seen before. Colonies and their people were looked at as mere resources that colonisers could use to foster economic growth in their home countries. No mercy was shown as people and resources were exploited. Wealth was taxed through an unfair and rigorous system that overburdened the poor while making the rich even richer. Any voice of dissent was eliminated (Tharoor 2016, 47).

As the land of the people was exploited, it soon became infertile. Colonisers moved from one patch of land to another, leaving behind several barren patches. There was enormous-scale destruction. Ignorance and dissociation led to large-scale violence and destruction and consequently started the Anthropocene (Voskoboynik 2018, 41). When colonies regained self-determination, they had to rebuild from scratch. A chosen few gained an opportunity for education and developed scientific and liberal temperaments to fight retrograde traditions Colonialism eventually led to an economically prosperous world but one with a high level of inequality. Also, such a level of economic prosperity was only made possible through the exploitation of natural resources and people.

Apart from agriculture, gold and silver mining across colonies destroyed local terrains. For instance, extensive silver mining in the Andes and the Sierra Madre mountain ranges in South America and Mexico irreparably damaged local ecosystems. Colonisers undertook extensive deforestation to make way for fuel furnaces. This practice resulted in a significant decline in soil fertility, also triggering soil erosion and flooding (Gioda et al. 2002; Moore 2007). British merchants cleared India's Malabar Coast of teak forests. Then, they moved to Burma to clear the Tenasserim Forest and finished their task in just two decades (Ponting 2007). Within a few years, Marquesas Islands, Fiji and Hawaii were cleared of their sandalwood forests (Adams 2003, 35).

Extensive deforestation led to Java losing half of its forest in just one century (Boomgaard 1996). Between 1895–1925, agricultural expansion and logging led to Madagascar losing 70% of its primary forests (Jarosz 1993). Such extensive deforestation also led to wildlife losing their natural habitats, which eventually caused the extinction of several species. Moreover, demand for fine furs in the elite European circle led to overhunting in Siberia and the Americas. Consequently, the American Fur Company founder, John Astor, became the first American multimillionaire in history (Richards 2003). Overhunting also reached marine life (Roberts 2007).

Environmental damage and species extinction were accompanied by ethnocide. People were subjugated, exploited and massacred to make way for industrial production. Some of the tribes of people who were eradicated are the Beothuk, Charrúa, Guanches and Lucayas (Voskoboynik 2018). Aimé Césaire (2018), a Martinican author, notes that between 'colonizer and colonized, there is room only for forced labour, intimidation, pressure, the police, taxation, theft, rape, compulsory crops, contempt, mistrust, arrogance, self-complacency, swinishness, brainless elites, degraded masses.' Human slavery flourished in the US and Europe, which provided enormous benefits to these regions.

Lack of sanitation and malnutrition, along with the presence of exploitation, cramped spaces and enslavement, compromised the health of large share of the native population, which had become more vulnerable to germs and diseases (McBrian 2016). It led to populations being wiped out by diseases they had no immunity against. It resulted in a disastrous decimation of life. In the Americas, enslavement, famine, overwork, wards and epidemic diseases brought about the decimation of nine-tenths of the native population (McNeill 2010). Demographic research shows the level of compromised immunity in the indigenous population (Bacci 2008). The British empire forced India to export more food than it could, which led to one of the worst famines in the country's history. Tens of millions of people vanished in India, and if their bodies were laid head to toe, it would cover the entire area of England 85 times over. (Hickel 2015). In Congo, 10 million people died while they were deployed for the extraction of ivory and rubber (Hochschild 1999).

Colonialism remains one of the most destructive phases of human history. Along with the destruction of native people, cultures and traditions, it stole away the idea of different understandings of the world. People in different regions lived differently with nature. However, in the quest of gaining and accumulating wealth and power in just a few corridors of the world, the rest of the planet was forced to engage only in those activities that would help in the colonial pursuits. Numerous ecological traditions, languages, identities, cosmologies and possibilities vanished.

The morally tainted benefits of colonialism formed the basis on which the wealth of the colonial power rests. Precious metals like gold and silver made European merchants and banks extremely rich. The wealth accumulated from plantations, mining and slave trading formed the backbone of colonial economies and made the Industrial Revolution a success story (Hickel 2017). People from the colonies also became consumers of the colonial powers, i.e., they were obliged to purchase products from European industries. As a result, in the late-nineteenth century, more than 50% of the revenues of the British government came from its colonies (Inikori 2002).

Consequently, the world economy was reconfigured by colonialism. In its pre-colonial era, the Indian economy accounted for the 27% of the global economy. It was reduced to 3% when the British left. Similarly, the Chinese economy accounted for the 35% of the global economy, which eventually was reduced to 7%. A completely different trend emerged in Europe. Europe's contribution to the global economy increased from 20% to 70% in the same period. The tables turned in terms of development too. In the eighteenth century, income inequality around the world was minimal. There is evidence that Europeans had lower average standards of living than the rest of the world (Davis 2000).

Around the eighteenth century, the districts around the Bay of Bengal and the Yangtze Delta were manufacturing hubs with artisan workshops. These workshops could easily compete with the workshops of Florence. Prasannan Parthasarathi, an Indian historian, argues that 'there is compelling evidence that South Indian labourers had higher earnings than their British counterparts in the 18th century and lived lives of greater financial security'. In the nineteenth century, North Africa and Latin American people survived on higher calories than an average European (Montgomery 2012). But decades of destruction and exploitation changed the global picture.

There have been attempts to downplay the adverse effects of the era of colonialism. However, the history of climate change would be grossly incomplete without connecting the dots. The era of globally divergent economic growth is the result of extensive destruction of ecosystems and suppression and exploitation of communities, which consequently resulted in a significant increase in emissions. For instance, between 1835 and 1885, land-use change in the US proved to be the most significant contributor to global CO2 emissions (Brooke 2014, 482).

Colonialism transformed the way economic growth was perceived, and it altered the scale, degree and rate of ecological destruction. In Medieval Picardy, it took 200 years to deforest 12,000 hectares of land. However, in 1650, during the Brazilian sugar boom, it only took a year for an equivalent amount land to be deforested (Hickel 2017). Extensive alterations in marine and terrestrial ecosystems were normalised by colonial powers and justified only on the basis of economic growth. Political ecologist Jason Moore (2017) argues that 'the rise of capitalist civilization after 1450, with its audacious strategies of global conquest, endless commodification, and relentless rationalization', marked 'a turning point in the history of humanity's relationship with the rest of nature, greater than any watershed since the rise of agriculture and the first cities'.

2. Colonial Climate Legacies

When formal colonialism largely came to an end in the nineteenth and twentieth centuries, the world had already been divided into developed and developing countries. Colonialism, moreover, defined the very concept of development; a concept that largely ignored the social and environmental externalities created by the development of the western world. As colonies gained their freedom, they kept on living under the legacy of colonialism. Several newly formed nation-states followed the colonial development track and continued to treat ecosystems as mere instruments for economic growth. The programmes were still designed to evict and displace large communities under the banners of development.

For example, India gained independence in 1947. Between 1947–2000, 24 million Adivasis were evicted from their land in the name of development projects. The construction of Narmada Dam alone displaced close to 100,000 people. Vast areas of the Amazon Rainforest have been destroyed by military and non-military governments of Brazil. Hosni Mubarak's government in Egypt transferred control of lands from small farmers to big landowners and justified this in the name of development (Voskoboynik 2018, 61).

Many former colonies started large deforestation projects. Over two decades, between 1960–1980, timber exports in Indonesia increased 200-fold. In Côte d'Ivoire, extensive deforestation resulted in just one-fortieth of the forest remaining today. Côte d'Ivoire increased its timber exports from 42,000 tonnes to 1.6 million tonnes between 1913–1980 (Ponting 2007, 189). Almost 50% of forests were abated in the global south between 1900–2010. The colonial-based model of development led to these results, and people who resisted it were faced with extrajudicial violence and extreme repression (Voskoboynik 2018, 61). Even today, hundreds of community leaders, environmental and social activists are frequently killed in the world because they resist large-scale economic projects that favour a tiny minority. In Honduras alone, at least 124 land and environmental activists were killed between 2010–2016 (Spanne 2016).

Any story remains incomplete if we simply focus on the winners and neglect the losers. Even though the full impact of colonialism is still to be reckoned in its fullness, it is evident that it altered cultures, philosophies, landscapes and international relations, resulting in the exploitative economic model that thrives on natural and human domination. Colonialism also led to the disappearance of ancient and important cultural traditions that could have helped to contain the devastation of nature that is at the root of climate change.

The world is more than 1.1°C warmer than pre-industrial levels, and human activities have caused this warming (Tollefson 2021). The year 2021 has already seen a series of extreme climate events (Plimmer 2021). These events are the result of decades of greenhouse gas emissions. We are witnessing the warming caused by the exploitative practices of the nineteenth and twentieth centuries. According to the IPCC's Sixth Assessment Report, if we continue on the present trajectory, the world will be 1.5°C warmer on average by mid-century than it was in pre-industrial times (IPCC 2021). A warmer world implies that we cannot escape the misery that climate change will cause. However, where have these human activities, which have been mainly considered essential and indispensable, brought us?

To better understand the link between human development, on the one hand, and colonial exploitation of nature and human beings, on the other, we need to focus on some data. Historical emissions have mainly been unequal across the globe. Roughly 40% of the world's emission debt is carried by US (Matthews 2016). However, there are 'carbon creditor' states like Bangladesh, India, China and Nepal, which have a larger share of the global population compared to their share of emissions (Matthews 2016).

The effect of colonialism on emissions is evident in the fact that, in 1825, Britain was responsible for 80% of global carbon emissions from fossil fuels (Malm 2005). The wealthiest countries in the world are responsible for 80% of historical global emissions, and yet their population share is just 20%. Until 2000, the US was responsible for 27.6% of historical emissions. In contrast, Brazil was responsible for 0.9%, and Nigeria was responsible for 0.2% (Malm 2016). The per capita emissions of El Salvador are 45-times lower than the average emissions of a Qatari national and 15-times lower than the average emission of an American citizen (World Bank 2021a).

The inequity that the global economic system has caused is also evident from the emissions data of private corporations (see also Grasso in this volume). Since 1750, two-thirds of all emissions have been caused by just 90 corporations (Heede 2014). Half of these emissions occurred after 1988. Between 1751–2010, ExxonMobil has been responsible for 3.22% of all emissions (Heede 2014). The point to be stressed here is that by 1988, humanity had enough evidence to show that we were altering the earth system (Heede 2014).

According to Oxfam, 50% of global emissions are caused by just 10% of the world population (Gore 2015). At the same time, the poorest 50% of the world population is responsible for just 10% of global emissions. Historically, the wealthiest 1% of the world population has emitted 175 times more greenhouse gases than the poorest 10% (Voskoboynik 2018, 115). The richest 1% of Saudi Arabians have a carbon footprint 2,000-times higher than the poorest Malians (Malm 2016). Inequality aggravates emissions (Kusumawardani and Dewi 2020). It is evident that higher inequality within a country leads to higher emissions (Dorling 2017). Inequality in society is usually the result of the unjust accumulation of wealth and income by a few and of the dearth of opportunities for the many. The wealthiest people have excess money to spend on non-essentials, whereas many others struggle for basics. Therefore, the wealthiest people in the world use more energy, consume more goods, fly more and heat larger homes. The 20% of the global human population consumes 80% of the world's resources (Activesustainability 2021).

The accumulation of wealth and power in the hands of the few means that those most responsible for climate change are likely to adapt to changing conditions and, therefore, are more likely to remain unaffected. Poor people, by contrast, who contributed the least to ecological destruction will most likely be unable to adapt and endure the most substantial impacts and pay the highest costs. Global capitalism is normally perceived as the only means for economic growth and social development. Yet, 30 years ago, over 50 nations were more prosperous than they are today (Matthews 2016). The number of Europeans with severe material deprivation increased from 7 million to 50 million between 2009–2013 (Malm 2016). In the last 10 years, the number of people living in extreme poverty has tripled in Italy (Malm 2016). In recent decades, income inequality has increased in almost every country (World Bank 2021b).

Global inequality is evident because today, five (male) persons control as much wealth as 50% of the global population (Heede 2014, 229–241). About 70% of global income goes to the wealthiest 20% of the world's population, while the poorest 20% of the global population receive just 2% of global income (Starr 2016). Approximately 73% of India's new wealth accrues to the top 1% of the population (Heede 2014). Bill Gates's net worth is greater than the total gross domestic product (GDP) of Haiti over the past 30 years, while a single American family, the Waltons, own more wealth than the US's poorest 42%. There have been multiple studies showing a significant relationship between social wellbeing and equality (Wilkinson and Pickett 2010). Societies with less inequality have lower crime rates, less violence, less divorce, less addiction, healthier populations and lower infant mortality rates (Dorling 2017).

Global prosperity is generally measured in terms of GDP. Neoliberal economics suggests that GDP should keep increasing, which would eventually lead to a more prosperous world. However, after years of sustained global GDP growth, the present-day world is more unequal than it ever has been (OECD 2021). Years of environmental and human exploitation have resulted in the accumulation of power and wealth in the hands of few. Decades of exploitation of the global south have resulted in vulnerable social and environmental structures. Madagascar, for example, is on the verge of facing the first climate change-induced famine, and the inhabitants of the country who will suffer the most have contributed almost nothing to climate change (Harding 2021). Crises like this risk undermining the human development goals achieved so far in developing countries. For instance, climate change-induced drought increases the rate of human trafficking in India (Dutta 2020). School dropout rates increase, and gender inequality deepens when families face financial crises induced by climate events (Sharma 2018).

There is enough evidence to show that local communities have always known the techniques to live in harmony with nature. We also know very well that local communities have developed expert knowledge about their ecosystems and have learnt to thrive in them (Voskoboynik 2018, 82). For instance, the Moru community in Sudan has mastered techniques like crop rotation, waste recycling and soil classification to maintain the fertility of the soil. The Chepang community of the Himalayas maintain the fertility of the soil by practising 'Khoriya', a crop rotation technique. Similarly, there are endless examples of expert knowledge of local ecosystems that people have built over the years (Prudencio 1993). The greed to derive wealth from every human activity has resulted in the destruction of precious traditional knowledge.

The social cost of carbon in the global south is higher than in the global north (Brock 2012). The key reason is that societies in the global south are still reeling under the historical havoc caused by the countries which today are developed. Climate change is a time-bound problem, and the emergency has already arrived. Countries in the global south do not have time to evolve into societies with fewer infrastructural and social fault lines that would withstand the destruction caused by climate change.

Conclusions

During the current climate emergency, millions of people risk losing their lives, homes and identities. In an unequal world, one life is more valuable than another. While different countries will adopt different strategies for coping with various facets of climate change, it is essential to implement adaptation and mitigation strategies that are inclusive. The poor have been systematically left behind by the post-colonial capitalist economic system. Moreover, the wealth created could be directly linked to ecological destruction. Oxfam's report on inequality post-Covid-19 suggests that the Covid-19 pandemic undid the progress made in terms of eradicating poverty and inequality. The world will likely experience several such crises in the coming future that will leave the most vulnerable people worse off. Wealthy countries and individuals who have accumulated wealth from the exploitation share a moral obligation to restore communities globally and help build a more resilient world against climate change.

References

Active sustainability. 2021 "Natural Resources Deficit". https://www.activesustainability.com/environment/natural-resources-deficit/

Adams, William M. 2003. "Nature And the Colonial Mind". In *Decarbonising Nature*, edited by William (Bill) Adams and Martin Mulligan, 2–35. London: Routledge.

Apter, Andrew. 1991. "MUDIMBE, V. Y., The Invention Of Africa: Gnosis, Philosophy, And The Order Of Knowledge, , 241 Pp., 0 253 33126 9". *Journal Of Religion In Africa*, 21 (2): 172–174. https://doi.org/10.1163/157006691x00285

Bacci, Massima Livi. 2008. *Conquest: The Destruction Of The American Indios*. Polity.

Boomgaard, Peter. 1996. "Environmental Impact Of The European Presence In Southeast Asia, 17Th-19Th Centuries". *Island and Empires* 1 (28): 22–35.

Brock, Hannah. 2012. "Climate Change: Drivers Of Insecurity And The Global South." https://reliefweb.int/report/world/climate-change-drivers-insecurity-and-global-south

Brooke, John L. 2014. *Climate Change And The Course Of Global History*. Cambridge: Cambridge University Press.

Brown, Kendall W. 2012. "A History Of Mining In Latin America From The Colonial Era To The Presentby Kendall W. Brown". *The Latin Americanist* 56 (3): 104–107. https://doi.org/10.1111/j.1557-203x.2012.01162_3.x

Cesaire, Aime. 2018. *Discourse on Colonialism*. AAKAR Books.

Crate, Susan. 2009. "Viliui Sakha Of Subartic Russia And Their Struggle For Environmental Justice". In *Environmental Justice And Sustainability In The Former Soviet Union*, edited by Julian Agyeman and Yelena Ogneva-Himmelberger. Cambridge: MIT Press.

Davis, Mike. 2000. *The Origin Of The Third World*. Antipode.

Diaz, Junot. 2011. "Apocalypse: What Disasters Reveal". *Boston Review*. https://bostonreview.net/articles/junot-diaz-apocalypse-haiti-earthquake/

Dorling, Danny. 2017. *The Equality Effect*. New Internationalist, 504. https://newint.org/features/2017/07/01/equality-effect

Dutta, Soumik. 2020. "In India, Climate Change Is Increasing Refugees & Human Trafficking - India". Inter Press Service, July 9. http://www.ipsnews.net/2020/07/india-climate-change-increasing-refugees-human-trafficking/?utm_source=rss&utm_medium=rss&utm_campaign=india-climate-change-increasing-refugees-human-trafficking

Gioda, Alain, Carlos Serrano, and Ana Forenza. 2002. "Dam collapses in the world : a new estimation of the Potosi disaster (1626, Bolivia)". *La Houille Blanche* 88 (4–5): 165–170.

Gore, Timothy. 2015. *Extreme Carbon Inequality*. Oxfam. https://www-cdn.oxfam.org/s3fs-public/file_attachments/mb-extreme-carbon-inequality-021215-en.pdf

Harding, Adrew. 2021. "Madagascar On The Brink Of Climate Change-Induced Famine". *BBC News*, August 25. https://www.bbc.com/news/world-africa-58303792

Harvard T.H. Chan School Of Public Health. 2021 "Coronavirus And Climate Change". *C-CHANGE* | https://www.hsph.harvard.edu/c-change/subtopics/coronavirus-and-climate-change/

Heede, Richard. 2014. "Tracing Anthropogenic Carbon Dioxide And Methane Emissions To Fossil Fuel And Cement Producers, 1854–2010". *Climatic Change* 122 (1–2): 229-241.

Hickel, Jason. 2015. "Enough Of Aid – Let'S Talk Reparations". *The Guardian*. Nov 27. https://www.theguardian.com/global-development-professionals-network/2015/nov/27/enough-of-aid-lets-talk-reparations

Hickel, Jason. 2017. *The Divide*. William Heinemann.

Hochschild, Adam. 1999. *King Leopold's Ghost*. London: Papermac.

Inikori, J. E. 2002. *Africans And the Industrial Revolution In England*. Cambridge: Cambridge University Press.

IPCC. 2021. "Climate Change Widespread, Rapid, And Intensifying – IPCC". 2021. *Ipcc.Ch*. https://www.ipcc.ch/2021/08/09/ar6-wg1-20210809-pr/

Jackson, Stephen. 2005. *Understanding Katrine; Perspectives From The Social Sciences*.

Jarosz, Lucy. 1993. "Defining and Explaining Tropical Deforestation". *Economic Geography* 69 (4): 366–379.

Kusumawardani, Deni, and Ajeng Kartiko Dewi. 2020. "The Effect Of Income Inequality On Carbon Dioxide Emissions: A Case Study Of Indonesia". *Heliyon* 6 (8): e04772. https://doi.org/10.1016/j.heliyon.2020.e04772

Malm, Andreas. 2005. *Searching For The Origins Of The Fossil Economy*. Vero Blogs.

Malm, Andreas. 2016. *Who Lit This Fire? Critical Historical Studies* 3 (2): 215–248. https://doi.org/10.1086/688347

Matta, Gagan. 2020. "Science Communication As A Preventative Tool In The COVID19 Pandemic". *Humanities And Social Sciences Communications* 7 (159): 1–14. https://doi.org/10.1057/s41599-020-00645-1

Matthews, H. Damon. 2016. "Quantifying Historical Carbon And Climate Debts Among Nations". *Nature Climate Change* 6 (1): 60–64. https://doi.org/10.1038/nclimate2774

McBrian, Justin. 2016. *Anthropocene Or Capitalocene?* Oakland: PM Press.

McNeill, John Robert. 2010. *Mosquito Empires*. Cambridge University Press.

Montgomery, David R. 2012. *Dirt; The Erosion Of Civilisations*. University Of California Press.

Moore, Jason W. 2007. "Silver, Ecology And The Origins Of The Modern World, 1450–1640". In Rethinking Environmental History, edited by Alf Hornborg and J. R. McNeill.

Moore, Jason W. 2017. "The Capitalocene". *The Journal of Peasant Studies* 44 (3): 1–38. http://dx.doi.org/10.1080/03066150.2016.1235036

OECD. 2021 "Inequality – OECD". https://www.oecd.org/social/inequality.htm

Oxfam International. 2021. "The Inequality Virus | Oxfam International". https://www.oxfam.org/en/research/inequality-virus

Paliath, Shreehari. 2021. "A Year After Exodus, No Reliable Data Or Policy On Migrant Workers". *Indiaspend.Com*. https://www.indiaspend.com/governance/migrant-workers-no-reliable-data-or-policy-737499

Palmer, Tim, and Bjorn Stevens. 2019. "The Scientific Challenge Of Understanding And Estimating Climate Change". *Proceedings Of The National Academy Of Sciences* 116 (49): 24390–24395. https://doi.org/10.1073/pnas.1906691116

Plimmer, Joe. 2021. "Climate Crisis: 50 Photos Of Extreme Weather Around The World – In Pictures". *The Guardian*. Jul 19, 2021. https://www.theguardian.com/world/gallery/2021/jul/19/climate-crisis-weather-around-world-in-pictures-wildfires-floods-winds

Ponting, Clive. 2007. *A New Green History Of The World: The Environment and the Collapse of Great Civilizations*. London: Penguin Books.

Project Drawdown. 2021. "Table Of Solutions | Project Drawdown". https://drawdown.org/solutions/table-of-solutions

Prudencio, Coffi Y. 1993. "Ring Management Of Soils And Crops In The West African Semi-Arid Tropics: The Case Of The Mossi Farming System In Burkina Faso". *Agriculture, Ecosystems & Environment* 47 (3): 237–264. https://doi.org/10.1016/0167-8809(93)90125-9

Richards, John F. 2003. *The Unending Frontier*. Berkeley: University of California Press.

Roberts, Callum. 2007. *The Unnatural History Of The Sea*. London: Island Press.

Ross, Corey. 2014. "The Plantation Paradigm". *Journal Of Global History* 9 (1): 49–71.

Sharma, Amrit P. 2018. "School Dropout Rate Of Children In A Changing Climate". https://www.researchgate.net/publication/348163575_SCHOOL_DROPOUT_RATE_OF_CHILDREN_IN_A_CHANGING_CLIMATE

Soni, Paroma. 2020. "Why India Needs To See Climate Change As Urgent Political Issue". *Downtoearth.Org.In*. https://www.downtoearth.org.in/blog/climate-change/why-india-needs-to-see-climate-change-as-urgent-political-issue-69783

Spanne, Autumn. 2016. "Why Is Honduras The World's Deadliest Country For Environmentalists?". *The Guardian*, April 7. https://www.theguardian.com/environment/2016/apr/07/honduras-environment-killing-human-rights-berta-caceres-flores-murder

Starr, Douglas. 2016. "Just 90 Companies Are To Blame For Most Climate Change, This 'Carbon Accountant' Says". *Science*, August 25. https://doi.org/10.1126/science.aah7222

Tharoor, Shashi. 2016. *An Era Of Darkness The British Empire In India*. Aleph Book Company.

Tollefson, Jeff. 2021. "IPCC Climate Report: Earth Is Warmer Than It's Been In 125,000 Years". *Nature* 596 (7871): 171–172. https://doi.org/10.1038/d41586-021-02179-1

Voskoboynik, Daniel Macmillen. 2018. *The Memory We Could Be*. New Society Publishers.

Weizmen, Eyal, and Fazal Sheikh. 2015. "Colonization As Climate Change In The Negev Desert". *Steidl*. https://steidl.de/Books/The-Conflict-Shoreline-Colonization-as-Climate-Change-in-the-Negev-Desert-0821232758.html

Wilkinson, Richard G, and Kate Pickett. 2010. *The Spirit Level*. Penguin.

World Bank. 2021a. "CO2 Emissions (Metric Tons Per Capita) | Data". https://data.worldbank.org/indicator/EN.ATM.CO2E.PC

World Bank. 2021b. "Poverty and Shared Prosperity 2020: Reversals Of Fortune". https://www.worldbank.org/en/publication/poverty-and-shared-prosperity .

World Bank. 2021c."Social Dimensions of Climate Change". https://www.worldbank.org/en/topic/social-dimensions-of-climate-change

3

From Rio to Paris: International Climate Change Treaties Between Consensus and Efficacy

SILVIA BACCHETTA

Combatting climate change requires a global coordinated effort. For this reason, international politics is the crucial setting to delineate a strategy that, on the one hand, involves as many countries as possible and, on the other hand, promotes effective measures to contain climate change. These two aspects represent two dimensions of international diplomacy: consensus (i.e., the willingness to be part of an international agreement) and efficacy (i.e., the need to devise norms that will be functional in tackling climate change). This chapter proposes to reconstruct how the consensus/efficacy trade-off has been balanced in the treaties that have progressively dictated international climate change policymaking.

Introduction

In 2021, the Intergovernmental Panel on Climate Change (IPCC) published its Sixth Assessment Report with updated data about climate change, stating that 'unless there are immediate, rapid and large-scale reductions in greenhouse gas emissions, limiting warming to close to 1.5°C or even 2° C will be beyond reach' (IPCC 2021). The report highlighted that our climate system is irreparably damaged, but it also stated that there is still time to slow down climate change by limiting warming through forward-looking climate policies. Combatting climate change requires a global coordinated effort that calls for a solution at the international level. In this sense, international diplomacy has tackled climate change by developing a global institutional framework (namely, a climate regime) to coordinate countries into developing apt pro-environmental policies (Lahn 2020; Gupta 2010; Randalls 2010; Mayer 2018). Each major climate regime (i.e., the Kyoto Protocol, the Copenhagen Accord and the Paris Agreement) applies a different approach in dealing with climate change, even though the treaties are composed by the same basic building blocks combined into three different yet organic regimes (Held and Roger 2018, 528).

To be successful, a climate regime ought to fulfil two core requirements: *consensus* and *efficacy*. According to this first requirement, countries are expected to be part of an international climate change regime. For this reason, obtaining a country's consensus is crucial for a successful international agreement because it signals that a country is willing to do something about climate change. The second requirement, instead, deals with the kind of policies that are promoted by a climate regime. If the end-goal is to contain climate change, then it is necessary to promote measures that are conducive to effectively hit the target of climate change mitigation or adaptation set by climate science. However, balancing these two requirements is difficult and each climate regime tried to modulate the interplay between consensus and efficacy. Analysing how the consensus/efficacy interplay shapes the design of international climate change policymaking provides an insightful angle for reconstructing the development of climate negotiations (Mayer 2018). Efficacy, on the one hand, is important because it directs political action towards a specific target. Insofar a climate agreement, once implemented, provides a good chance of reaching a given climate target (e.g., limiting average temperatures to between 1.5°C and 2°C above pre-industrial levels) then it can be considered effective, regardless of its specific policy plan – as s there can be different ways to allocate mitigation and adaptation burdens to reach a certain target (Gao, Gao and Zhang 2017). Consensus, instead, is related to a genuinely political dimension. It signals the willingness of a country to be part of an international agreement, assuming that compliance with the agreement's prescriptions will follow. Clearly, the more a climate regime accommodates countries' preferences, the more willing participants there will be. However, effectively combatting climate change requires to follow a strict schedule of (often demanding) environmental reforms, which can hinder widespread participation. This means that ultimately there is a trade-off between consensus and efficacy (Mayer 2018).

This chapter proposes a reconstruction of the history of international climate regimes by overviewing how the consensus/efficacy trade-off has been modulated during the last 20 years of international climate change policymaking. The first section will focus on the early years of climate change negotiations, which spans from the Rio Declaration to the development of the first major climate regime, the Kyoto Protocol. On the one hand, the Rio Declaration (1992) is a cornerstone document which provides guidelines for combatting climate change, including both consensus and efficacy among the guiding principles for international climate diplomacy. The Kyoto Protocol (1997) is the first major climate regime modulating the consensus/efficacy trade-off by prioritising efficacy over consensus. The second section will focus on the post-2000 phase of climate regimes, which includes the Copenhagen Accord and the current climate regime, the Paris Agreement. This second stage is characterised by a shift towards consensus after the Kyoto Protocol's failure. The Copenhagen Accord (2009) tried to push the consensus dimension while expecting countries to make voluntary pledges in line with the ambitious efficacy goals of the Kyoto Protocol. The false start of the Copenhagen Accord prompted a reassessment of the institutional mechanisms at play. The result of this reassessment is the current climate regime, the Paris Agreement (2015), which balances efficacy with consensus within a novel normative framework. The final section proposes a commentary on the problem of non-compliance that affects climate change policymaking. Despite having perfectioned its institutional mechanisms, we are still far from effectively combat climate change because, in the end, countries do not implement the necessary environmental policies. To solve the non-compliance issue, there are two possible ways: a conservative view, which tries to correct the flaws of the current climate regime, and a reformist view, which instead argues that the climate regime ought to be radically changed in order to prompt compliance. In conclusion, a brief appendix discusses the outcomes of the COP26 held in Glasgow in 2021.

1. The Early Years: From Rio to Kyoto (1992–2001)

The first phase of international treaties on climate change begins with the Rio Declaration and ends with the failure of the Kyoto Protocol. The early years of climate negotiations are crucial for understanding the guiding principles and the inner workings of international cooperation to combat climate change. The Rio Declaration set out the cornerstone principles upon which international cooperation was to be built (United Nations 1992a). These principles, including consensus and efficacy, still provide a charter for climate action, and they form the basis of every climate change regime. The Kyoto Protocol was the first major international agreement to try to address climate change (IISD 1997). Despite being initially considered a diplomatic achievement, many countries defected from the Kyoto Protocol over the years, especially developed countries which were expected to bear most of the burdens of greenhouse gas (GHG) emissions reductions. The downfall of the Kyoto Protocol was due to the fact that the regime's prescriptions clashed with the interests of well-off countries. In addition, the Kyoto Protocol could not adequately balance consensus and efficacy, as it focused on efficacy, pushing concerns regarding the role of consensus to the background.

1.1 The Rio Declaration

To foster international awareness about climate change, the IPCC published its first scientific report about climate change in the early 1990s. The IPCC's reports stated that the concentration of GHGs in the earth's atmosphere was increasing rapidly, posing a danger for the stability of the planet's climate. The data provided by the IPCC's first report prompted the development of a stable institutional framework for establishing international climate policy (Böhringer 2003, 457).

In this initial phase, the main task was to pinpoint the key elements that would guide the international climate regime for years to come. The result of this effort was the Rio Declaration, which translated various values for environmental protection into a systematic document listing 27 cornerstone principles (United Nations 1992a; Viñuales 2015). These principles tackle many issues that soon became familiar for those interested in climate justice and environmental ethics. For example, the Rio Declaration mentions the right to sustainable development (Moellendorf 2014; 2011), the problem of intergenerational justice (Brandstedt 2015; McKinnon 2012) and a formulation of the precautionary principle (Gardiner 2006; Steel 2015; Hartzell-Nichols 2017).

For this chapter's purpose, the most relevant principles are the twelfth and the ninth, which discuss respectively consensus and efficacy. The twelfth principle declares that 'states should cooperate to promote a supportive and open international economic system' (United Nations 1992a) to address in a more conducive way problems of environmental disruption. Most importantly, it also declares that 'environmental measures addressing transboundary or global environmental problems should, as far as possible, be based on an international consensus' (United Nations 1992a). In this way, the Rio Declaration sets consensus as one of the main requirements to guide international policymaking regarding environmental issues. International climate agreements should, in other words, seek a consensus among those who participate, trying to accommodate the needs of each country to achieve cooperation. Additionally, the Rio Declaration implicitly discusses the efficacy dimension. The idea that a climate regime has to promote effective measures can be tracked down in the ninth principle which states that cooperation among countries should also include the improvement of scientific understanding, in order to deepen our 'scientific and technologic knowledge' (United Nations 1992a).

The importance of the Rio Declaration lies especially in its normative dimension, as it provides a value-laden paradigm for international climate regimes, even though it does not mention how to embed its principles in the inner workings of international diplomacy. To fill this void, it was drafted another cornerstone document, the United Nations Framework on Climate Change Convention (UNFCCC), which deals with the organisational structure that should be adopted by climate change regimes (United Nations 1992b). The UNFCCC also established some of the familiar (and controversial) requirements for climate policy implementation, such as the idea of legally binding mitigation commitments (Bodansky and Rajamani 2018) and the differentiated mitigation responsibilities between developed and developing countries (Gupta 2010, 640). Both the values and the procedures of international environmental diplomacy were then applied to the first UN climate regime, the Kyoto Protocol, which tried to incorporate the values listed in the Rio Declaration with an effective institutional, procedural, and legal framework to prompt climate action.

1.2 The Kyoto Protocol

The Kyoto Protocol was the first attempt to deal with the problem of curbing GHG emissions through a coordinated global effort. It is, for all intents and purposes, 'the climate regime's first systematic approach to addressing the problem of global warming' (Held and Roger 2018, 529). The Kyoto Protocol was the result of a widespread and far-reaching diplomatic effort, as this climate regime tried to involve many countries and tried to address most issues connected with climate change. On the downside, it also showed that the traditional tools of diplomacy were not adequate for 'crafting deals that actually make a difference' (Victor 2011, xxviii). More specifically, it applied to climate change the strategy of 'finding agreement where agreement is feasible and pushing other issues into the future' (Victor 2011, xxix) to get the treaty up and running, focusing on 'symbolic goals, such as limiting global warming to 2 degrees, while largely ignoring the more important practical need to set goals that governments can actually honor' (Victor 2011, xxviii). In this sense, the Kyoto Protocol can be considered both a success and a failure. It was undoubtedly a win for international diplomacy, as the behind-the-scenes negotiations led to the ratification of a viable and functioning climate regime. Yet, the implementation of the climate regime's prescriptions was very difficult to achieve, as effective measures to contain climate change would have mostly impacted the 'economic competitiveness' (Victor 2011, xxx) of participating countries – especially the well-off ones.

The Kyoto Protocol was initially drawn up following the blueprint of the UNFCCC, according to 'key parameters defining the kind of agreement states would pursue, including its legal form (it would be binding) and the parties it was intended to apply (namely, wealthier 'Annex I' states)' (Held and Roger 2018, 528). It divided participating countries into developed and developing countries – respectively called Annex I and non-Annex I countries (IISD 1997). The GHG emissions reduction requirements were assigned according to each country's level of development, therefore assigning the bulk of mitigation and adaptation burdens to the wealthier Annex I countries (Gupta 2010). More specifically, it established an overall GHG emissions reduction target (5% per cent below 1990 levels to be achieved by 2012) with additional individual targets that, once negotiated, countries were legally beholden to meet. The treaty allowed some flexibility for the specific policy strategies each country could implement to fulfil its commitments, though specifying that the parties were considered 'responsible under international law for meeting specific emission outcomes' (Held and Roger 2018, 529).

Despite being considered a game-changer treaty, it was clear that the lack of certain mechanisms could be a hindrance for the implementation of The Kyoto Protocol. For example, the Kyoto Protocol did not allow developing countries to make voluntary commitments. Additionally, the distinction between Annex I and non-Annex I countries led to some issues for the allocation of mitigation and adaptation burdens, as some of the top polluting countries (e.g., China, India and Brazil) had no mitigation commitments to meet because they were listed as non-Annex I countries. Moreover, the Kyoto Protocol lacked an effective compliance mechanism. So, if a country failed to meet its targets, it could simply withdraw from the agreement without suffering any penalty (Held and Roger 2018, 529). These internal contradictions of the Kyoto Protocol contributed to the climate regime's downfall in the subsequent years, when many Annex-I countries defected. The most significant defection happened in 2001 when the United States withdrew from the Kyoto Protocol because, according to the US government, the requirements of GHG emissions reduction were too costly and too demanding for the country's economy. (Böhringer 2003; Gupta 2010). After 2008, many countries, including Japan and Russia, refused to settle on new targets (Savaresi 2016, 2), bringing the negotiations to a gridlock that resulted into leaving behind the Kyoto Protocol in order to draw up a plan for an alternative climate regime.

If we analyse the Kyoto Protocol using the two dimensions of consensus and efficacy, we can see that it leaned more on the efficacy side, especially when compared to subsequent regimes. The Kyoto Protocol is efficacy-led as the mitigation targets are applied top-down to countries. Moreover, making the Kyoto Protocol a binding agreement might be seen as a way to enforce compliance with the regime's commitments as much as possible. The measures devised by the Kyoto Protocol were thought to be high-reaching and many Annex I countries considered the implementation of such measures to be too costly. Imposing these demanding mitigation burdens without considering how many resources each Annex I country was willing to actually allocate to climate change mitigation or adaptation led many countries to withdraw from the agreement. For what concerns consensus, there is little to say. The limited flexibility of the Kyoto Protocol did not give much room for manoeuvre to set more country-specific mitigation or adaptation goals. With its top-down architecture, the Kyoto climate regime is concerned with consensus insofar as it brings a country's diplomats to the negotiating table. Once the regime was ratified, every decision about goals, targets or the allocation of mitigation or adaptation burdens was imposed top-down on countries. With this little flexibility, if a country did not meet the top-down requirements, it had only one alternative option: shirking from the agreement.

The consensus/efficacy trade-off instantiated by the Kyoto Protocol suggests that an imbalance between the elaboration of efficient mitigation policies and states' willingness to be part of an agreement leads to an ineffective agreement. Imposing demanding obligations on countries without considering their available resources for climate change mitigation and adaptation or without giving them some freedom in setting their targets is a failing strategy. Indeed, many countries, rather than damaging their economy, preferred to renege on the climate regime, leaving only few countries to abide by the Kyoto Protocol. For an agreement that had the ambition to create a global, comprehensive and effective solution to combat climate change, the Kyoto Protocol was an institutional failure (Rosen 2015). Yet, the Kyoto Protocol represented a learning experience for international climate politics. It introduced in the global climate regime some institutional mechanisms that, with some improvements, were 'a valuable starting point for shaping climate policies in the future' (Böhringer 2003, 451).

2. The Post-2000 Breakthrough: From Copenhagen to Paris

The Kyoto Protocol's failure kick-started a new phase for the global climate regime. The Kyoto Protocol demonstrated that pursuing efficacy at the expense of consensus would produce a regime with few countries joining it, so the efforts of international diplomacy shifted towards incorporating mechanisms that would increase the number of participating countries. The Kyoto Protocol's experience had taught that countries should be part of the decision process regarding the allocation of mitigation and adaptation burdens. Therefore, it became important devising a more flexible climate regime to accommodate the specific needs and demands of each participating country. The first attempt to apply this new model of international climate politics is the Copenhagen Accord, a short-lived climate regime ratified in 2009 (IISD 2009). Initially, the Copenhagen Accord aimed at the creation of a bottom-up, consensus-based agreement. Despite the good intentions, it was clear since the beginning that reaching any kind of substantial agreement would be very difficult. In contrast the Kyoto Protocol's successful negotiations, international diplomacy could not find a common point of agreement in order to resolve each country's objection. Eventually, the Copenhagen Accord became just a *pro forma* agreement that did not bring forward the environmental agenda to contain climate change. The second attempt to create a novel climate regime came in 2015 with the Paris Agreement (IISD 2015), which is still the operating climate regime. After the Copenhagen Accord's demise, diplomats learned their lesson and they addressed most of the institutional, factual, and political problems that had hindered the previous climate regimes. The Paris Agreement's architecture and institutional mechanisms, indeed, are thought to elicit widespread countries' participation, and to effectively combat climate change, therefore keeping together both concerns about consensus and efficacy.

2.1 The Copenhagen Accord

The intention behind the Copenhagen Accord was to promote an agreement that purported 'a shared vision, adaptation measures, mitigation measures, technology development and transfer, and financial assistance and investment to be adopted in Copenhagen 2009' (Gupta 2010, 646). Indeed, the common understanding was that countries had to be more involved in negotiating climate targets, therefore, instead of imposing top-down mitigation burdens and legally binding commitments to each participating country, the new climate regime had to apply a more participatory structure, giving countries more flexibility to model climate change policies according to their own needs and resources (Savaresi 2016). Even though the classification between Annex I and non-Annex I countries still remained, it became less relevant in allocating mitigation and adaptation burdens (Rajamani 2010, 831–32). Under the Copenhagen climate regime, the Kyoto Protocol's binding commitments became *nationally determined contributions* (IISD 2009). The key feature of NDCs was that each country could decide how to contribute to combatting climate change, making participation more appealing – therefore focusing on consensus rather than on efficacy.

The Copenhagen Accord's negotiations began with the expectation of building a bottom-up agreement rooted in international cooperation and national participation, focusing on specific opportunities to cut emissions through domestic policies (Levi 2010). This expectation was largely misguided, as from the beginning of negotiations it was clear that there would be underwhelming results: negotiations were difficult, many countries vetoed the proposed measures (namely GHG emissions cuts), and there was general dissatisfaction with the results of the meetings. For example, the first draft of the Copenhagen Accord proposed an 80% reduction in GHG emissions, which was lowered to 50% during negotiations. The participating countries could only agree on maintaining the average global temperature below 2° C above pre-industrial levels, without adopting any more specific short- or medium-term targets (Gupta 2010; Randalls 2010). In the end, due to countries' reluctance to accept binding but efficient short- or medium-term targets, the Copenhagen Accord became just a *pro forma* agreement, despite the few persisting elements it introduced – such as the 2°C target and the idea of creating a climate regime 'open to all' (Held and Roger 2018, 530–31).

The Copenhagen Accord promoted 'a model of global climate governance that would operate, at bottom, according to strictly voluntary governance logic' (Held and Roger 2018, 530). By applying a bottom-up strategy, the Copenhagen Accord focused entirely on the dimension of consensus. Yet, this bottom-up strategy soon backfired as it was difficult to impose effective mitigation targets within this fully voluntary institutional framework. The result was an underwhelming treaty, with many participating countries but no bite in terms of combatting climate change, which is the bottom-line of any climate regime (Vidal, Stratton and Goldenberg 2009). Despite being a failure, in the big picture of the history of international climate change treaties, the Copenhagen Accord paved the road towards the Paris Agreement. The Copenhagen Accord's proceedings signalled that the winning strategy to have a cooperative agreement was to enhance consensus through the creation of a flexible regime. However, a climate regime should also deliver on the efficacy dimension, namely it should promote mitigation and adaptation targets to reduce GHG emissions effectively. In this sense, the Copenhagen Accord could not deliver any significant goals for GHG emissions reduction.

2.2 The Paris Agreement

The Kyoto Protocol applied a top-down strategy to combat climate change, whereas the Copenhagen Accord tried to punt into place a bottom-up mechanism. Both architectures offered some advantages, but in the end the disadvantages were greater, leading both climate regimes to fail. Nevertheless, these previous regimes were useful because they paved the road to build a novel climate regime, namely the Paris Agreement, that focused both on consensus and efficacy. The key task of the Paris Agreement is to grant some flexibility to foster broad agreement and widespread participation without sacrificing the promotion of viable and effective mitigation and adaptation goals – that should possibly be set at each country's terms (Dimitrov 2016), while addressing also pressing matters of global distributive justice (Okereke and Coventry 2016; Okereke 2010; Moellendorf 2012).

With the development of the Paris Agreement, policymakers tried to pursue these objectives through a novel climate regime that integrates both a bottom-up and a top-down mechanism – albeit prioritising the former. As a matter of fact, the Paris Agreement consists in a binding agreement with mandatory provisions regarding mitigation and adaptation goals that each country can adapt to its own needs (Höhne et al. 2017; Rajamani 2016b). For this reason, the tool of *nationally determined contributions*, introduced in the Copenhagen Accord, was enriched with a Kyoto-like strategy that consists in setting mandatory long-term goals (IISD 2015). The key concept for the Paris climate regime is flexibility, so as to encourage international cooperation without vexing national governments with top-down impositions. This means that developed and developing countries can tailor their goals to their own needs, provided they stay on track with the overall mandatory goals set by the Paris Agreement (Streck, Keenlyside and von Unger 2016; Bodansky 2016). To guarantee this flexibility, the Paris Agreement contained a revised version of Copenhagen's nationally determined contributions, namely *intended nationally determined contributions* (Victor 2015). The parties can pledge their own mitigation or adaptation goals that had to be reached by a certain medium-term deadline. Such pledges 'are not up for negotiation' (Held and Roger 2018, 532) and they had to be 'taken as is' (Held and Roger 2018, 532). This means that the international regime has to accept each country's pledge with no interference. In this way, each country can adapt its policy strategy to its own needs: some countries can make ambitious pledges, while others can play it safer and pledge more modest targets.

The introduction of *intended nationally determined contributions* is a great concession in favour of the dimension of consensus. According to each party's ability, any country can be part of an international climate regime (with all the side perks, such as international recognition, participation in international talks, geopolitical weight, and potentially fruitful economic connections) without the hassle of meeting top-down commitments. Yet, the Paris Agreement was also able to promote the efficacy dimension. Indeed, even though each country is free to pledge whatever mitigation or adaption goal it deems appropriate, these pledges are integrated into a 'legally binding framework that builds around them a range of important procedural obligations' (Held and Roger 2018, 532; Rajamani 2016a), including periodical checks on each country's progress in implementing domestic policies to reach its pledged target. Moreover, the Paris Agreement aims to promote a more ambitious overall mitigation target than the previous treaties. Even though it still maintains the goal of keeping the temperatures rise below 2°C, it also encourages efforts to keep the rise below 1.5°C. Thanks to these mechanisms, within the Paris climate regime it is possible to combine a consensus-based architecture with the promotion of effective mitigation and adaptation goals to reduce GHG emissions, overcoming the main flaws of both the Kyoto Protocol and the Copenhagen Accord.

Nevertheless, there are also some weaknesses in the Paris climate regime as well. Perhaps its most important flaw is the lack of a punitive mechanism intervening when mandatory goals are not met (Streck, Keenlyside and von Unger 2016; Geden 2016; Spash 2016). Indeed, the only form of control mentioned in the Paris Agreement is a (still) undetermined mechanism involving a committee that will be 'non-adversarial and non-punitive, which means that it has no teeth and can do nothing about non-compliance' (Spash 2016, 3). Furthermore, from 2017 to 2021, history repeated itself: during the Trump administration the United States decided to withdraw from the Paris Agreement, reminding everyone that what was achieved with much effort by a certain political administration could be easily undone by the next one. That turn of events highlighted the fact that any kind of plan regarding climate change policymaking will be ineffective if there is not an overarching and consistent long-term planning. Indeed, the US withdrawal reminded that climate policy requires more than making representatives from all over the world sit at the same table and agree on a set of measures, it needs a lasting political commitment to continue to be part of a climate regime, and to implement the necessary policies to combat climate change.

The Paris Agreement is the result of years of difficult negotiations in which diplomats were able to find a compromise between the urgency of effective policies and the need to involve as many countries as possible to produce a coordinated global effort to combat climate change. From the point of view of diplomacy, the Paris Agreement was an undeniable success that stemmed from the previous climate regimes by perfectioning the mechanisms that worked, pruning those that did not work and changing what could be modified to make it work (Allan et al. 2021). The Paris climate regime seems to have found the right way to balance the trade-off between consensus and efficacy, but some problems related to combatting climate change persist, even within an operating, well-balanced climate regime. Besides the more technical aspects, the main problem of each international climate regime – from the Kyoto Protocol to the Paris Agreement – is that it is still difficult to prompt countries to implement the expected mitigation or adaptation policies. Even with an accommodating climate regime such as the Paris Agreement, the transition from pledges to policy is not always sufficient to produce results. Indeed, very few countries are respecting and staying on track with their initial pledges (Victor 2016).This lack of practical results for GHG emissions reductions remains a climate regime's issue that needs to be solved.

Concluding Remarks: The Non-Compliance Issue and the Future of the Climate Regime

This chapter proposes a reconstruction of the history of the international climate treaties through the evolution of the consensus/efficacy trade-off, showing how each regime tried to balance the need to produce results in terms of climate change adaptation and mitigation (efficacy) and the need to involve as many countries as possible (consensus). This overview skims almost 30 years of climate regimes, pinpointing the different ways in which efficacy and consensus have been balanced: initially, the Kyoto Protocol prioritised efficacy over consensus, then the Copenhagen Accord tried unsuccessfully to combine consensus and efficacy on the misguided assumption that in a fully voluntary climate regime, countries would have set ambitious mitigation and adaptation goals for themselves. Eventually, the Paris Agreement was able to integrate specific mechanisms to accommodate both consensus and efficacy. Apart from the specificity of each climate regime, throughout the years international diplomacy has always found a way to overcome obstacles and enact an operative climate regime. Yet, these climate regimes did not produce any significant progress in terms of combatting climate change.

Probably, the major problem of dealing with climate change consists in translating pledges into actual environmental policies to be implement in each country. It is not difficult to bring countries with different interests and needs to the table and make them agree on a common institutional framework. Rather, the real challenge is to make countries follow up their international commitments with actual domestic policies. Instead, pledges mostly become empty promises as countries systematically fail to implement mitigation or adaptation policies: climate change policymaking is indeed affected by a systematic problem of non-compliance. The problem of non-compliance, which includes both the problem of implementation and free-riding, adds a further challenge to the already complex issue of climate change. The problem is twofold: on the one hand, there is the need to prompt countries to implement climate change policies to reduce GHG emissions and curb climate change, while on the other hand, it is important to discourage free-riding, so as to make each country contribute to combatting climate change. There are two main proposals to solve the problem of non-compliance: on one side, there is a more conservative view that proposes the integration of some mechanisms to prompt compliance without any alteration to the current climate regime; on the other side there is a reformist view holding that the climate regime should be thoroughly reformed. More specifically, the conservative view holds that to solve the problem of compliance, it would be enough to think of possible compliance-enhancing strategies that are consistent with the current climate regime's structure. According to the reformist view, instead, it is necessary to radically 'reconceptualize climate agreements' (Nordhaus 2021) because the persistent problem of compliance derives from the systematically 'flawed architecture' (Nordhaus 2021) of climate regimes.

According to the conservative view, the problem of compliance is related to each country's domestic circumstances that hold back the implementation climate change measures. Indeed, each country has limited resources to distribute over many issues, and climate change is only one of the many problems to tackle. Ultimately, it depends on how each country prioritises climate change over other issues. If a country has more urgent issues to solve, the implementation of mitigation and adaptation policies can be put on hold. Alternatively, climate change could simply not be a priority for a country's political agenda, so the implementation of climate change policies is delayed or downsized. Additionally, a government may decide to participate to international climate change negotiations purely for instrumental reasons unrelated to climate change, such as to improve a country's international credibility or to forge new alliances. If the problem is that governments do not implement mitigation and adaptation policies because they do not deem climate change as a primary concern, the solution could be to promote bottom-up actions that would push climate change up in a government's political agenda signalling climate change as a relevant issue. Such bottom-up methods might include cultivating green virtues (Peeters, Diependaele and Sterckx 2019, 442) or pushing climate action through strikes, demonstrations or actions of civil disobedience that might contribute to change a country's political agenda (Martin 1996; Delmas 2018; Welchman 2001).

According to the reformist view, the problem of non-compliance derives from the climate regime's inherently faulty architecture that relies on 'voluntary arrangements, which induce free-riding that undermines any agreement' (Nordhaus 2021). Therefore, it is necessary to radically change the climate regime 'from a voluntary agreement to one with strong incentives to participate' (Nordhaus 2021). The proposed solution is to think of the climate regime as a 'climate club' (Victor 2011; Nordhaus 2015; 2021). This proposal builds on creating a club system, namely, establishing a voluntary group in which there are mutual benefits derived from sharing the costs of a shared good or service, adapted to climate change. Instead of thinking about an overarching agreement, it would be better to create small groups of countries (i.e., climate clubs) to develop more manageable policies and to 'make it easier for club members to concentrate on the benefits of cooperation' (Victor 2011, xxx). Indeed, a climate club would provide rules for membership, prompting countries to agree to undertake emissions reduction policies to meet a goal (such as the 1.5°C temperature limit). At the same time, it would include penalties for nonparticipants, such as the imposition of penalty tariffs for non-club members. In this way, the climate club would be a cooperative system that is advantageous for its members and that would grant a stable membership with limited incentives to defect or free ride (Nordhaus 2021), therefore solving the problem of compliance at its roots.

At the moment, we are far from reaching a satisfactory goal in combatting climate change, notwithstanding the years of negotiations, the ever-changing climate regimes, and the work to highlight problems and propose solutions to correct them. The bottom line is that, since 1995, we are still worrying about the same issues – rising temperatures, excessive GHG emissions, extreme weather events, and ocean pollution – with no significant progress, as the most recent international conference has once again proved.

Appendix: The COP26 in Glasgow

The twenty-sixth Conference of the Parties (COP26) was held in Glasgow, Scotland in October 2021. The main purpose of COP26 was to confirm the Paris Agreement's commitments but the results of the conference were unsatisfactory: in the final document there were many lukewarm decisions that matter-of-factly delayed any effective policy strategy to combat climate change. For example, COP26 had to deliver important policy guidelines regarding the use of fossil fuels, especially coal: the expectation was to make countries pledge to phase out coal to reduce emissions; instead in the final document, many countries settled for a more modest pledge to phase down coal (United Nations 2022). Moreover, COP26 confirmed the goal to keep warming below 1.5°C above pre-industrial levels, even though this goal becomes more and more unattainable, especially considering that there are no impactful policies in place to achieve this target (Vidal 2021). In Glasgow, there was indeed a lack of political will to overcome the difficulties of combatting climate change, and it is a reiteration of the same lack of political will that we have diagnosed in the previous climate treaties. In the current situation, combatting climate change still seems an uphill struggle as it was in 1995.

References

Allan, Jen Iris, Charles B. Roger, Thomas N. Hale, Steven Bernstein, Yves Tiberghien, and Richard Balme. 2021. "Making the Paris Agreement: Historical Processes and the Drivers of Institutional Design." *Political Studies*, online first. https://doi.org/10.1177/00323217211049294

Bodansky, Daniel. 2016. "The Legal Character of the Paris Agreement." *Review of European, Comparative & International Environmental Law* 25 (2): 142–50. https://doi.org/10.1111/reel.12154

Bodansky, Daniel, and Lavanya Rajamani. 2018. "The Issues That Never Die." *Carbon & Climate Law Review* 12 (3): 184–90. https://doi.org/10.21552/cclr/2018/3/4

Böhringer, Christoph. 2003. "The Kyoto Protocol: A Review and Perspectives." *Oxford Review of Economic Policy* 19 (3): 451–66. https://doi.org/10.1093/oxrep/19.3.451

Brandstedt, Eric. 2015. "The Circumstances of Intergenerational Justice." *Moral Philosophy and Politics* 2 (1): 33–55. https://doi.org/10.1515/mopp-2014-0018

Delmas, Candice. 2018. *A Duty to Resist: When Disobedience Should Be Uncivil*. Oxford: Oxford University Press.

Dimitrov, Radoslav S. 2016. "The Paris Agreement on Climate Change: Behind Closed Doors." *Global Environmental Politics* 16 (3): 1–11. https://doi.org/10.1162/GLEP_a_00361

Gao, Yun, Xiang Gao, and Xiaohua Zhang. 2017. "The 2 °C Global Temperature Target and the Evolution of the Long-Term Goal of Addressing Climate Change—From the United Nations Framework Convention on Climate Change to the Paris Agreement." *Engineering* 3 (2): 272–78. https://doi.org/10.1016/J.ENG.2017.01.022

Gardiner, Stephen M. 2006. "A Core Precautionary Principle." *Journal of Political Philosophy* 14 (1): 33–60. https://doi.org/10.1111/j.1467-9760.2006.00237.x

Geden, Oliver. 2016. "The Paris Agreement and the Inherent Inconsistency of Climate Policymaking." *Wiley Interdisciplinary Reviews: Climate Change* 7 (6): 790–97. https://doi.org/10.1002/wcc.427

Gupta, Joyeeta. 2010. "A History of International Climate Change Policy." *Wiley Interdisciplinary Reviews: Climate Change* 1 (5): 636–53. https://doi.org/10.1002/wcc.67

Hartzell-Nichols, Lauren. 2017. *A Climate of Risk. Precautionary Principles, Catastrophes, and Climate Change*. New York: Routledge.

Held, David, and Charles Roger. 2018. "Three Models of Global Climate Governance: From Kyoto to Paris and Beyond." *Global Policy* 9 (4): 527–37. https://doi.org/10.1111/1758-5899.12617

Höhne, Niklas, Takeshi Kuramochi, Carsten Warnecke, Frauke Röser, Hanna

Fekete, Markus Hagemann, Thomas Day, et al. 2017. "The Paris Agreement: Resolving the Inconsistency between Global Goals and National Contributions." *Climate Policy* 17 (1): 16–32. https://doi.org/10.1080/1469306 2.2016.1218320

IISD. 1997. "Summary of the Kyoto Climate Conference." *Earth Negotiations Bulletin* 12 (76): 1–16.

IISD. 2009. "Summary of the Copenhagen Climate Conference." *Earth Negotiations Bulletin* 12 (459): 1–30.

IISD. 2015. "Summary of the Paris Climate Conference." *Earth Negotiations Bulletin* 12 (663): 1–47.

IPCC. 2021. "Climate Change Widespread, Rapid, and Intensifying – IPCC." https://www.ipcc.ch/2021/08/09/ar6-wg1-20210809-pr/

Lahn, Bård. 2020. "A History of the Global Carbon Budget." *Wiley Interdisciplinary Reviews: Climate Change*, 11(3): 1–9. https://doi. org/10.1002/wcc.636

Levi, Michael. 2010. "Beyond Copenhagen." *Foreign Affairs*, November. https://www.foreignaffairs.com/articles/commons/2010-02-22/beyond-copenhagen

Martin, Michael. 1996. "Ecosabotage and Civil Disobedience." *Environmental Ethics* 12 (4): 291–310.

Mayer, Benoit. 2018. *The International Law on Climate Change*. Cambridge: Cambridge University Press.

McKinnon, Catriona. 2012. *Climate Change and Future Justice: Precaution, Compensation, and Triage*. New York: Routledge.

Moellendorf, Darrel. 2011. "A Right to Sustainable Development." *The Monist* 94 (3): 433–52. https://www.jstor.org/stable/23039153

Moellendorf, Darrel. 2012. "Climate Change and Global Justice." *Wiley Interdisciplinary Reviews: Climate Change* 3 (2): 131–43. https://doi. org/10.1002/wcc.158

Moellendorf, Darrel. 2014. *The Moral Challenge of Dangerous Climate Change. Values, Poverty, and Policy.* New York, NY: Cambridge University Press.

Nordhaus, William. 2015. "Climate Clubs: Overcoming Free-Riding in International Climate Policy." *American Economic Review* 105 (4): 1339–70. https://doi.org/10.1257/aer.15000001

Nordhaus, William. 2021. "The Climate Club." *Foreign Affairs*, December. https://www.foreignaffairs.com/articles/united-states/2020-04-10/climate-club

Okereke, Chukwumerije. 2010. "Climate Justice and the International Regime." *Wiley Interdisciplinary Reviews: Climate Change* 1 (3): 462–74. https://doi.org/10.1002/wcc.52

Okereke, Chukwumerije, and Philip Coventry. 2016. "Climate Justice and the International Regime: Before, during, and after Paris." *Wiley Interdisciplinary Reviews: Climate Change* 7 (6): 834–51. https://doi.org/https://doi.org/10.1002/wcc.419

Peeters, Wouter, Lisa Diependaele, and Sigrid Sterckx. 2019. "Moral Disengagement and the Motivational Gap in Climate Change." *Ethical Theory and Moral Practice* 22 (2): 425–47. https://doi.org/10.1007/s10677-019-09995-5

Rajamani, Lavanya. 2010. "The Making and Unmaking of the Copenhagen Accord." *International and Comparative Law Quarterly* 59 (3): 824–43. https://doi.org/10.1017/S0020589310000400

Rajamani, Lavanya. 2016a. "The 2015 Paris Agreement: Interplay Between Hard, Soft and Non-Obligations." *Journal of Environmental Law* 28: 337–58.

Rajamani, Lavanya. 2016b. "Ambition and Differentiation in the 2015 Paris Agreement: Interpretative Possibilities and Underlying Politics." *International and Comparative Law Quarterly* 65 (2): 493–514. https://doi.org/10.1017/S0020589316000130

Randalls, Samuel. 2010. "History of the 2° Climate Target." *Wiley Interdisciplinary Reviews: Climate Change* 1 (4): 598–605. https://doi.org/10.1002/wcc.62

Rosen, Amanda M. 2015. "The Wrong Solution at the Right Time: The Failure of the Kyoto Protocol on Climate Change." *Politics & Policy* 43 (1): 30–58. https://doi.org/10.1111/polp.12105

Savaresi, Annalisa. 2016. "The Paris Agreement: A New Beginning?" *Journal of Energy & Natural Resources Law* 34 (1): 16–26. https://doi.org/10.1080/02646811.2016.1133983

Spash, Clive L. 2016. "This Changes Nothing: The Paris Agreement to Ignore Reality." *Globalizations* 13 (6): 928–33. https://doi.org/10.1080/14747731.2016.1161119

Steel, Daniel. 2015. *Philosophy and the Precautionary Principle. Science, Evidence, and Environmental Policy*. Cambridge: Cambridge University Press.

Streck, Charlotte, Paul Keenlyside, and Moritz von Unger. 2016. "The Paris Agreement: A New Beginning." *Journal for European Environmental & Planning Law* 13 (1): 3–29. https://doi.org/10.1163/18760104-01301002

United Nations. 1992a. "Rio Declaration on Environment and Development." https://www.un.org/en/development/desa/population/migration/generalassembly/docs/globalcompact/A_CONF.151_26_Vol.I_Declaration.pdf

United Nations. 1992b. "United Nations Framework Convention on Climate Change," http://unfccc.int/files/essential_background/background_publications_htmlpdf/application/pdf/conveng.pdf

United Nations. 2022. "Report of the Conference of the Parties serving as the meeting of the Parties to the Paris Agreement on its third session, held in Glasgow from 31 October to 13 November 2021" https://unfccc.int/sites/default/files/resource/cma2021_10_add1_adv.pdf

Victor, David G. 2011. *Global Warming Gridlock. Creating More Effective Strategies for Protecting the Planet*. Cambridge: Cambridge University Press.

Victor, David G. 2016. "Making the Promise of Paris a Reality." In *The Paris Agreement and Beyond: International Climate Change Policy Post-2020*, edited by Robert N. Stavins and Robert C. Stowe. Cambridge, MA: Harvard Project on Climate Agreements.

Victor, David G. 2015. "Why Paris Worked: A Different Approach to Climate Diplomacy." *Yale Environment 360*. https://e360.yale.edu/features/why_paris_worked_a_different_approach_to_climate_diplomacy

Vidal, John. 2021. "It Could Have Been Worse, but Our Leaders Failed Us at Cop26. That's the Truth of It." *The Guardian*, December 20, 2021. https://www.theguardian.com/environment/2021/nov/13/heres-the-truth-our-leaders-at-cop26-have-failed-us-the-rest-is-spin

Vidal, John, Allegra Stratton, and Suzanne Goldenberg. 2009. "Low Targets, Goals Dropped: Copenhagen Ends in Failure." *The Guardian*, October 12, 2009. https://www.theguardian.com/environment/2009/dec/18/copenhagen-deal

Viñuales, Jorge E. 2015. *The Rio Declaration on Environment and Development: A Commentary.* Oxford: Oxford University Press.

Welchman, Jennifer. 2001. "Is Ecosabotage Civil Disobedience?" *Philosophy & Geography* 4 (1): 97–107. https://doi.org/10.1080/10903770124815

4

Has Climate Change Ended Nature?

ELENA CASETTA

In his 1989 book *The End of Nature*, Bill McKibben claims that, because of large-scale climate change produced by human technologies, no place on Earth can be considered natural anymore. In 2000, at a conference in Cuernavaca, Mexico, Paul J. Crutzen proclaimed that we live in the Anthropocene, a new phase in the history of the planet in which humankind has imposed itself as a decisive influence on the global ecology, interfering with its fundamental systems. Is nature truly over? And, if so, are we left with nothing more to do than mourning its end? In this chapter, I reconstruct how humans have allegedly ended nature (Section 1); hence I analyse two different possible readings of the 'end of nature' claim, namely the ontological and the epistemological readings, showing that the first one is either false or unfounded, while the second one is possibly true (Section 2). Finally, I show how the analysis previously conducted can help in better focusing the (ontological) target of our conservation actions and the (epistemological) tools at our disposal (Section 3).

1. How Humans Are Supposed to Have Ended Nature

The view that human activities may have an impact on climate is not a novelty. The German naturalist and explorer Alexander von Humboldt (1769–1859) was probably the first person to theorise the existence of a link between human activities and climate, pointing out three ways in which the firsts affect the second: deforestation, reckless irrigation and the 'great masses of steam and gas' produced by industrial centres (Wulf 2015, 326). The American scientist Eunice Foote (1818–1888) was, by comparison, the first person who tested experimentally the hypothesis that atmospheric gases affect Earth's temperature, or what we call today 'the greenhouse effect'. In her 1856 article, 'Circumstances affecting the heat of the sun's rays', Foote related changes in the types and quantities of atmospheric gases – including carbon dioxide (CO_2) – to earth's temperature, concluding that 'an atmosphere of that gas [CO_2] would give our earth a high temperature' (Foote 1856, 383).

The phenomena underlined by Humboldt were destined to escalate together with the greenhouse effect tested by Foot. The main reasons for this were the massive increase of human population, on the one hand, and energy consumption and related anthropogenic emissions of greenhouse gases on the other hand.

Starting with population growth, around 1800 there were just one billion human beings on Earth, and it had taken many thousands of years to reach that number. By 1930, the human population had doubled. Then, in the span of one human lifetime, the global population tripled from 2.3 billion in 1945 to 7.2 billion in 2015. Concerning energy, in the late eighteenth century, humankind shifted from an organic energy regime, based on human and animal force for power, and wood and biomasses for heat, to a fossil-based energy regime. According to John McNeill and Peter Engelke (2014, 9), coal became the world's primary fuel at the end of the nineteenth century, and then oil took up the position in the mid-1960s. At the same time, energy use increased: by 1870, human beings were using more fossil fuel energy each year than the annual global production from all photosynthesis; today they use about 32 billion barrels of oil each year: 'the burgeoning rate of energy use in modern history makes our time wildly different from anything in the human past' (McNeill and Engelke 2014, 10). Burning fossil fuels and deforestation are the two main ways through which humans add carbon to the atmosphere: as of 7 June 2021, the Research News Webpage of the United States National Oceanic & Atmospheric Administration (NOOA), announced that the atmospheric carbon dioxide measured at NOAA's Mauna Loa Atmospheric Baseline Observatory reached a monthly average of 419 parts per million (ppm) – compared with the 280 ppm pre-industrial baseline (McNeill and Engelke 2014, 64); this is the highest level registered since accurate measurements began.

At first, the increasing human impact on the biosphere was gradual, but since the mid-twentieth century, the so-called 'Great Acceleration' began (McNeill and Engelke 2014; Steffen et al. 2015). The human impact escalated fast, and human actions began to interfere significantly with crucial biogeochemical cycles (the interconnected processes through which the elemental components of organic matter are cycled through the biosphere, such as the carbon, sulphur, nitrogen and water cycles), affecting their capacity for self-adjustment and hence altering global climate. Humans can hence be compared, according to some, to a 'great force of nature' (Ellis 2018, 2) that is reshaping the planet, putting an end to the relatively stable conditions that characterized the Holocene – the post-glacial geological epoch that started approximately 11,000 years ago.

The magnitude and the pervasiveness of the phenomena described above have led some authors to decree the end of nature by the hand of humans. In particular, the American environmentalist and journalist Bill McKibben – in an early popular book about climate change, entitled *The End of Nature* (1989) – denounced the anthropogenic changes to nature that were affecting the entire planet, to the point that 'we are at the end of nature' (McKibben 1989; rev. ed. 2003, 7). What makes the difference, according to him, is that even though human activities like deforestation and pollution have had an impact on the environment for a long time, they used to occur on a local scale (they altered the places in which they occurred but not those in which they did not), while climate change is a global phenomenon that involves the entire planet, including places not inhabited by human beings.

> Short of widescale nuclear war, global warming represents the largest imaginable such alteration: by changing the very temperature of the planet, we inexorably affect its flora, its fauna, its rainfall and evaporation, the decomposition of its soils. Every inch of the planet is different; indeed, the physics of climate means the most extreme changes are going on at the north and south poles, farthest from human beings (McKibben 1989; rev. ed. 2003, xv).

In 2000, at a conference in Cuernavaca, Mexico, Nobel-prize winning atmospheric chemist Paul J. Crutzen stood up in frustration towards his colleagues still referring to our epoch as the Holocene, exclaiming, 'We are in the Anthropocene!' (Ellis 2018, 1). As illustrated in a brief note published the same year in the 'Global Change News Letter' (Crutzen and Stoermer 2000), during the Holocene human activities 'gradually grew into a significant geological, morphological force'. The global effect of these human activities had become evident in the latter part of the eighteenth century, the conventional starting date of the Anthropocene. And, unless major catastrophes occur, humankind seems destined, according to Crutzen and Stoermer, to remain a major geological force for millennia, maybe millions of years. (Note, however, that the Anthropocene has not yet been formally recognized by geologists and that several criticisms have been raised against the idea – see, for instance, Santana 2019).

If we take seriously the claim that anthropogenic climate change has ended nature, then – as political scientist Steven Vogel argues in *Thinking like a Mall: Environmental Philosophy after the End of Nature* (2015) – we would be left with nothing more to do than mourn nature's end, because, once destroyed, nature cannot be restored. In fact, to restore nature, human intervention would be required, but – by definition – human productions are artifacts, and a newly planted forest, for instance, would be an artifact as well. Accordingly, and paradoxically, restoring nature would result in increasing the number of artifacts on the planet (Katz 1992). The end of nature challenges traditional environmentalism, focused mainly on nature conservation and ecology, studying the workings of intact ecosystems rather than ways to manage them (Editorial 2008). However, such a challenge can be constructive, making us rethink nature and the place of our species in it, and hence helping to better focus our conservation targets and our epistemological tools, bringing to light new possible paths for action.

2. Two Ways of Understanding the Alleged End of Nature

Does the advent of the age of humans mean the end of nature? Is nature *actually* over? The first step in answering these questions is to clarify what 'nature', and hence its end, means. The claim that nature has ended can be understood in both an ontological and epistemological way. While ontology has to do with the world, its entities and processes, epistemology has to do with our knowledge and especially our justified beliefs about the world, its entities and processes.

The ontological reading of the end of nature suggests that once there was something that we called 'nature', and that it does not exist anymore because of our activities, especially anthropogenic climate change. But what was it, that *something*? Limiting our analysis to the western use of the word, there are two main possibilities. The first one is that *nature* is what is opposed to the *supernatural*. According to Aristotle (384–322 BC), who first defined 'nature' (Owens 1968; Lammer 2016), the technical, philosophical meaning of the term has to be limited to things that change. More precisely, 'natural' things have an inner principle of change and being at rest, namely all the – living and not living – things of the visible and tangible universe, as opposed to abstract things, like theorems and numbers, and the unmoved motor. In this sense, since artifacts are not supernatural things, they seem to belong to the ontological domain of natural entities (however – as we are going to see – they do not possess the principle of change in themselves). Similarly, John Stuart Mill (1865–1868) defined 'nature' in its technical meaning, as 'the sum of all phenomena, together with the causes which produce them' (Mill 1874; 2009, 66); in other words, the sum of the phenomena and their causes that inhabit the non-supernatural world.

If we endorse such a scientific understanding of 'nature', Ellis's claim – that humans have become a great force of nature that is reshaping the planet – makes perfect sense. Around 2.4 billion years ago, in the so-called 'Great Oxidation Event', cyanobacteria started changing Earth's atmosphere from a mainly CO_2-based one to an oxygen-based one, reshaping the earth and causing the extinction of organisms unsuited to the new atmosphere, allowing the evolution of life as we know it today. Under the Aristotelian-Millian technical understanding of 'nature', the fact that human beings are changing Earth's atmosphere would be a perfectly natural event – just like the Great Oxidation. While fully coherent from a scientific point of view, it is evident that such an understanding does not leave, at least *prima facie*, much room for manoeuvre: human beings and their activities are as natural as every other non-supernatural entity and process; the claim that anthropogenic climate change has ended nature is simply false; and if we want nature to run its course, we should just let it be. I argue below that this 'let it be' attitude does not necessarily follow from the technical meaning of 'nature'.

Both Aristotle and Mill recognize a second meaning of 'nature', however. Aristotle contrasts natural entities with those entities that are not supernatural, but that do not have in themselves the principle of change, namely artifacts. While a horse embryo has the principle of change, for instance – i.e., the active principle that allows the embryo to become an adult organism – a statue does not: a certain portion of matter becomes a statue only by virtue of a human maker. Mill also introduces a distinction between natural and hand-made entities, recognizing that the technical definition of 'nature' conflicts with 'the common form of speech by which Nature is opposed to Art, and natural to artificial':

> For in the sense of the word Nature which has just been defined, and which is the true scientific sense, Art is as much Nature as anything else; and everything which is artificial is natural ... Art is but the employment of the powers of Nature for an end. Phenomena produced by human agency, no less than those which as far as we are concerned are spontaneous, depend of the properties of the elementary forces, or of the elementary substances and their compounds. The united powers of the whole human race could not create a new property of matter (Mill 1874; 2009, 67).

Hence, for Mill, a non-scientific, non-technical sense of 'nature' must be recognized that opposes natural entities and phenomena to other entities and phenomena that take place by virtue of human agency. If everything that takes place by virtue of the – intentional or unintentional – activity of humans is considered artificial, then it is true that, because of climate change, no nature remains, because no place on Earth remains untouched by human-caused global warming.

What is in play in the ontological reading of the claim that nature has ended is the ontological status of human beings: either human beings are natural entities, together with the products of their activities (as in the first meaning of nature), or they are separate from nature (as in the second meaning of nature). The roots of the alleged divide between human beings and nature can be traced back once again to Aristotle. Humans, for Aristotle, enjoy a sort of twofold status (Owens 1968). On the one hand, being composed of matter (the body) and form (the soul), they belong to the domain of natural entities. However, one part of the human soul, the intellect, is twofold, consisting of passive intellect, which is perishable, and active intellect, which is separate and imperishable. Just like the unmoved motor, the active intellect does not change. The presence of the active intellect calls into question the belonging of human beings to the domain of perishable, changing things, namely natural entities. Despite the serious interpretive difficulties Aristotelian psychological theory has engendered (Shields 2020), the idea that human beings enjoy a peculiar status compared to 'mere' natural entities – either by virtue of a special relationship with some supernatural beings (think of religious narratives like the Hebrew Genesis), or of their minds (think of the Cartesian *res cogitans*, which distinguishes human beings from other animals) – is part of the culture of our species. However, on the basis of Charles Darwin's theory of evolution by natural selection, humans' peculiar ontological status can be questioned. Indeed, the Darwinian theory allows to explain the presence of human beings on Earth together with their mind without invoking a supernatural Creator. Both are the product of million years of natural selection conserving random adaptive mutations and discharging non-adaptive ones. From the point of view of evolutionary biology, we are natural entities just like the cyanobacteria, just different branches of the same tree stemming from a common root.

Two objections can be raised to the claim that human beings do not enjoy any peculiar status compared to 'mere' natural entities, and while the first one misses the target, I argue that the second one can help in finding our way out of the nihilistic outcome that the alleged end of nature seems to engender.

The first objection is that it is possible to find a biologically sound foundation of the difference between humans' and other organisms' *activities*. According to American philosopher Eric Katz, humans are natural entities, but their activities can be both natural and unnatural. They are unnatural when they go 'beyond our biological and evolutionary capacities' (Katz 1995, 95). For instance, 'natural' childbirth is a human activity, yet it can be considered 'natural' because it lies within the scope of humans' biological make-up. On the contrary, medicalized childbirth should be considered unnatural because it manipulates a natural biological process (Katz 1995, 95). While *prima facie* appealing, this way of distinguishing natural from unnatural human activities faces two limits. First, where to trace the boundary? If a mother learns a position that helps her in giving birth, would her childbirth be unnatural? We would probably answer negatively, but we would probably answer in the positive if the mother were given an epidural. Yet, both the new position and the epidural are the product of human beings learning something – of human culture, so to speak – and hence, why should we distinguish the first from the second? And here comes the second limit: any way of tracing the boundary seems to presuppose a distinction between the products of culture and the products of nature. But why should human culture and the products of it not be within the scope of human biology, the result of human evolution and, hence, perfectly within our evolutionary capacity? Just as a beaver has evolved its capability of building dams, we have evolved ours. To put it in another way, how could we do something which is not within our evolutionary capacity? Unless we are supernatural beings, the only possible answer is that we cannot. In Elliott Sober words, 'If we are part of nature, then everything we do is part of nature' (Sober 1986, 180, emphasis in the original).

The second objection is that, unlike cyanobacteria, we have developed the capability of self-reflection and choice. Like cyanobacteria, humans and human activities are perfectly natural but, while cyanobacteria could not help but emit oxygen, we can *decide* to do something to limit CO_2 emissions. We can decide not to end nature. Provided, of course, that nature has not yet ended. But, as we have seen, in the scientific understanding of 'nature', this claim is false; while in the 'common form of speech' pointed out by Mill, it is unfounded.

It is our – evolved – capability of self-reflection and choice that grants that the 'let it be' attitude mentioned before does not follow necessarily from the first meaning of 'nature'. The fact that human beings and their activities are natural entities and processes does not necessarily imply that humans cannot change the course of their actions. On the contrary, recognising that climate change is anthropogenic means recognising who is factually responsible and who has the means to act to counteract that change. We shall return to this idea in the last Section of the chapter; first, the other possible – epistemological – meaning of the claim that nature has ended has to be clarified.

The epistemological reading of the claim that nature has ended because of anthropogenic climate change is that our beliefs about nature have come to an end because they no longer reflect what nature has become.

> When I say 'nature,' I mean a certain set of human ideas about the world and our place in it. But the death of those ideas begins with concrete changes in the reality around us – changes that scientists can measure and enumerate. More and more frequently, these changes will clash with our perceptions, until, finally, our sense of nature as eternal and separate is washed away, and we will see all too clearly what we have done' (McKibben 1989; rev. ed. 2003, 7).

What is that 'set of human ideas' about which McKibben is writing? According to him (1989; rev. ed. 2003, 61), 'our view' is that nature is separate and independent from human beings: 'Nature's independence *is* its meaning; without it there is nothing but us'. In other words, our view is that nature is wilderness, namely 'pristine places, places substantially *unaltered* by man' (McKibben 1989; rev. ed. 2003, 56, emphasis in the original). This is the view of nature that anthropogenic climate change ended.

Wilderness was defined in the 1964 US Wilderness Act (1964) as follows,

> A wilderness, in contrast with those areas where man and his own works dominate the landscape, is hereby recognized as an area where the earth and its community of life are untrammeled by man, where man himself is a visitor who does not remain.

According to the Act, to fall under the definition of wilderness, an area must satisfy the following criteria: bearing unnoticeable human imprint; offering 'outstanding opportunities for solitude or a primitive and unconfined type of recreation'; being at least 5,000 acres of land; and containing 'ecological, geological, or other features of scientific, educational, scenic, or historical value' (Wilderness Act 1964).

Several serious criticisms have been raised against the idea of wilderness (Callicott and Nelson 2008; Merchant 2003). Here I shall briefly present one of them, namely that the idea of wilderness is a myth (Sarkar 2005).

By myth, philosopher of science Sahotra Sarkar (2005, 28) means 'a general story with normative implications, parts of which are known to be false, or at least implausible, but which is nevertheless useful in analysing other, more veridical stories that share some crucial aspects with it'. In the case of wilderness, the lack of veridicality of the narrative lies in the claim that the places designated as wilderness by the Wilderness Act, or the places that we perceive today as wilderness, have not been occupied by humans – they are pristine. These claims are generally false. To give just one paradigmatic example, when European colonisation began, tens of millions of people were living in North and Central America, with a long history of interaction with the land. Then, mainly because of European-originated diseases, up to 90% of the Indigenous population died: 'Arguably it was only because of this massive depopulation that it even became possible to view the North American continent as a pure wilderness: the land seemed unpopulated by humans simply because they had died' (Vogel 2015, 5). For colonists, the 'wilderness condition' of North America was a negative one, to be eradicated and replaced by neat and tidy farms and cities. Its human inhabitants, the savages, also had to be domesticated and civilised (Standing Bear 1933). When Yosemite in 1864 and then Yellowstone in 1872 were declared national parks, Indigenous people were excluded from them and their land was proclaimed 'to have been unoccupied by humans from the beginning of time' (Sarkar 2005, 40). Today, wilderness is generally loaded with a positive connotation that reflects the needs and desires of the new inhabitants, such as the need for 'solitude or a primitive and unconfined type of recreation' mentioned in the Wilderness Act (1964).

In the introduction to their book on the wilderness debate, environmental philosophers J. Baird Callicott and Michael P. Nelson (2008) explain how the wilderness idea has been accused of being 'a conversation of the West', that is the 'Americanized Western civilization', or the Euro-American one, which is true. However, I would add that a further difference should be considered when speaking of human beliefs concerning nature. As a matter of fact, even limiting our reflections to the present western world, the European view of nature might be quite different from the American one, especially because in Europe nature has been anthropised for much longer than in America. If we look, for instance, at European conservation policies, we can realise that wilderness in Europe started receiving attention only in the 1990s and that an agreement for a common understanding and interpretation of what 'wilderness' means in the European context was only created in 2012, according to the European Wilderness Society website. (For a review of the meaning of 'nature' in European languages, see Ducarme and Couvet 2020).

If the view of nature ended by climate change is a myth rather than an epistemologically well-founded belief, then claiming that climate change ended the idea/myth of wilderness would just imply the end of an unfounded belief. Recognising this fact may turn the epistemological 'end of nature' into an incentive to proceed towards both a better, unbiased knowledge of nature itself and a better understanding of the relation between human beings and the rest of nature.

Let us now sum up the results of the analysis so far. Our starting question was whether anthropogenic climate change had ended nature, as has been claimed. We distinguished two ways of understanding this claim, an ontological one and an epistemological one. Under the ontological reading, the claim is either false or unfounded. Under the epistemological reading, it is true and possibly beneficial. This is good news. From an ontological point of view, nature has not ended – hence, there is more to do than just mourning its end. From an epistemological point of view, some of our beliefs about nature have been proven to be false – hence, we are in the position of replacing them with better ones and rethinking our relationship with other natural entities.

3. Nature Has Not Ended, Action Is Needed

That nature has not ended means that we still have time to act. The ontological and epistemological analysis previously conducted may now help in better focusing the target of our conservation actions and the tools at our disposal.

From the ontological analysis conducted above, it follows both that the focus of our environmental concerns, and more specifically of climate policies, is not nature *per se*. What is it, then? Let us go back to the technical-scientific meaning of 'nature', namely nature as the non-supernatural world. Everyone may agree that our environmental concerns do not involve the entirety of nature (for instance, we are not worried about the state of other planets, which are a part of nature at the same time as the earth is), but just a part of it, namely the biosphere –Earth's 'zone of life' extending from a few kilometres into the atmosphere to the oceans' deep-sea vents, which originated between 3.5–3.8 billion years ago.

According to the founder of the modern concept of the biosphere – the Russian scientist Vladimir Ivanovich Vernadsky (1863–1945) – the biosphere originated together with life: life is a geological force that can – and did – change Earth's landforms, climate and atmosphere. It has transformed the earth from a rocky place with shallow oceans and an atmosphere made of toxic gases into the planet we know and inhabit today. 'Between its inorganic "lifeless" and living parts, inhabiting it, exists continuous exchange of matter and energy, expressed by atomic movement caused by the living matter' (Vernadsky, quoted in Svirezhev and Svirejva-Hopkins 2008, 468). Moreover, our environmental concerns do not involve the biosphere *per se*: for instance, we would not want the biosphere to become as it was about 20,000 years ago, when permanent summer ice covered about 25% percent of the land area, a large part of the world was dry and inhospitable and the atmosphere

93   *Global Climate Justice: Theory and Practice*

was laden with dust. Rather, we are concerned with a particular state of the biosphere, that can be roughly identified with the world as we know it, so to speak, namely the state of the biosphere during the Holocene. It is a relatively warm and cosy state that started when the last ice age ended, the temperature increased by 6°C, sea levels rose by 120 metres and CO_2 in the atmosphere increased by one-third (Maslin 2014, 3-4). This is what we care about; these are the conditions that allow a good life to us and the organisms that we know and cherish.

The previously conducted ontological analysis allows us to recognize that our environmental concerns are not aimed at nature, but rather at a specific state of the biosphere. This state of the biosphere is changing. In 1990, the first report of the Intergovernmental Panel on Climate Change (IPCC) concluded that anthropogenic climate change would soon become evident, but it was not able to confirm that it was already happening. Today, as the last IPCC report reads (IPCC 2021, FAQs Section, 6).

> [the] evidence is overwhelming that the climate has indeed changed since the pre-industrial era and that human activities are the principal cause of that change... the main human causes of climate change are greenhouse gases released by fossil fuel combustion, deforestation, and agriculture.

Recognising human beings as the driver of climate change is the first, necessary step, to cope and possibly control and limit that change.

We clarified how humans and their activities are a part of nature – if 'nature' is understood in its technical sense. If the biosphere is the sum and interaction of all its biotic and abiotic components, considering humans as separate from the rest of nature runs the risk of reflecting an incomplete view of the biosphere. Two caveats, however, are in order.

First, it does not follow from the claim that human beings are natural entities that human activities cannot damage other natural entities and processes, as well as humans themselves. A useful theory to show this point is the Niche Construction Theory (NCT). Organisms, through their metabolisms, activities and choices alter their local environments, 'constructing' their own niches – think of the construction of nests and burrows, but also the alteration of the soil and more generally of local climate by plants. In doing so, they modify the sources of natural selection, generating a form of feedback in evolution (Odling-Smee et al. 1996). Normally, niche construction is beneficial to the constructor, since it counteracts natural selection's pressures on it; however, it can be deleterious for other species, for instance, when the niche constructor is an invasive species. Sometimes it can even decrease the fitness of the constructor itself, for instance, when in building their niches, 'organisms also partly destroy their habitats, through stripping them of valuable resources or building up detritus' (Laland et al. 2000). This last case is the so-called 'negative niche construction process'. Human beings' activities can be read, like those of any other organism, through the lenses of niche construction theory: their activities aim at improving their own fitness (adaptive niche construction), but – as a side-effect – they can decrease the fitness of other organisms and even their own fitness. However, as mentioned, unlike other organisms, human beings can – in principle – control their niche construction activity to limit its possible negative effects.

Since human beings must be considered a constitutive part of the biosphere, niche construction theory can reveal a useful instrument to proceed towards a better understanding of human interactions with the environment. A recent study has suggested, for instance, that anthropogenic climate change can be understood as a 'monumental niche construction process' (Meneganzin et al. 2020) that is putting present and future generations' at risk. However, while some attention has already been paid to the possible environmental applications of NCT from environmental studies (Ellis 2016) and conservation biology (Boogert et al. 2006), a systematic incorporation of NCT in environmental and climate policies is, to my knowledge, still missing.

The second caveat is that, while recognising that human beings and their activities are a part of nature allows us to consider the biosphere as a whole, it does not imply that a distinction between natural entities and artifacts can't be traced, as John Stuart Mill already recognised.

I suggested elsewhere (Casetta 2020) that the natural/artificial distinction can be *operationally* maintained if conveniently reframed. Following Sarkar,

> Even if humans are conceptualized as part of nature, we can coherently distinguish between humans and the rest of nature. There is at least an *operational* distinction; that is, one that we can straightforwardly make in practical contexts. We can distinguish between anthropogenic features (those largely brought about by human action) and non-anthropogenic ones' (Sarkar 2012, 19, emphasis in the original).

Operationally, the categorical and fixed distinction between natural and artificial entities and processes may be reframed in a more dynamic way that recognizes that naturalness – and artificialness – are matters of degree and that they are relative to time. Different – and context-sensitive criteria – may then be employed to assess and monitor the naturalness of a place or an ecosystem, such as for instance the degree of change expected if humans were removed, the degree of sustained control and the extent and abruptness of change following the cessation of human activities (Angermeier 2000). To give but one concrete example, from this view it follows that a newly planted forest – a so-called artificial object, since it has been planted by humans – may become, through time, a natural one if its persistence ceases to depend on human support. In such a framework, the concept of wilderness can be rethought as well and its narrative freed from non-veridical elements.

As said, recognizing human beings as the driver of climate change is the first, necessary step to cope and possibly control and limit that change; and acknowledging that they are part of nature allows us to consider the biosphere as a whole, taking into account all its components and processes. In such a perspective, on the one hand, NCT can prove to be a useful descriptive and predictive tool for anthropogenic climate change; on the other hand, reframing the natural/artificial distinction in a more dynamic way can help in facing the challenges that 'the end of nature' poses to both ecology and environmentalism.

The second step consists in knowing how the biosphere works and how it is expected to work in the future, i.e., recognizing and studying the relations between its biotic and abiotic components, and developing reliable climate models. With the second step, we enter the field of epistemology. Once the ontological focus has moved from wilderness to the biosphere, at least two questions may be asked, namely which justified beliefs do we have on the actual state of the biosphere? And which is the best scientific approach to the study of the biosphere as a whole?

Concerning the first question, the last IPCC report – published between August 2021 and April 2022 – maintains that several new instruments have been deployed to collect and integrate data. To give but one example, when the IPCC started in 1990, very little was known about the consequences of climate change on the deep ocean, while today it is known that oceans soak up most of the surplus energy captured by greenhouse gases in the atmosphere and that even the deep ocean is warming. It is known which human activities have the most impact on climate (i.e., greenhouse gases released by fossil fuels, deforestation, agriculture and aerosols from burning coal) and more and better observations of their impact are available. More sophisticated climate models allowing the prediction of patterns of change have been elaborated. For instance, while old-generation models – climate modelling started in the 1950s – focused mainly on the atmosphere, considering oceans and land surfaces only marginally, today models include detailed considerations of many other variables (such as oceans, ice, snow and vegetation). Models can then now simulate complex interactions between different entities and processes of the biosphere, such as, for instance, the interaction between clouds and air pollutants. As the IPCC (FAQ Section, 20) states, 'Developments in the latest generation of climate models, including new and better representation of physical, chemical and biological processes, as well as higher resolution, have improved the simulation of many aspects of the Earth system'.

The second question stems from considering human beings and their activity as a constitutive part of the biosphere. Studying the functioning of the biosphere requires the cooperation of several sciences – from physics to chemistry, from climatology to Earth System Science, from ecology to evolutionary biology, and so on. Moreover, if human beings are genuinely considered part and parcel of the biosphere in the same way as other natural entities, the humanities come into play. In the light of this, I suggest that a transdisciplinary approach (Klein 2004) that includes natural and social sciences is required. Two kinds of transdisciplinarity can be distinguished. 'Deep' transdisciplinarity aims at building up a total system of knowledge with no constraints at all on the type of knowledge in play: shamanistic practices can be considered together with scientific knowledge (Max-Neef 2005). However, deep transdisciplinarity is an epistemologically risky enterprise because the epistemic status of the resulting discipline would be strongly questionable (Marques da Silva and Casetta 2015). 'Shallow' transdisciplinarity is a more cautious approach, epistemologically, that calls for 'trans-sector problem solving' where the focus of research is a certain global problem such as – in our case – anthropogenic climate change that requires collaboration among a mix of actors from different disciplines, professions and sectors of society.

Conclusions

When McKibben denounced 'the end of nature', his main intention was to shake public opinion, to denounce the consequences of climate change and its anthropogenic nature. It was 1989, the IPCC had just been founded and climate scepticism was widespread. Still in 2003, leading Republican consultant Frank Luntz advised the Bush administration to replace talking about 'global warming' with 'climate change', because the latter phrase was considered less frightening and because 'change' avoids implying human agency (Heink and Jax 2014). The memo also suggested that politicians endorse the view that there was no scientific consensus on the dangers of greenhouse gases (Burkeman 2003). Several things have since changed. Despite any political communication agenda, 'climate change' turned out to be no less frightening than 'global warming', and evidence that points at the human agency cannot be questioned anymore, as the last IPCC assessment report documents.

McKibben's operation may probably be ascribed to what German philosopher Hans Jonas (1903–1993) called a 'heuristic of fear' (Jonas 1984, x). According to him, while fear seems to be, at least *prima facie*, a negative emotion, it can instead serve as a guide if fear is understood as an 'imaginative-anticipatory' heuristic. In other words, to avert those negative scenarios that we fear, we may be urged to act or re-evaluate our current course of action. While there can be some effectiveness in a heuristic of fear, it may also sort the opposite effect, like the one denounced by Steven Vogel, i.e., mourning instead of acting. For instance, fearing the end of nature understood as wilderness might result in underestimating the importance of the management of our everyday environment, made mainly of cities (that host 55% of the population), which already are local hotspots of global warming (they are generally warmer than their surroundings). More generally, considering human beings as separate from the rest of nature might lead to both an incomplete and partial view of the current functioning of the biosphere and the misrepresentation of the human role in such a functioning.

References

Angermeier, Paul. 2000. "The Natural Imperative for Biological Conservation," *Conservation Biology*, 12 (2): 373–381.

Boogert, Neeltje J., David M. Paterson, and Kevin N. Laland. 2006. "The Implications of Niche Construction and Ecosystem Engineering for Conservation Biology." *BioScience* 56 (7): 570578. https://doi.org/10.1641/0006-3568(2006)56[570:TIONCA]2.0.CO;2

Burkeman, Oliver. 2003. "Memo Exposes Bush's New Green Strategy." *The Guardian*, March 4. https://www.theguardian.com/environment/2003/mar/04/usnews.climatechange.

Callicott, J. Baird and Michael P. Nelson. 2008. "Introduction". In *The Great New Wilderness Debate*, edited by Idd., Athens (GA): University of Georgia Press, 1–20.

Casetta, Elena. 2020. "Making Sense of Nature Conservation After the end of Nature." *History and Philosophy of the Life Sciences* 42: 18. https://doi.org/10.1007/s40656-020-00312-3.

Crutzen, Paul J. and Eugene F. Stoermer. 2000. "The 'Anthropocene'." *Global Change News Letter* 4: 17–18.

Ducarme, Frédéric and Denis Couvet. 2020. "What Does 'Nature' Mean?" *Palgrave Commun* 6 (14): 1–8. https://doi.org/10.1057/s41599-020-0390-y.

Editorial. (2008). "Handle with Care." *Nature* 455: 263–264. https://doi.org/10.1038/455263b

Ellis, Erle. 2018. *Anthropocene: A Very Short Introduction*. Oxford: Oxford University Press.

Ellis, Erle. 2016. "Why Is Human Niche Construction Transforming Planet Earth?" In: *Molding the Planet: Human Niche Construction at Work*, edited by Maurits W. Ertsen, Christof Mauch, and Edmund Russell, *RCC Perspectives: Transformations in Environment and Society*, 5, 63–70. https://doi.org/10.5282/rcc/7733

Foote, Eunice. 1856. "Circumstances Affecting the Heat of the Sun's Rays." *American Journal of Science and Arts* 22: 382–383.

Heink, Ulrich and Kurt Jax. 2014. "Framing Biodiversity." In *Concepts and Values in Biodiversity*, edited by Dirk Lanzerath and Minou Friele. London: Routledge, 73–98.

IPCC Sixth Assessment Report. 2021. *Frequently Asked Question*, https://www.ipcc.ch/report/ar6/wg1/downloads/faqs/IPCC_AR6_WGI_FAQs_Compiled.pdf

Katz, Eric. 1992. "The Big Lie: Human Restoration of Nature." *Research in Philosophy and Technology* 12: 231–241.

Katz, Eric. 1995. "Restoration and Redesign: The Ethical Significance of Human Intervention in Nature." *Restoration & Management Notes* 9 (2): 90–96.

Klein, Julie T. 2004. "Prospects for Transdisciplinarity." *Futures* 36: 515–526.

Jonas, Hans 1984. *The Imperative of Responsibility: In Search of an Ethics for the Technological Age*. Chicago: The University of Chicago Press.

Laland, Kevin N., John Odling-Smee, and Marcus W. Feldman. 2000. "Niche Construction, Biological Evolution, and Cultural Change" *Behavioral and Brain Sciences* 23: 131–175.

Lammer, Andreas. 2016. "Defining Nature: From Aristotle to Philoponus to Avicenna." In *Aristotle and the Arabic Tradition*, edited by Ahmed Alwishah and Josh Hayes, 121–142. Cambridge: Cambridge University Press.

Marques da Silva, Jorge and Elena Casetta. 2015. "The Evolutionary Stages of Plant Physiology and a Plea for Transdisciplinarity." *Axiomathes* 25 (2): 205–215. https://doi.org/10.1007/s10516-014-9257-4

Maslin, Mark. 2014. *Climate Change. A Very Short Introduction.* Oxford: Oxford University Press, 1st ed. 2004.

Max-Neef, Manfred. 2005. "Foundations of Transdisciplinarity." *Ecological Economics* 53 (1): 5–16. https://doi.org/10.1016/j.ecolecon.2005.01.014

Meneganzin, Andra, Telmo Pievani and Stefano Caserini. 2020. Anthropogenic climate change as a monumental niche construction process: background and philosophical aspects. *Biol Philos* 35: 38. https://doi.org/10.1007/s10539-020-09754-2

Merchant, Carolyn. 2003. *Reinventing Eden: The Fate of Nature in Western Culture.* New York and London: Routledge.

Mill, John Stuart. 1874. "Nature." In Id. *Three Essays on Religion,* edited by L.J. Matz. Peterborough, ON: Broadview Edition, 65-104.

McKibben, Bill. 1989. *The End of Nature.* New York: Penguin Random House. Revised and updated edition: 2003. London: Bloomsbury.

McNeill, John R. and Peter Engelke. 2014. *The Great Acceleration. An Environmental History of the Anthropocene since 1945.* Cambridge (MA): Harvard University Press.

Odling-Smee, John, Kevin N. Laland and Marcus W. Feldman. 1996. "Niche Construction." *The American Naturalist* 147 (4): 641–648.

Owens, Joseph. 1968. "Teleology of Nature in Aristotle." *The Monist* 52 (2): 159–173.

Santana, Carlos. 2019. "Waiting for the Anthropocene." *British Journal for the Philosophy of Science* 70: 1073–1096.

Sarkar, Sahotra. 2005. *Biodiversity and Environmental Philosophy. An Introduction.* Cambridge: Cambridge University Press.

Sarkar, Sahotra. 2012. *Environmental Philosophy: From Theory to Practice.* Oxford: Wiley-Blackwell.

Shields, Christopher. 2020. "Aristotle's Psychology." In *The Stanford Encyclopedia of Philosophy* (Winter 2020 Edition), edited by Edward N. Zalta. https://plato.stanford.edu/archives/win2020/entries/aristotle-psychology/

Sober, Elliott. 1986. "Philosophical Problems for Environmentalism." In *The Preservation of Species*, edited by Bryan B. Norton. Princeton: Princeton University Press, 2014, 173–194.

Standing Bear, Luther. 1933. *Land of the Spotted Eagle.* Lincoln and London: University of Nebraska Press, 2006, 192–225.

Steffen, Will, Wendy Broadgate, Lisa Deutsch, Owen Gaffney, and Cornelia Ludwig. 2015. "The Trajectory of the Anthropocene: The Great Acceleration." *The Anthropocene Review* 2 (1): 81–98. https://doi.org/10.1177/2053019614564785

Svirezhev, Yuri M. and Anastasia Svirejeva-Hopkins. 2008. "Biosphere: Vernadsky's Concept." In *Encyclopedia of Ecology*, edited by Sven Erik Jorgensen and Brian D. Fath, Oxford: Elsevier, 467–471.

Vogel, Steven. 2015. *Thinking like a Mall. Environmental Philosophy after the End of Nature.* Cambridge (MA): MIT Press.

Wilderness Act. 1964. https://www.fsa.usda.gov/Assets/USDA-FSA-Public/usdafiles/Environ-Cultural/wilderness_act.pdf

Wulf, Andrea 2015. *The Invention of Nature: Alexander von Humboldt's New World.* New York: Knopf.

SECTION TWO

THE GLOBAL DISTRIBUTIVE CHALLENGE
OF CLIMATE CHANGE

5

Libertarianism and Climate Ethics

ELIAS MOSER

Libertarians defend a minimal state that is only allowed to raise taxes in order to secure and enforce individual rights. Any redistributive aim is regarded as illegitimate. Therefore, it is not common to apply a libertarian political theory to justify policies to mitigate or adapt to climate change, since they require large-scale redistribution and taxation. Nevertheless, my aim in this chapter is to outline a libertarian framework for reflecting on the moral problem of human-caused climate change. I distinguish between two potential libertarian arguments. First, climate change can be said to threaten many rights of currently living and future individuals. The libertarian argument of rights-infringements defends governmental actions to prevent and to exact compensation for conduct that leads to a violation of individual rights. I discuss the conditions under which this line of reasoning justifies governmental actions to mitigate climate change. I conclude, however, that the argument is based on highly restrictive assumptions. Second, the use of the atmosphere can be considered an appropriation of a commonly owned resource. Applying a Lockean theory of just appropriation, libertarianism allows for climate action by showing that the excessive use of the atmosphere is unjust. Therefore, there is a case for justified redistribution. I defend this latter argument as more promising alternative to justify duties of climate change prevention. I will outline the normative implications by distinguishing them from those of more common accounts to climate justice.

Introduction

Commonly, libertarians are defendants of a minimal state that is only allowed to impose taxes in order to secure basic rights from interference. The primary purpose of the state is to protect and enforce strong property rights. Any redistributive aim, e.g., social insurance or the provision of public goods is considered illegitimate. The outcome of free markets based on decentralized, voluntary interactions is defended as a just allocation of income and capital among members of society. Yet, the transition from a fossil-fuel-based economy to a carbon-neutral economy cannot be achieved without enormous governmental investments in infrastructure, subsidies of sustainable energy use, and legal interventions in the free market, such as taxes, regulations and prohibitions. Thus, a libertarian political theory seems incapable of justifying policies to combat climate change.

As I will show, however, libertarian political philosophy does provide a highly interesting approach to thinking about the problem of climate change. It has two potential arguments that may support climate action. The first argument is based on an understanding of excessive greenhouse gas (GHG) emissions and the causation of global warming as a transgression of rights (Zwolinski 2014; Torpman 2021). The rights of those negatively affected by climate change are violated by those who emit too much. According to libertarian thought, rights-encroachments can only be justified if rights-holders consent to them. However, climate change leads to severe rights-violations of individuals who cannot possibly consent – such as children and future generations. Thus, a libertarian might argue that polluters have a duty to either drastically reduce emissions or to pay compensation for their rights-infringements.

The second argument draws on the idea of justified acquisition of property, which is a core element of any libertarian defence of strong property rights. Libertarians often refer to John Locke's (1980, chap. V) account of just appropriation, which maintains that property is generated not only by the use of one's labour and just transfer, but also the use and acquisition of external goods that, originally, do not belong to any individual. Worldly resources, Locke assumes, belong to everyone alike. He therefore introduces restrictions to the legitimate appropriation of these resources (the so-called 'provisos'). People, he concedes, are only allowed to use resources up to a point at which there is still 'enough, and as good, left' for others to use. Now, climate change can be construed as a problem of unjust acquisition of property. The atmosphere can be considered a finite resource and the current overuse of it as a sink does not seem to leave 'enough, or as good' for our children and future generations. Therefore, the Lockean proviso provides a powerful argument for the restriction of the use of the atmosphere and for duties of climate change mitigation.

Because libertarianism may conceptualize climate change as a problem of illegitimate property and because it asks the question of historically unjust use of commonly owned worldly resources it offers a unique rights-based account on climate justice that allocates duties of omission and of compensation according to legitimate claims. In this chapter, my aim is to elucidate and discuss these two arguments and to outline the conditions under which libertarianism accepts restrictions on individual liberties and property in order to combat climate change. I argue that, to justify duties of climate change mitigation, the second argument is more promising than the first.

The next section introduces the libertarian argument of individual rights-infringements and critically reflects its assumptions and implications. The third section distinguishes different possible views on just acquisition of natural resources according to the Lockean proviso. I introduce Robert Nozick's (1974, 167–174) 'weak interpretation' of the proviso and contrast it with stronger versions, such as left-libertarian accounts (Steiner 1994, 234–236; Otsuka 1998; Vallenthyne 2007). Depending on the strength of the interpretation of the proviso, different normative conclusions can be drawn from the libertarian account. Furthermore, I compare the implications of the libertarian argument from unjust acquisition of property with common accounts of climate justice. The last section outlines some conclusions.

1. The Argument of Rights-Infringement

Although most libertarians are against redistribution, they do not oppose the existence of the state as such. A minimal state that secures individuals' rights to life, liberty and property is generally accepted as justified (Brennan 2012, 57–59). Justified governmental institutions can restrict rights-infringements by means of a system of sanctions, law enforcement, and legal courts. If individual rights are infringed upon, the state determines the amount of compensation, exacts it from the wrongdoers and allocates it to the victims. Climate change can be said to threaten some of the most fundamental rights of individuals. Natural catastrophes threaten lives, water scarcity causes severe health problems, and sea-level rise and desertification force people to migrate (IPCC 2021, 15–31). Herein lies the libertarian argument for climate action (Singer 2004, ch. 2). Causing climate change can be seen as an act of illegitimate rights-infringement and the perpetrators should either be prohibited from causing it or owe compensation to affected individuals.

A few clarifications are required to understand this argument. First, the responsibility for climate change is different from paradigm cases of rights-infringements, such as theft, bodily injuries and homicide. Not every CO_2-emitting activity ultimately results in the violation of other people's rights. The concentration of GHGs in the atmosphere can cause a problem to humans only if emission levels exceed the earth's absorptive capacity. And even if that is the case, the overuse of the atmosphere could still be so small that humans have enough time to adapt without having their life, health or liberty threatened. It is, therefore, wrong to think about the problem in terms of categorical prohibitions (as with theft or killing). But I will come to that later.

Second, an individual's action alone cannot be said to effectuate a rights-infringement. Climate change is an outcome of collective action – the aggregate of a large number of individual excess emissions. Nevertheless, even if through 'miniscule' and 'imperceptible' actions, individuals cause harm to other people by causing global warming (Broome 2012, 56). The low magnitude of one's contribution does not preclude legitimate restrictions on her actions (Vallentyne and van der Vossen 2014, Sect. 3; cf. Railton 2003, 191).

Third, rights-infringements do not necessarily occur even if CO_2 emissions exceed some critical level. Environmental problems may occur in one way or another. There is no certainty as to how people will be affected by climate change or even whether they will be affected at all. Thus, the joint causation of an increase in global temperatures is not necessarily a direct violation of rights. It is an imposition of risk upon those potentially affected (cf. Nozick 1974, 73–77; Railton 2003, 193). The negative effect of risk can be conceptualized as a so-called 'expectation value' – the negative value of an outcome multiplied by the probability of its incidence. The worse the outcome and/or the higher the probability, the greater the harm for the individual.

Even if bad outcomes do not ultimately occur, imposing risk is an action that libertarians agree must be prohibited or compensated for (e.g., Shahar 2009, 228). Let me show this with an example. If a person is speeding, she is unjustifiably increasing the risk of harm to other road users. Even if the potential harm does not materialise, driving too fast has a negative effect on other individuals. They enjoy fewer freedoms because they have to be more careful on the road, or they have to pay more to insure themselves against harm. Thus, risk has an undesired effect on the potential enjoyment of individual rights. Libertarian approaches consider this negative effect an unjustified curtailment of rights and, thus (under certain circumstances) a prohibition of the risk-increasing action as justified (Nozick 1974, 65–71, 73–74).

1.1 Duties of Compensation or Duties of Prevention?

What exactly does the argument of rights-infringement justify? Which governmental actions are justified to combat climate change? To answer these questions, we need to understand the libertarian argument for the state. As mentioned above, libertarians regard a minimal state as justified to secure individual rights. The first five chapters of Robert Nozick's *Anarchy, State and*

Utopia outline why and how people would hypothetically consent to a centralised institution that restricts conduct that violates or endangers individual rights and that implements legal procedures to exact compensation in cases of rights-infringement.

The starting point of the argument is a conception of a state of nature in which people find themselves without governmental institutions. Individuals in the state of nature, Nozick assumes, possess rights to life, liberty, physical integrity and property. However, since some people would frequently infringe upon the rights of others, people face a constant risk. Individuals are, therefore, ready to partially abandon some of their rights and authorise a state to take measures to protect and enforce some of their other rights. Nozick and other libertarians thus assume that individuals would hypothetically consent to restrictions on their liberties and the authorisation of a minimal state.

A similar hypothetical history can be imagined in order to justify restrictions on CO_2 emissions. Present and future individuals hold rights that might potentially be infringed upon as a result of climate change. This expectation leads them to accept taxes, laws, and regulations in order to protect their rights in the future. Policies to prevent or adapt to climate change can be justified if individuals are willing to abandon some of their rights in order to reduce the risk of being killed or suffering severe impairments to their health and property.

However, Nozick (1974, 58–59, 71–73) and other libertarians would not argue that every rights-infringement justifies a prohibition of certain conduct. Rights can be protected either with prohibitions or with compensation for their violation. For example, if a person cannot or will not fulfil her contractual obligations, then she usually owes damages to the infringed party. Breach of contract is not prohibited, but the loss must be compensated. If rights-infringements can also be compensated for, one must ask, why ever prohibit conduct? In the context of climate ethics, this question translates into the question of how we should react to the challenges of climate change: should high-emitting countries set up funds to cover adaptation costs (i.e., compensate) or should they reduce CO_2 emissions and thus increase our effort to mitigate global warming (i.e., prohibit) or both?

Libertarians can easily argue for compensation for risks imposed on those potentially affected by climate change – that is compensation for the negative expectation value. Every person's excessive emissions increase the likelihood of people suffering from global warming. Within a country whose carbon footprint is above a sustainable level, everyone is partly responsible for increased risk. The state could, therefore, exact compensation for that risk from all excessively emitting individuals by taxing them. The raised funds in turn would be used to cover the adaptation costs of present and future people who are negatively affected by climate change.

It should be noted, however, that such a tax is not justified with reference to distributive justice. The redistribution that results from a duty of compensation does not infringe strong property rights. Rather it is a rectificatory means to reinstall justice in reaction to a prior injustice. Thus, there is a libertarian argument in favour of compensatory measures to cover adaption costs that is compatible with the libertarian rejection of distributive justice. With regard to climate change mitigation, however, the libertarian argument is more intricate.

1.2 Intergenerational Compensation for Prevention

Nozick is concerned with the question of how a minimal state can be justified given that it restricts certain conduct. Some individuals, he admits, do not benefit from such restrictions (such as mitigation efforts) and, therefore, in a hypothetical state of nature, would be unwilling to subject to a legal system that restricts their liberties. He asks why these people would accept being restricted by the state (Nozick 1974, 51) and he introduces what he calls a 'principle of compensation' (Nozick 1974, 78): individuals that do not benefit from being subjected to restrictions should be compensated for the restriction of their liberties.

This principle does not apply to cases of classical rights-infringements. Everybody can be assumed to benefit from the restriction of theft, violence, and killing since all members of society are potential victims. Presumably everyone would voluntarily subject herself to the legal system that restricts this conduct and would restrain herself from engaging in it, without claiming amends. However, not all risks of rights-violations apply to all equally. A state that prevents excessive emissions only benefits some while others carry the burden without ever being at risk of suffering from hazardous effects of climate change. In order to ensure anonymous, hypothetical consent, libertarians therefore need to argue for a principle of compensation for those who are not at risk but nevertheless urged to reduce emissions.

The major difference in the risk distribution exists between the living people and the future generations. The latter face far greater risks of severe impairments on their quality of life. Thus, reducing CO_2 emissions benefits them more than present generations. Moreover, measures to prevent climate change must be taken now. So, the costs of reducing emissions are borne by individuals in the present. From this perspective, living people would most probably agree to prevent climate change only if they were compensated for their costs of doing so since they only benefit little and bear all the costs.

At first glance, this line of thought seems to run contrary to any conception of climate justice. It seems to imply that we are allowed to charge the potential victims for the omission of our wrongdoing. Intuitively, it is not future people who owe something to us. Rather, we owe them. Nevertheless, a highly interesting proposal made by John Broome (2012, 43–48) goes in the same direction. It is not based on reflections of justice, but rather applies efficiency as a normative standard. He argues that an efficient hypothetical bargain between present and future people would result in the commitment to emissions reduction policies (cf. Posner and Sunstein 2008, 169–70). But he maintains that the future people would have to compensate us by paying public debts our generation has to take on to finance the restructuring of the economy. Broome argues for the legitimacy of charging future generations with the costs of mitigation with pragmatic reasons. In a state of crisis, we should not 'encumber the task of fixing climate change with the much broader task of improving the distribution of resource' (Broome 2012, 47). However, for libertarians, compensation is not something that is instrumentally reasonable. It is something that is indeed required by justice.

This libertarian proposal of an 'intergenerational contract' based on a Nozickian principle of compensation might be appropriate from the perspective of rectificatory justice, but it would have dramatic consequences from the perspective of distributive justice. An imbalance of burdens and benefits (risk reduction) of climate action exists not only between generations, but also between different countries. Certain states are fortunate to be relatively unaffected by the negative consequences of climate change, whereas for others it will have a catastrophic impact. As a matter of fact, wealthy regions of the world, such as the United States, will be less affected by climate change, while poorer countries, such as India, will be affected significantly (Sunstein 2007, 10–17).

The libertarian proposal would thus imply that the poor compensate the rich for preventing climate change. Hence, the libertarian argument for climate change mitigation yields distributive consequences that few scholars in climate ethics would be willing to accept. The argument, therefore, leads to *a reductio ad absurdum*. But not only are the conclusions of the argument susceptible to criticism. There are also critical assumptions behind it.

1.3 Critical Assumptions behind the Argument

To justify climate action either as prevention of or compensation for rights-infringements only allows for very little governmental interference if it is assumed that future people cannot possess rights. The hazardous effects of climate change, such as sea-level rise, desertification and natural catastrophes will, of course, affect living people. But the true extent of the crisis, with all its detrimental effects to quality of life, will mostly affect future generations. In order to be a viable argument in climate ethics, the rights-based libertarian claim therefore needs to assume that future human beings have claims vis-à-vis present people to prevent or compensate for the harm caused by excessive emissions. The assumption that future people possess rights, however, is restrictive and philosophically challenging.

First, it is restrictive in the sense that it excludes a conception of rights that is defended by adherents of the 'will theory of rights' (Hart 1982, 183–185; Steiner 1994, 55–107). According to this conception, a rights-holder is necessarily vested with control over enforcement, the possibility of waiving the right and forgoing a claim to compensation in the case of rights-violations. Thus, rights grant the individuals moral and legal powers to exercise their will. Yet, future people cannot possibly possess powers to demand enforcement or waive their rights. Therefore, a will theorist cannot, without contradiction, conceive of future people holding rights (Steiner 1994, 249–261).

In particular, with respect to the powers of waiver, libertarians are usually sympathetic with the will theory of rights (e.g., Steiner 1994). Their non-paternalistic stance is supported by a conception of rights that ensures the possibility of a rights-holder's consent to encroachments (cf. Nozick 1974, 58). I do not believe that libertarians are necessarily committed to a will theory of rights (Steiner and Vallentyne 2009, 57). The rights of future people might also be conceptualised within an 'interest theory' (Lyons 1969, 176; Raz 1986, 165–186; Kramer 1998, 60–100) that conceives of the core function of rights as the protection of individual interests, such as satisfaction of needs, attainment of well-being or development of capabilities. Nevertheless, the assumption of rights of future people is restrictive in that it excludes certain ideas of the notion of rights.

Second, the assumption that future people possess rights is philosophically challenging because harms created by excessive GHG emissions are not imposed on a specific individual. The actions we take today, may influence whether future people exist or not. Also, the actions we take today may have an effect on the circumstances under which future people will grow up and live and therefore may have an impact on people's identities. Lastly, they may have an effect on the number of people that will live in the future. If we decide not to reduce GHG emissions today, this changes the way people will live in the future. The people living in a world that is significantly warmer would not be the same people as those living in a world that has more or less the same temperature as ours. Hence, we cannot say that we made someone in particular worse off (than that person would have been otherwise). This problem is discussed under the term 'non-identity problem' (Parfit 1984, 351–377).

Concerning the question of whether excessive CO_2 emissions violate the rights of future people, the non-identity problem is significant. In a 'strict sense', rights imply duties directed to specific rights-holders (Hohfeld 2001, 13). If the duties to prevent climate change constitute rights in the strict sense, they need to be owed to a particular person or group of people. In the case of actions affecting future people, however, a person's compliance or non-compliance with a duty might make the right holder non-existent. The libertarian argument therefore has to abandon this strong 'person-affecting' conception of rights-infringements (Meyer 2021, Sect. 2).

Actions that constitute a transgression do not cause a specific person to have her rights infringed upon. In a future that occurs when we take no climate action, there would be no person that could possibly claim she would have her rights better realized if previous generations had taken climate action. But if no person is worse off than she would have been otherwise, how can this constitute a rights-infringement? One possible answer is to introduce a certain threshold of well-being, such as, for example, some basic needs that must be satisfied (Meyer 2003, 147–149). Whenever a person is born and finds herself below that level, her rights are infringed upon. Thus, those living today must ensure that people born in the future will not be worse off than they ought to be according to a critical threshold.

Now, both of these conceptual understandings of rights are themselves not uncontroversial and require further investigation, which I cannot accomplish within the scope of this chapter. It should be noted that the exclusion of particular theories of rights as well as the weak person-affecting conception of rights-infringements render the libertarian argument less 'robust'. On the one hand, libertarians who do not accept either of these assumptions might conclude that the only thing that needs to be prevented or compensated for are the hazardous effects of climate change on living people. With this restriction in mind, the problem of climate change as rights-infringement is similar to a case of, for example, air pollution (Rothbard 1973, 301–321; Nozick 1974, 77). But the moral challenge of climate change is not sufficiently described as a problem between contemporaries. Action is required to secure the liveability of the planet for future generations. On the other hand, non-libertarians who are not convinced that the libertarian account can accept either of these two implications, tend to refrain from endorsing a libertarian account of climate ethics altogether. Thus, the argument so far is based on a weak footing. In the next section, I discuss a more promising approach to justifying climate action from a libertarian perspective.

2. The Argument of Unjust Appropriation

Libertarianism places a strong emphasis on the moral significance of property rights. In principle, the state and society have no authority to take away and redistribute something an individual has rightfully acquired, were it for social welfare, health insurance, pension funds or the provision of public goods – except, of course, for the maintenance of a minimal state; e.g., law enforcement institutions and courts. Nozick's (1974, 151) 'entitlement theory of justice' captures this core idea: If a person acquires property either through her own labour or through voluntary transaction, she is entitled to dispose of it. Others (the state included) have no claim to deprive her of this property. Redistribution is justified only when either property was unjustly acquired, or transactions were coercive or involuntary.

The claim for strong ownership rights can be grounded in a Kantian principle of respect for persons (Nozick 1974, 33; Kymlicka 2002, 107–108). One should treat other individuals as an end in themselves and not as means to other ends. Libertarians believe the fruits of a person's labour originates in the self and, therefore, everything a person produces belongs to herself exclusively (van der Vossen 2010, Sect. 1). Depriving someone of her property is an infringement of individuals rights and, consequently, a violation of the principle of a Kantian ideal of self-ownership.

2.1 Lockean Justification of Property

Individuals can create legitimate property through their labour – that is, the activities that belong to the individual herself. However, their acquisition of property at some point necessarily includes the use of goods external to the self – that is, worldly resources. In order to produce something of economic value, individuals have to take land and plant crops, mine ore and forge tools, etc. Libertarians therefore have to employ a theory that explains how people can legitimately appropriate these natural resources in the first place. Commonly, they do so by referring to John Locke's (1980, chap. V) theory of just appropriation. According to him, individuals can legitimately acquire private ownership over commonly owned worldly resources. By exerting effort to increase the economic value of natural goods, individuals are able to appropriate them. He describes this process as 'mixing oneself' with external resources. Locke, however, concedes that there are limits to the legitimate acquisition of property. For example, a single person cannot simply claim all the natural resources for herself. He thus introduces restrictions – the two

Lockean provisos. First, private ownership can only be justified as long as there is 'enough, and as good, left' for others to use. Second, a person should only have a legitimate claim on ownership as long as she does not waste it (Locke 1980, Sect. 7, 33).

The first proviso is of particular importance when it comes to the libertarian argument for climate action. The atmosphere can be conceived of as a commonly owned resource. It is a sink we use when we emit CO_2 to produce goods and create economic value. Now, ownership over this economic value is, according to the Lockean theory of just acquisition, only legitimate if the atmosphere is not overused – in his words, there is 'enough, and as good, left' for others. So, the starting point of the libertarian argument for climate action is the understanding of the atmosphere as a common resource. This resource is only renewable to a certain extent – the planet's flora is capable of absorbing only a certain amount of CO_2. Emissions above the level of the earth's absorptive capacity lead to global warming and therefore potentially infringe the second Lockean proviso.

For simplicity, philosophers who conceptualise emissions from a perspective of Locke's theory often speak of acquiring partial 'ownership of the atmosphere' (e.g., Bovens 2011, 132). This expression is somewhat misleading. The atmosphere, as a sink, is a resource but no property is acquired by using it. Simply put, you do not gain ownership of a trash can by dumping waste into it. By using the atmosphere, the resource emitter generates economic value, which she produces with the help of the commonly owned sink. Hence, the idea of original appropriation in the case of GHG emissions is different from, e.g., appropriation of land (where not only the harvest but also the very piece of land belongs to the owner). Thus, to address the issue of climate change and the overuse of the atmosphere, Locke's proviso must be interpreted as a constraint on the use of resources rather than a constraint on appropriation (Mack 1995, 216–218).

Now, if the use of resource is and has been beyond the Lockean limits, the acquisition of property of the produced economic value is invalid. This leads to the conclusion that the material wealth generated with the help of excessive GHG emissions cannot be considered a libertarian property right that is worthy of protection against governmental interference. Thus, from the libertarian point of view, there is an argument for redistribution of the economic gains that have been created with unjustly high emissions.

This libertarian argument from unjust overuse of the atmosphere is backward-looking. Since property has been unjustly acquired, it must be returned to those people who do not have enough, and as good, resources left at their avail. Historically high-emitting countries should give reimbursement of adaptation costs as compensation for their overuse of the atmosphere. But there is also a forward-looking argument. Since future overuse of the atmosphere would imply an unjustified generation of property rights, the state is entitled to take preventive measures. It can impose restrictions as a mitigation strategy just as it is justified in preventing other forms of unjust appropriation such as theft, fraud, or robbery.

2.2 Different Degrees of the Proviso

Some problems with this argument occur when we need to define the particular point at which the atmosphere can be said to be overused. Unlike fishing grounds or forests, there is no specific threshold at which a clearly identifiable maximum is reached. On the one hand, the state of overexploitation does not occur only when one can no longer live on the earth. On the other hand, it is debatable whether even a minimal increase in the CO_2 concentration in the atmosphere, which leads to a slow, continuous global warming, is already a problem. Most scholars share the (empirically informed) moral conviction that today the atmosphere is being overused. The overall 'climate budget' is significantly smaller than the one our current standard of living requires. But when was the specific point in time, henceforth called t^*, at which the use of the atmosphere was so high that the economic value generated was no longer legitimately appropriated?

Libertarian philosophers disagree over the strength of the Lockean proviso and, therefore, over t^*. Some argue that the proviso only restricts the legitimate use of resources in case rights of non-owners are violated (Nozick 1974, 174–181; Mack 1990; Narveson 1999), whereas others believe that the Lockean condition is by far more demanding. They claim that the proviso involves egalitarian principles of original distribution of resources (Steiner 1994, 235; cf. Steiner 1987, 64–68; Otsuka 1998; Vallentyne 2007, 200). So, there is a continuum of different interpretations of Locke's theory (cf. Wendt 2017, 169). Weak interpretations regard much of the use of resources as justified, whereas strong versions demand a more equal distribution of resources. This strength relates to the potential libertarian case for governmental action to exact compensation for and prevention of illegitimate appropriation.

2.2.a Right-libertarian proviso

All libertarians would subscribe to the view that the use of resources should not threaten the individual rights to life, liberty, or property of non-users. If, e.g., a chemical factory uses a nearby river as a sink for toxic waste, which causes health problems to the inhabitants of the adjacent village, libertarians agree that both the chemical factory infringes the rights to bodily integrity and that the Lockean proviso is violated. The normative requirement of the proviso is, thus understood, already included in the libertarian case for the protection of individuals from non-consensual rights-infringements. So-called 'right-libertarian' thinkers state that the Lockean proviso amounts to little more than the restriction of rights-violations by appropriating resources.

However, the libertarian claim for climate action arising out of illegitimate appropriation goes beyond the argument from rights-infringement (Torpman 2016, 33–34). Even if no rights are infringed upon, appropriation may violate the Lockean proviso. Nozick states that a person's appropriation of resources should not put anyone else in a worse-off position than she would have been had the resource not been appropriated (Nozick 1974, 177). Like Locke and many libertarians, Nozick believes that a property rights regime generally has beneficial effects on society. The fact that people can appropriate something enables them to create great economic value. Compared to the *status quo ante* (e.g., in a hunter-gatherer society), living in a property-owning society is better for all members in almost all respects. So, arguably, Nozick's right-libertarian proviso is rarely unfulfilled. However, it is subject to interpretation as to what it means not to be worse off and where the baseline is drawn between people benefiting and people being made worse off by other people's appropriation (Wündisch 2013, 206–207).

In a world economy where all individuals emit CO_2 below the earth's absorptive capacity, individuals do not infringe the proviso. Only above this threshold, the excessive use of the atmosphere becomes an issue. There are two potential interpretations of the weak proviso here. On the one hand, it could be claimed that, whenever GHG emissions are above the sustainable level, the global temperature rises, and this could be said to make people worse off *ceteris paribus* than if the atmosphere were not overused. On the other hand, one needs to concede that by overusing the atmosphere the world economy creates economic value. Considering per capita growth in gross domestic product (GDP) over the last two centuries, it is clear that past and present generations have vastly benefited from past emissions. This economic gain may outweigh the negative impact of global warming such that no one is actually worse off. The weak proviso, therefore, would come into play only when current or future generations suffer so much from the negative effects of global warming that material wealth cannot compensate for the damage. The distinction made here dovetails with the distinction between weak and strong conceptions of 'sustainability' (Beckerman 1995; Neumayer 2010): 'strong sustainability' demands that the same basic stock of natural resources should be available to future generations; 'weak sustainability', on the other hand, simply demands that future generations should have sufficient resources to achieve a similar level of prosperity as the people living today, whereby the loss of resources can be compensated for by man-made capital.

Drawing on Locke, Nozick and other libertarians would rather follow the second interpretation of the weak proviso: The benefits of economic growth can, to a large extent, outweigh the harm caused by climate change and even an overuse of the atmosphere does not make anyone worse off. A weak interpretation of the Lockean proviso therefore leaves open the space for a right-libertarian argument against government action to foster climate change mitigation. A more demanding understanding of 'making someone worse off' would have to be taken as a basis for arguing for justified state intervention and taxation.

2.2.b Left-libertarian proviso

At the other end of the spectrum, the so-called left-libertarians harbour a strong notion of the Lockean proviso. Such an account is, e.g., provided by Michael Otsuka (1998, 79) who proposes that if an appropriator (user) creates value from the use of commonly shared resources, she should leave enough to enable other people to acquire the same level of well-being. The legitimate use and appropriation of common resources is only ensured if everyone has an equal opportunity to obtain welfare. If a person uses more than that, she becomes subject to redistributive claims against her. Others may legitimately demand compensation up to the point at which they enjoy the possibility of attaining the same level of well-being (this account can also be called an 'egalitarian proviso'; cf. Steiner and Vallentyne 2009).

The left-libertarian account is a reaction to the problem raised above – namely, that in order to consider a specific person worse off than before the appropriation (use) of a resource, one must define a specific baseline (Otsuka 1998, 78). Any proviso, Otsuka correctly observes, needs to refer to some standard below which a person can be said to experience a disadvantage. He criticises the Nozickian proviso because it allows that a single person to consume the entire atmosphere without having to share more than what is necessary for others to survive. The proviso, thus conceived, legitimises a monopolistic assumption of all resources by a single appropriator.

A weak proviso that defines legitimate constraints merely in terms of potential rights-infringements is insufficient. As a baseline, left-libertarians therefore propose an egalitarian distribution of claims to commonly owned resources. When using external resources, a person should leave enough of them, such that others may acquire the same amount and quality of resources (Steiner 1987; 1994) or opportunities to acquire well-being (Otsuka 1998; Vallentyne

2007). Invoking such a left-libertarian proviso in an argument for climate action has strong policy implications. On the one hand, this argument demonstrates that there is no legitimate ownership over the wealth accumulated through past overuse of the atmosphere; whereas 'overuse' is already present when future generations do not have the same opportunities to obtain welfare. On the other hand, this argument authorises the state to take preventive measures against future excessive use and to forestall illegitimate appropriation.

However, there is a broad spectrum of possibilities for interpreting the proviso that lies between a weak and an egalitarian proviso (for a so-called 'sufficiency proviso' see, e.g., Wendt 2017). In this enquiry, I do not intend to give precedence to or defend any specific interpretation. What is important to see is that, once one moves away from a right-libertarian weak constraint, libertarian political theory provides its own argument justifying redistribution to cover adaptation and mitigation costs. The following subsection aims to explain some of the most important aspects of this libertarian argument for climate action. I will show how the libertarian approach differs from other conceptions of climate justice.

2.3 Relation to Accounts of Climate Justice

Based on the argument of unjust appropriation, libertarians may claim that rich industrialised countries unjustly inherited their wealth because their ancestors strained the atmosphere well beyond their legitimate share. In order to reinstate a just state of affairs, rich countries, on the one hand, have a duty to redistribute the portion of their wealth that has been accumulated as a result of excessive emissions. Call this the *backward-looking aspect* of the argument of unjust appropriation. On the other hand, if they want to possess legitimate property in the economic value that is produced with GHG emissions, they have a duty to stop overusing the atmosphere as a sink. Hence, they have to re-establish a state in which emissions are down to a point at which the disadvantaged have 'enough, and as good left' for them to use. Call this the *forward-looking aspect* of the argument of unjust appropriation.

2.3.a Congruence with 'Polluter Pays Principle'

One principle that is vividly discussed in the debate on climate justice is the so-called 'polluter pays principle' (PPP) (cf. Shue 1999; Caney 2006). It states that the actors responsible for the largest amount of CO_2 in the atmosphere should also bear the largest share of the costs for adaptation and mitigation. The principle grounds its normative force in the idea of rectificatory justice.

Those who cause the problem should also pay for the damage. The normative implications of the principle are mostly congruent with those of the libertarian argument of unjust appropriation.

The forward-looking argument from unjust appropriation requires today's excessive polluters to reduce their carbon footprint to a sustainable level. Whereas the libertarian argument sees this as a precondition for justified acquisition of property in the future, the PPP demands a reduction in emissions because, as long as the footprint is above the sustainable level, the emitter is considered the originator of the problem and therefore the holder of duties of mitigation.

There is a difference, however, between the argument of unjust appropriation and the PPP with respect to the backward-looking argument. The libertarian argument for owing compensation for unjust appropriation is immune from the 'objection of excusable ignorance'. This objection holds that, before there was a scientifically and politically established consensus behind the existence of human-caused climate change, excessive GHG emitters did not knowingly commit a wrong (Gosseries 2004, 39–41; Caney 2006; Page 2008; Meyer and Roser 2010). Thus, high-emitters before that time (e.g., before the first IPCC report in 1990 or the Rio Summit in 1992) cannot be held *morally responsible* for the hazardous effects of their activities on climate. So, the PPP has no normative foundation to claim that earlier emissions also need to be compensated for.

The libertarian claim for redistribution, however, does not rely on an idea of rectificatory justice. It is not based on the assumption that past actions have caused morally blameworthy damage. Thus, a possible excuse for causing a negative state of affairs does not exempt actors from redistributive duties. In fact, an excuse is not needed. Previously high-emitting states do not owe compensation for imposing harm on others. Redistribution of wealth in favour of climate action is justified because the ownership of the economic gains produced by high emissions was unjustly acquired. Historically high-emitting countries have claimed property they could not legitimately have acquired. Such unjust enrichment needs to be remedied irrespective of the fact that it has been produced unknowingly or unwillingly.

To explain this, one can draw an analogy to receiving stolen goods. Imagine you ignorantly buy a painting from an art thief. Although you are not morally culpable, you must return the painting to its rightful owner, since you never acquired legitimate ownership of it. Here, a principle of corrective justice comes into play that is independent of retributive justice. The argument of unjust appropriation has the same structure. It is valid regardless of the guilt of the appropriators.

2.3.b Difference from the 'Beneficiary Pays Principle'

Another prominent proposal in climate justice is the so-called 'beneficiary pays principle' (BPP). It holds that states that benefited from excess emissions in the past have a duty to compensate other states for adaptation and mitigation (Caney 2006; Butt 2007; Meyer and Roser 2013). It is a plain fact that today's rich countries are historically responsible for the lion's share of excess CO_2 in the atmosphere and that these countries are financially capable of bearing the costs of both adaptation to and mitigation of climate change. The BPP is, therefore, an interesting proposal that combines principles of rectificatory and distributive justice. Those responsible for the problem are mostly those that have benefitted from causing it; and they are also those in an economically privileged position to solve it.

The argument of unjust acquisition yields a normative conclusion similar to the implications of BPP. If we assume that the Lockean proviso (the requirement that there is enough and as good left) has been infringed upon by today's wealthy countries in that they have used too big a share of the atmosphere, some portion of the accumulated wealth has been unjustly acquired. The libertarian would thus demand compensation for the unjustified enrichment.

But the coincidence of normative conclusions from the libertarian argument with those from the BPP does not necessarily prevail. An original acquisition and transfer may be unjust even if no one benefits from it. Not all of the historically high-emitting countries happen to be rich. Consider, e.g., former members of the Soviet Union. Because of their industrialisation in the twentieth century, these states are responsible for a large portion of the GHG concentration in the atmosphere. But they lag far behind western countries with respect to per capita GDP. Since the citizens of these states are not as well off as those of other countries, they can be considered less of a beneficiary and, thus according to BPP, owe less adaptation and mitigation repayment to the rest of the world. In contrast, the libertarian argument of unjust acquisition is insensitive to any difference in wealth distribution among countries today. The fact that the atmosphere has been overused is sufficient to justify duties of redistribution of unjustly acquired property.

2.3.c Defence of 'Grandfathering'?

By referring to Locke's theory of appropriation, the libertarian account defends the principle of distribution of emission rights that is discussed under the infamous term 'grandfathering'. Grandfathering involves a policy that distributes rights, power, or material benefits in proportion to a state of distribution before the implementation of the policy. With respect to climate justice, a grandfathering principle of distribution of burdens to adapt and mitigate climate change grants greater rights to emit to those countries that previously emitted at a higher level. In turn, states with historically low emissions would receive fewer emission rights.

To some extent, such a principle would unjustly reinforce the *status quo ante* by the distribution of rights. Among moral philosophers, grandfathering principles are, therefore, rarely defended (Caney 2009, 128). Following the Lockean account, however, Luc Bovens (2011) defends grandfathering. He argues that making excessive use of the atmosphere as a sink today is not justified anymore since it does not leave enough and as good for others. However, if one believes that the use of the atmosphere is subject to the Lockean conception of just appropriation, there must have been some point in time t^* at which states and individuals were justified to make use of the resource. Now, according to those shares in the legitimate use of the atmosphere before t^*, states should be able to continue to use the atmosphere.

Thus, Bovens defends a preservation of claims before t^*. Such a conception of legitimate use of resources, he argues, correlates with the convictions of justice we have concerning the legal regimes regulating the use of other commons, such as fishing grounds or forests. Existing users receive a state-guaranteed quota for further use, which is larger than that of other potential users. Thus, a grandfathering principle for the distribution of mitigation and adaptation costs might be considered compatible with common-sense morality. For simplicity, I do not discuss Bovens' limitations of the normative conclusion from the Lockean account, such as redistribution of emissions rights for humanitarian reasons or in emergency cases.

In applying the Lockean theory of appropriation, the libertarian is inclined to come to the same conclusion. However, one needs to keep in mind that a historical emitter of GHGs did not acquire property of the atmosphere. She only acquired property in the economic value that resulted from the just use of the resource. So, a forward-looking principle that assigns emission rights does not follow from the principle of just appropriation (Schuessler 2017, 148–149). The atmosphere differs from, e.g., a piece of land. By emitting, no property is acquired in the atmosphere, and therefore, future emission rights cannot be legitimised by past emissions with the help of the Lockean theory of just appropriation. According to the argument of unjust appropriation, however, rich countries with past records of high emissions do not owe compensation for their making use of the atmosphere before t^*. Therefore, rather than defending a principle of grandfathering, the libertarian account may provide a backward-looking excuse. But a forward-looking justification for excessive emissions by rich countries is not implied by the libertarian argument.

Conclusion

This chapter aimed to show that libertarian political theory provides an interesting approach to climate justice. This is the case even though libertarian theories are decidedly against claims of distributive justice.

One starting point for libertarian theories to justify state intervention to combat climate change lies in the argument that climate change is responsible for a variety of rights-infringements of currently living and future people. On the basis of this argument, obligations to compensate for the damage caused by excessive emissions can be justified. However, when it comes to duties to prevent climate change, the argument has undesirable distributive consequences. Furthermore, it presupposes the existence of rights of future generations. A libertarian theory can make such an assumption without contradiction, but the assumption renders the theory less attractive. If one is unwilling to accept it, the libertarian argument only provides reasons to justifiably compensate those affected by climate change today, but no reasons to reduce global warming for future generations.

A more promising libertarian argument refers to a Lockean theory of just appropriation of commonly owned resources to generate property. It is based on the claim that, depending on the interpretation of the restrictions on the justified use of resources, there is no legitimate ownership over the economic gains created by excessive use of the atmosphere. So redistribution for the purpose of covering adaptation and mitigation costs is justified. The normative implications of the argument are distinct from those of the more familiar approaches to climate justice, the 'polluter pays principle' and the 'beneficiary pays principle.' Thus, libertarian political theory provides an original approach to thinking about the moral problem of anthropogenic climate change.

References

Beckerman, Wilfred. "How Would you Like your 'Sustainability', Sir? Weak or Strong? A Reply to my Critics." *Environmental Values* 4, no. 2 (1995): 167–179. https://doi.org/10.3197/096327195776679574

Bovens, Luc. "A Lockean Defense of Grandfathering Emission Rights." In *The Ethics of Global Climate Change*, edited by Denis Arnold, 124–144. Cambridge: Cambridge University Press, 2011.

Brennan, Jason. *Libertarianism: What everyone Needs to Know*. Oxford: Oxford University Press, 2012.

Broome, John. *Climate Matters: Ethics in a Warming World*. New York: WW Norton & Company, 2012.

Butt, Daniel. "On Benefiting from Injustice." *Canadian Journal of Philosophy* 37, no. 1 (2007): 129–152. https://doi.org/10.1353/cjp.2007.0010

Caney, Simon. "Environmental Degradation, Reparations, and the Moral Significance of History." *Journal of Social Philosophy* 37, no. 3 (2006): 464–482. https://doi.org/10.1111/j.1467-9833.2006.00348.x

Caney, Simon. "Justice and the Distribution of Greenhouse Gas Emissions." *Journal of Global Ethics* 5, no. 2 (2009): 125–146. https://doi.org/10.1080/17449620903110300

Gosseries, Axel. "Historical Emissions and Free-Riding." *Ethical Perspectives* 11, no. 1 (2004): 36–60.

Hart, H. L. A. "Legal Rights." In *Essays on Bentham: Studies in Jurisprudence and Social Philosophy*, edited by H. L. A. Hart, 162–193. New York: Clarendon Press, 1982.

Hohfeld, Wesley. *Fundamental Legal Conceptions as Applied to Judicial Reasoning*. Edited by David Campbell, and Peter Thomas. London: Routledge, 2001.

IPCC. *Sixth Assessment Report: Summary for Policymakers*. https://www.ipcc.ch/report/ar6/wg1/downloads/report/IPCC_AR6_WGI_SPM.pdf

Kramer, Matthew. "Rights Without Trimmings." In *A debate over rights*, edited by Matthew Kramer, Nigel Simmonds, and Hillel Steiner, 7–111. Oxford: Clarendon Press, 1998.

Kymlicka, Will. *Contemporary Political Philosophy: An Introduction*. Oxford: Oxford University Press, 2002.

Locke, John. *Second Treatise of Government*. Edited by Crawford MacPherson. London: Hackett, 1980.

Lyons, David. "Rights, Claimants, and Beneficiaries." *American Philosophical Quarterly* 6, no. 3 (1969): 173–185. https://www.jstor.org/stable/20009306

Mack, Eric. "Self-Ownership and The Right of Property." *Monist* 73, no. 4 (1990): 519–43. https://www.jstor.org/stable/27903208

Mack, Eric. "The Self-Ownership Proviso: A New and Improved Lockean Proviso." *Social Philosophy and Policy* 12, no. 1 (1995): 186–218. https://doi.org/10.1017/S0265052500004611

Meyer, Lukas. "Past and Future: The Case for a Threshold Notion of Harm." In *Rights, Culture, and the Law: Themes from the Legal and Political Philosophy of Joseph Raz*, edited by Lukas Meyer, Stanley Paulson, and Thomas Pogge, 143–160. Oxford: Oxford University Press, 2003.

Meyer, Lukas. "Intergenerational Justice." In *The Stanford Encyclopedia of Philosophy*, edited by Edward Zalta, 2021. https://plato.stanford.edu/entries/justice-intergenerational/#pagetopright

Meyer, Lukas, and Dominic Roser. "Climate Justice and Historical Emissions." *Critical Review of International Social and Political Philosophy* 13, no. 1 (2010): 229–253. https://doi.org/10.1080/13698230903326349

Meyer, Lukas, and Dominic Roser. "Climate Justice: Past Emissions and the Present Allocation of Emission Rights." In *Spheres of Global Justice*, edited by Jean-Christophe Merle, 705–712. Dordrecht: Springer, 2013.

Narveson, Jan. "Property Rights: Original Acquisition and Lockean Provisos." *Public Affairs Quarterly* 13, no. 3 (1999): 205–227. https://www.jstor.org/stable/40441228

Neumayer, Eric. *Weak versus Strong Sustainability: Exploring the Limits of Two Opposing Paradigms*, 3rd ed. Cheltenham: Edward Elgar, 2010.

Nozick, Robert. *Anarchy, State, and Utopia*. New York: Basic Books, 1974.

Otsuka, Michael. "Self-Ownership and Equality: A Lockean Reconciliation." *Philosophy & Public Affairs* 27, no. 1 (1998): 65–92. https://doi.org/10.1111/j.1088-4963.1998.tb00061.x

Page, Edward. "Distributing the Burdens of Climate Change." *Environmental Politics* 17, no. 4 (2008): 556–575. https://doi.org/10.1080/09644010802193419

Parfit, Derek. *Reasons and Persons*. Oxford: Clarendon Press, 1984.

Posner, Eric, and Sunstein, Cass. "Climate Change Justice." *Georgetown Law Journal* 96, no. 5 (2008): 1565–1612. https://heinonline.org/HOL/Page?collection=journals&handle=hein.journals/glj96&id=1571&men_tab=srchresults

Railton, Peter. "Locke, Stock, and Peril: Natural Property Rights, Pollution, and Risk." In *Facts, Values, and Norms*, edited by Peter Railton, 187–225. Cambridge: Cambridge University Press, 2003. https://doi.org/10.1017/CBO9780511613982

Raz, Joseph. *The Morality of Freedom*. Oxford: Clarendon Press, 1986.

Rothbard, Murray. "For a New Liberty: The Libertarian Manifesto." Ludwig von Mises Institute, 1973.

Schuessler, Rudolf. "A Luck-Based Moral Defense of Grandfathering." In *Climate justice and historical emissions*, edited by Lukas Meyer, and Pranay Sanklecha, 141–164. Cambridge: Cambridge University Press, 2017. https://doi.org/10.1017/9781107706835.008

Shahar, Dan. "Justice and Climate Change: Toward a Libertarian Analysis." *The Independent Review* 14, no. 2 (2009): 219–237. https://www.jstor.org/stable/24562317

Shue, Henry. "Global Environment and International Inequality." *International Affairs* 75, no. 3 (1999): 531–45. https://doi.org/10.1111/1468-2346.00092

Singer, Peter. *One World Now*. New Haven: Yale University Press, 2015.

Steiner, Hillel. "Capitalism, Justice and Equal Starts." *Social Philosophy and Policy* 5, no. 1 (1987): 49–71. https://doi.org/10.1017/S0265052500001242

Steiner, Hillel. *An Essay on Rights*. Oxford: Blackwell, 1994.

Steiner, Hillel, and Vallentyne, Peter. "Libertarian Theories of Intergenerational Justice." In *Intergenerational Justice*, edited by Axel Gosseries, and Lukas Meyer, 50–76. Oxford: Oxford University Press, 2009.

Sunstein, Cass. "The Complex Climate Change Incentives of China and the United States." *University of Chicago Law & Economics*, Olin Working Paper 352 (2007). http://dx.doi.org/10.2139/ssrn.1008598

Torpman, Olle. "Libertarianism and Climate Change." PhD diss, Stockholm University, 2016. https://www.philosophy.su.se/polopoly_ fs/1.279738.1461161241!/menu/standard/file/Torpman%20%282016%29%20 Libertarianism%20and%20Climate%20Change%20%28dissertation%29.pdf

Torpman, Olle. "Libertarianism, Climate Change, and Individual Responsibility." *Res Publica* 28, no. 1 (2021): https://doi.org/10.1007/s11158-021-09514-3.

Vallentyne, Peter. "Libertarianism and the State." *Social Philosophy and Policy* 24, no. 1 (2007): 187–205. https://doi.org/10.1017/ S0265052507070082

van der Vossen, Bas. "Libertarianism." In *The Stanford Encyclopedia of Philosophy* edited by Edward Zalta, 2019. http://plato.stanford.edu/archives/ fall2014/entries/libertarianism

Wendt, Fabian. "The Sufficiency Proviso." In *The Routledge Handbook of Libertarianism*, edited by Jason Brennan, Bas van der Vossen, and David Schmitz, 169–183. London: Routledge, 2017.

Wündisch, Joachim. "Nozick's Proviso: Misunderstood and Misappropriated." *Rationality, Markets and Morals* 4 (2013): 205–220. https://core.ac.uk/ download/pdf/25534723.pdf

Zwolinski, Matt. "Libertarianism and Pollution." *Philosophy and Public Policy Quarterly* 32, no. 3 (2014): 9–21. https://ssrn.com/abstract=2443030

6

Beyond Rawls: The Principle of Transgenerational Equity

TIZIANA ANDINA

Climate change poses complex problems of justice, which are particularly difficult to address because the relationship between the actors involved in the justice relationship is not always symmetrical. In other words, climate change imposes a reflection on the importance of time in justice issues and underline, better than in other contexts, how transgenerationality is in fact a crucial matter for justice. The article has three aims. Firstly, starting from an analysis of selected points in Rawlsian justice theory, it illustrates the reasons why classical theories of justice are unsuitable for dealing with transgenerational issues. Secondly, within the framework of a general theory of transgenerationality, it offers useful arguments for the formulation of a principle of transgenerational equity. Finally, it discusses how the transgenerational equity principle can be applied to climate change.

Introduction

The issue of climate change is a particularly interesting problem of justice because of its exemplary nature. It illustrates above all that problems of justice are complex issues that cannot be addressed easily without taking into account time and the asymmetry between the parties that temporality determines. In circumstances where two parties are in an asymmetrical relationship, it is impossible to understand justice as a form of compensation (Plato and Waterfield 1993, 331E-332B), not only because, as Socrates notes, pure compensation is not always just, but above all because diachronic relationships involve complex relationships. The factual impossibility of reciprocity is further complicated by the power that one party is able to exercise over the other. In the specific case of transgenerational relations, what is at issue is the power that present generations can exercise over future entities. Whereas, when we are harmed by a friend, we are generally in a position to claim some form of compensation, in the case of relations between parties living in different time periods, not only is such compensation impossible, but the substantive content of the unjust acts cannot be erased, i.e., it is not possible to rewind the tape of time.

Let us take, for example, a state that has been releasing a large quantity of CO_2 into the environment for a certain period of time, and let us assume that it has done so for a noble purpose, namely to promote the development of its economic system and increase the wealth and improve the quality of life of its citizens. The side effect of this choice is a progressive and perhaps irreversible deterioration of both the environment and the climate in which future generations will live. In other words, the power of choice and action that certain generations have enjoyed and continue to enjoy in a certain historical time period may definitively compromise the power of choice and action of other generations in a subsequent time period. This situation, intuitively, seems unfair for at least two reasons. First, it appears unfair because, despite the data that scientists have amassed since the second half of the twentieth century, the exercise of power through the performance of utilitarian actions by some generations has not encountered limits to protect future generations. Second, it seems unfair because it seems plausible to believe that there are things – including the ecosystem – that represent the condition of possibility of life for everyone and that, for this very reason, they are not at the complete disposal of anyone.

Now, an imaginary reader, while agreeing with what we have argued, might not be willing to agree on two points. First, on the determination of this 'all'. He might, for example, be unwilling to consider future generations as an entity to be included among those we must consider as objects of justice, since they are in fact fictitious entities that do not yet exist in time and space. I have tried to respond to this objection elsewhere (Andina 2022). Here, however, I focus on the question of justice and develop both moral and political considerations. Assuming as valid the idea that utilitarianism is an ineffective theory for responding to climate justice and thus that it is an *a fortiori* unsuitable theory in a broader intergenerational perspective (Jamieson 2007), can we expect any help from neo-contractualist theories? That is, can we expect any significant help from the development of neo-contractualist positions that are not intended to be based on a maximization of individual profit? And if both utilitarianism and neo-contractualism prove to be ineffective theories for dealing with the challenges of transgenerationality, what other options do we have for promoting transgenerational equity? I argue that the neo-contractualist model is unsuitable for reasoning about transgenerational issues even independent of the assumption of a utilitarian alternative precisely because it excludes time from its modelling.

1. Justice is not enough: Rawlsian theory as the test of trans-generationality

The reflections of John Rawls in both *A Theory of Justice* (1971) and a series of subsequent writings (cf. Rawls 1999, 2005) concern themselves, among other things, with the problems of transgenerational justice, that is, with the form that justice between different generations takes. Before introducing Rawls's arguments, it is worth emphasising two aspects.

Firstly, although the American philosopher was among the first to realise the importance of the temporal relationship between the parties in a justice relationship, the issue of transgenerational justice remains rather marginal, perhaps also due to historical and contextual reasons. The second aspect concerns the tools available to a classical theory of justice, such as Rawls's theory, to deal with parties in an asymmetrical relationship. The asymmetrical relation is in fact one of the essential features of transgenerational relations. Rawlsian theory not only shows the inadequacy of cooperation-based theories of justice in the face of the problems raised by transgenerational relations in an exemplary way, but also how the concept of justice reveals its own inadequacy in dealing with transgenerational relations and their conditions of equity (see, at least, Pulcini 2020, Tronto 1993 and Held 2006). In other words, there are intrinsic limits to the concept of justice as it is treated by cooperation-based theories of justice that do not allow the main implications of transgenerational relationships to be addressed convincingly. Therefore, it seems reasonable to argue that if we want to talk about transgenerationality and its applications (including the issue of climate sustainability) we cannot limit ourselves to justice, but must refer to a broader set of principles, including the principle of transgenerational responsibility.

1.1 Transgenerationality in 'A Theory of Justice'

Rawls offers a normative theory of justice. This means that having started, as is the practice, from philosophical-anthropological considerations about human nature, Rawls describes relations of justice not as they are, but as they should be in an ideal society. In other words, *A Theory of Justice* intends to determine the principles of justice on which social relations should be based in order to give life to decent societies. The question is therefore, roughly, what can an ideal theory of justice tell us about asymmetrical relations? Not much, unfortunately.

The ideal social system described by Rawls (1999, 7–8, 216) displays certain characteristics. For one, it is closed, extended in time and isolated from other societies. A sort of bubble, in short. The bubble is populated by individuals that display certain characteristics. For example, they are willing to cooperate and are not subject to the pressures of basic needs, since the bubble enjoys a fairly favourable economic environment. In other words, resources are neither overabundant, nor equally available to all, nor they excessively scarce. This means that the inhabitants of the bubble do not struggle for survival. In addition, there is a neutral and plural value system. The inhabitants of Rawlsian society express different worldviews and beliefs, for example, regarding moral values. In this context, social and political equilibrium is roughly guaranteed by two elements. First, it is guaranteed by the characteristics of the individuals who inhabit the bubble: they are rational people that, in principle, make the most rationally advantageous choices available to them; and they are reasonable, i.e., they are willing, in certain circumstances, to give up their own immediate profit as long as the other inhabitants share the same attitude. This means they are willing to forego maximising profit.

A second element that guarantees political and social stability is the willingness and ability of individuals to cooperate, ideally also between different generations: 'Thus, justice as fairness starts within a certain political tradition and takes as its fundamental idea that of society as a fair system of cooperation over time, from one generation to the next' (Rawls 2005, 14). In order for the bubble to survive over time, i.e., for its inhabitants to live a life in peace and well-being, cooperation must also be thought of diachronically, between different generations.

In the Rawlsian context, just as in the case of Kantian morality, future generations play an eminently instrumental and regulatory role. We know that one day they will exist, be part of society and contribute to the deliberative processes, so if a society accepts the principles of justice upon which it has deliberated and citizens agree to apply them rigorously, future generations will participate in a process of justice implementation in which they will be protected. Each inhabitant of the bubble, in other words, would collaborate in the making of justice, cooperating in an inclusive process in which all individuals – in the present and future – can aspire to live in (greater) equity. Commitment to the realization of justice would therefore guarantee not only the present generations, but also future ones. All well, then? Not really, because of certain conditions that Rawls postulates in his ideal theory. Let us first try to understand in more detail what function – if any – future generations fulfil in the Rawlsian bubble.

> Each generation must not only preserve the gains of culture
> and civilisation and maintain intact those just institutions that
> have been established, but it must also put aside in each

period of time a suitable amount of real capital accumulation. This saving may take various forms from net investment in machinery and other means of production to investment in learning and education (Rawls 1999, 252).

In *A Theory of Justice*, we read that each generation has two tasks, one – so to speak – conservative, the other expansive. The first task is primarily concerned with the past, the second with how the future is to be oriented. Future generations must preserve what has been gained in terms of cultural progress and civilisation, i.e., they must strive to stabilise what has been acquired by previous generations and keep intact the institutions that have been created. This part of the task takes the form of capitalisation and consists of stabilising what is already available to a community. The second task, on the other hand, has an expansive character, so it does not take the form of capitalisation but of investment and, perhaps, of gift-giving, since each generation must set aside an adequate amount of accumulated capital to invest in goods that are not necessarily of immediate use. This provision can take various forms. It can be, for example, an investment in production tools or it can be an investment in knowledge and education. What is certain is that it seems difficult to interpret it solely in terms of justice – i.e., what one generation must pass on or give back to another. It seems rather reasonable that it should be about the logic of gift, i.e., what one generation gives to others with the aim of improving their future.

Now, within the Humean framework of the conditions of justice that Rawls accepts without substantial modification, both the conservative and expansive tasks are indeed problematic (cf. Brandstedt 2015). According to Hume, three circumstances must be present for justice: a moderate scarcity of resources, a moderate selfishness of the parties and relative equality between the parties (Rawls 1999, 109–110). These three conditions, taken together, should be conducive to the creation of conditions of justice. Indeed, we can expect that a moderate scarcity of resources will drive cooperation to optimise production and distribution; that moderate selfishness will make possible the reconfiguration of certain personal goals to more general goals; and finally, that the relative equality of the parties will make possible the formulation of equitable principles of justice.

However, when applied to asymmetrical relations, the third condition – relative equality between the parties – is problematic since people living at different times encounter different historical, cultural and environmental conditions. Therefore, admitting the fact that these conditions are usually relatively or substantially different, the more general condition of equality between the parties seems impossible to achieve. Let us assume, for example, that a certain generation G precedes a generation G_1 in time. Obviously, G will be able to make choices that have a significant impact on the quality of life of G_1, whereas the reverse situation does not arise, i.e., G_1 is not able to determine retroactively the quality of life of G. Thus, in many circumstances, there can be no reciprocity between generations, and this constitutes a problem for any theory of justice, including the theory of justice as fairness (Gosseries and Meyer 2009, 119–146).

There is a second argument concerning the thought experiment used by Rawls to determine the principles of justice derived from the 'original position' (for a more detailed discussion of Rawls's original position, see Hinton 2015). In the original position, people are contemporaries, meaning they all live in the same period of time and are also unaware of the characteristics of the generation to which they belong: they do not know whether it is rich or poor, whether previous generations set aside money for its welfare and so on (Rawls 1999, 121).

This assumption is certainly not accidental. Rawls could, in fact, have assumed different conditions. For example, he could have imagined that the bubble contains representatives of all generations or, perhaps, to make it a little less crowded, one representative for each group of contiguous generations. And yet he does not choose this solution. His motivation is, after all, quite simple and appeals to common sense: a thought experiment involving the coexistence of individuals from totally different eras would be difficult to conceive and would require too much effort. Therefore, he opts to leave out the temporal property (Hart, Hacker and Raz 1977, 278 e ff). This decision has an important consequence. If we assume that the generations are contemporary, we deprive ourselves of the possibility of intervening in the savings rate. This is because all generations enter the bubble at the same point in time x at which certain conditions are given, which are the same for all. In such a context, it makes no sense for the inhabitants to commit themselves to investing or setting aside resources for future generations – the situation would be different if, having no idea of the moment in history into which they find themselves parachuted, the inhabitants of the bubble discovered that it is possible for someone to be born at a stage in which previous generations have pursued their maximum profit. Thus, if Rawls had defined his bubble with the idea of temporality, he would have determined a condition in which the inhabitants would be expected to think about the future because, hypothetically, they could not rely on the equivalence of material conditions at all stages of life within the bubble. In this context, each individual in the original condition reasonably ends up acting in service of his maximum profit. In other words, he rationally chooses to follow rules that benefit him, without necessarily caring what will happen to future generations. In order to avoid this paradox, which indicates the weakness of Rawlsian neo-contractualism in the face of the problems of transgenerationality or, perhaps indicates how the ideal theory only accords well with presentist metaphysics, Rawls introduces an element that is, in effect, rather foreign to the theory as a whole. He suggests considering individuals in the original position not only as rational agents, but also as representatives of specific family lines – real heads of families who, as such, will have precise interests in protecting their descendants.

In a nutshell, what Rawls is trying to do is provide a minimum safeguard for transgenerationality, that is, the protection of primary transgenerationality through the recognition of biological linkages. Parents have affective reasons for looking after their children's future, and it is a fact that the protection of primary transgenerationality is part of the make-up of many species, not just human beings. If this is true, one can also imagine that each generation implements practices to protect biological transgenerationality and parental bonds. As for the rest, Rawls assumes that interventions in the savings rate depend roughly on the level of wealth of a society: societies or population groups in a certain social context that are in situations of relative poverty will be called upon to set aside less, whereas those who can do more will be called upon to do so.

Now, this reasoning presents at least two types of criticality. With regard to primary transgenerationality, it is clear that there is an element of obscurity or indeterminacy in the link between parents and their offspring. In other words, although it seems reasonable to think that in most cases fathers (or mothers) take care to protect their children while also protecting, as far as possible, the environment in which they will live, we know that this is not always the case. It is evident that Rawls considers transgenerationality a less important aspect than the general normativity he constructs through his general theory. The Rawlsian approach presents at least two fundamental problems. First, it entrusts the protection of transgenerational ties to the good will of each person and the exercise of a kind of virtue ethics. Second, it assumes that all parents love their children more than themselves, which is not true in all cases. Thus, parent-child love is not expendable in a theory of rational choice because it undermines the assumption of self-interest (cf. the argument developed by English 1977).

Rawls' arguments do not provide enough support for transgenerationality to be protected on a large scale and, above all, too little to think that this exercise works outside the parental circle. In other words, while it is possible – though certainly not inevitable – for parents to protect the life, health and future of their children, it is, on the other hand, quite utopian to think that a community will protect the life, health and future of the generations to come if there is no compulsion to act (or not act), that is, if there is no normative context to accompany and reinforce individual and institutional decisions. This is because primary transgenerationality generally has a spectrum of interest limited to children and grandchildren, rarely covering a longer time span. On the other hand, it must be stressed that the climate issue is a global problem that can only be tackled through global cooperation. Such cooperation can only extend beyond national interests. Thus, the motivation of parental or family ties can hardly justify fair savings on a transgenerational and cosmopolitan level. The most unexplored aspects of transgenerationality therefore concern the temporal dimension that belongs to those who will exist long after us.

In *Political Liberalism*, Rawls returns to the issue by proposing a different strategy. The idea remains that of preserving temporally undifferentiated access to the original position with the aim of strengthening the veil of ignorance. Therefore, he proposes that in the original position, an agreement is made to set aside, on the basis of a principle that can be followed by all generations, those that ideally preceded and those that will follow (Rawls 2005, 273-274). It is clear, however, that such a principle is extremely difficult to formulate precisely because the transgenerational question engages temporality not in an ahistorical sense, but as a historicised time in which human beings exercise their capacity for action.

2. The principle of transgenerational equity

If we believe there are good arguments for considering the Rawlsian approach ineffective and if, at the same time, we are of the opinion that transgenerationality raises urgent issues, then we should conclude at least three things. Firstly, there is an urgent need to think about the corrective measures to be applied to neo-contractualism, which, at least in the Rawlsian version, is unsuitable for dealing with transgenerational issues since it does not integrate temporality into its theoretical model. Secondly, utilitarianism is unsuitable for dealing with the problems raised by climate change, and this gives us good reason to believe that it is not the most suitable strategy for discussing transgenerational issues either.

Finally, we should also consider the possibility of reconfiguring political ontology in such a way that the categories of movement and deferral prevail over the more classical categories of permanence and substance. Secondary transgenerationality implies three things: the constant reference to a certain dimension of temporality, the future and the systematic recourse to a particular fictitious entity, namely future generations. Future generations are one of the entities that populate the ontology of future time; other entities could be mentioned, such as future climate or future biome. From an ontological point of view, and with regard to the question of temporality, rather than favouring the traditional division of time into distinct segments, such as present, past and future, it is a question of considering time as that which endures while understanding the future as that which exists despite not having a determined form. The future, therefore, exists in a different way than the past or the present.

Research and debate on climate change often focuses on the present and the past, i.e., on what is happening now and what has happened, roughly, during the last century, assuming that the actions that have damaged or are damaging the climate and the environment have been carried out by different agents over a fairly long period of time. This means that any analysis of the causes of climate change is *ipso facto* measured by the concept of responsibility over time.

This is why we speak of cumulative responsibilities in a twofold sense: both in the sense that these responsibilities must be ascribed to a multiplicity of public and private actors, and in the sense that they date back to different historical phases. This stratification needs to be taken into account by international institutions and policymakers to make the fairest possible decisions on climate change measures. Cumulative responsibilities generally relate to the past and have a good degree of complexity in themselves. However, the climate issue not only concerns that which has already been done, but also challenges us with regard to the future, i.e., with regard to the mitigation actions that need to be taken to halt the progressive deterioration of the climate and the environment.

The tool that international diplomacy has used most widely to reason about global actions and responsibilities is the principle of shared but differentiated responsibilities. This principle was explicitly formulated at the United Nations Conference on Environment and Development held in Rio in 1992 and known as the Earth Summit.

It is based on four premises: historical responsibility, equity, capacity and vulnerability. Historical responsibility is aimed at reconstructing the past and determining responsibilities as precisely as possible; equity is the general criterion aimed, so to speak, at balancing the scales, i.e., at finalising equitable actions and decisions; capacity is aimed at considering scientific and technological skills, i.e., the tools of knowledge and technology that can be used to limit climate deterioration. Finally, vulnerability encompasses the difficulties facing countries in a particularly fragile state of economic and cultural development and which, because of this, would not be able to cope with the burden of the costs of energy transition.

Historical responsibility, equity, capacity and vulnerability constitute the different aspects of the principle of shared but differentiated responsibilities, the aim of which is to provide a theoretical tool to support policymaking. To this end, there are two strategies: firstly, the examination of the present situation on a global scale (equity, capacity and vulnerability) and secondly, the analysis of responsibilities formulated on a longer time scale (historical responsibility). Historical responsibility is primarily concerned with understanding the past to determine as precisely as possible who is responsible for what has happened. In this sense, it is possible to regard the determination of historical responsibility as a necessary but not sufficient tool for guiding the future – scholars theorise both the non-existence of the future and its indeterminacy: the former hold that only what is or what has been can be said to exist in the sphere of temporality (McTaggart 1908), the latter (for example the so-called Growing Block Theory) hold that the future, despite its indeterminacy, does exist (Paul 2010).

In this context, we think it would be very useful to think about possible ways of broadening the criterion of vulnerability. Not only are some countries more vulnerable than others, but there are likely to be generations that are more vulnerable than others. Scientific evidence suggests that if actions to combat climate change are not both decisive and targeted, then future generations will be more vulnerable to climate change than past or present generations. Thus, there are good reasons to conclude that the principle of vulnerability must be extended to future generations. However, a reformulation of the vulnerability principle is not sufficient.

The broadening of the criterion of vulnerability must provide for the parallel reformulation of the criterion of historical responsibility. We attribute to states the historical responsibility for institutional actions, public actions and the actions of private citizens that fall within the perimeter of what a legal system allows. While it is true that responsibility is primarily individual (see, for example, Gilbert 2006), it is also true that companies, states and composite administrations (meta-states) are the only entities that can be attributed the task of guaranteeing the exercise of equity over time, and thus of elaborating and providing the instruments for exercising justice and equity in a diachronic perspective. What entities other than states and composite administrations could perform the task of keeping track of and possibly guaranteeing a form of intergenerational rebalancing?

It is precisely because of the implementation of intergenerational justice that actions dating back 20, 30 or even 40 years cannot fail to concern us. However, it is also appropriate to think about transgenerational responsibility. A first remark concerns the direction of responsibility: it cannot be limited to the past. We must ask whether it makes sense to believe that humankind, through the political entity to which we have delegated the task of preserving societies over time, namely the state, should not commit itself to directing the future by taking responsibility for it. Moreover, also in accordance with the recognition of the personal character of responsibility, we would argue that responsibility for the future is more cogent than responsibility for the past. We know that responsibility is personal; that is why we are responsible for the actions of those who have gone before us insofar as they are attributable to, or carried out by, institutional subjects that form the diachronic structure of the community.

Objections to these arguments are generally of two kinds. The first type concerns what we may call opacity concerning the 'consequences of action' (cf. Singer 1972, 2009). In other words, not all consequences of actions are predictable and not everything is predictable in the same way. Moreover, predictions are, by definition, uncertain. If this is true, it is not very reasonable for people to choose to limit their freedoms or not maximize their profits; nor is it reasonable for states to choose to do so. To stay with the case of climate change, it is clear that in the last century, the data available to the scientific community were qualitatively and quantitatively different from the data available to us today. Therefore, the issue of making reliable forecasts is not easy to solve because of the nature of scientific knowledge, which is in constant flux. However, it is important to emphasise that it is one thing to talk about forecasts, i.e., the possibility of predicting the future, and quite another to talk about steering the future in one direction or another. In the first case, it is a question of predicting what will happen; in the second, of orienting the present towards certain objectives that seem important to us and that, presumably, will only be realised at a certain temporal distance from us. In this sense, it is useful to emphasise that the consequences of an action or series of actions can only be understood at least with a sufficient degree of clarity by examining a fact or action 'from the future' (Danto 1965, 2007); that is, after a certain period of time has elapsed since the action was carried out, so that the most important consequences of that action can be considered. In certain matters, therefore, we must provide for the adoption of a principle of prudence that protects future generations with respect to worst-case scenarios.

The second objection generally involves a principle of prudence or, if we prefer, realism, and can be summed up as follows: how can we hope to interpret with any degree of certainty the tastes, expectations and wishes of those who will come after us (Zwarthoed 2015)? Above all, how can we do so in the case of those generations that will be at a considerable temporal distance from us? Supporters of this objection believe that the idea of more or less openly directing the future conceals an attempted exercise of power, since it imposes our vision of the world on those who will come after us. In reality, it is rather naive to believe that the most significant social actions (and sometimes even many of our individual actions) do not imply consequences in the long or short run. In other words, whether we take responsibility or not, the future is always conditioned by what we decide to do or not do. The central point, therefore, is to delimit the perimeter of responsibility and action, not to recognise that there is a responsibility for the orientation of the future.

If we assume that the future exists and is open, it is quite reasonable to assume that we should aim to steer it in some directions rather than others, such as, for example, towards actions that involve attention to sustainability and equity. In view of this, I suggest incorporating the principle of shared but differentiated responsibilities in such a way as to broaden the sphere of responsibility which, in the current formulation, primarily concerns the past. It seems desirable to broaden this to the future, first and foremost by considering the responsibility that present generations must assume towards future generations. This extension can be described by the principle of transgenerational responsibility, which can be articulated as follows:

1. Life, in most circumstances, is preferable to death;
2. Achieving a good quality of life is a goal that each generation legitimately sets for itself;
3. Transgenerational bonds bind generations across time, determining reciprocal rights and duties within the transgenerational chain;
4. Transgenerational actions have a particular structure that involves collaboration between generations in order for a specific action to be carried out (for a more on transgenerational action see Andina 2020); and,
5. Transgenerational social actions must respect transgenerational constraints and strive to orient the future in ways that do not undermine the right of future generations to have a good quality of life.

The principle of transgenerational responsibility recognises the existence of transgenerational bonds and, likewise, acknowledges the peculiar structure of transgenerational actions. Similarly, it contemplates the possibility of granting rights, as well as duties, to those abstract artefacts, future generations, which envisage a change of state, i.e., the passage from potency to act. The probable existence of future generations, together with the fact that these particular artefacts are called upon to make long-lasting and complex transgenerational actions practically feasible, leads to the conclusion that the principle of transgenerational responsibility must be extended to future entities with particular regard to future generations, also in view of the circumstance that preceding generations often derive a concrete advantage from postulating the existence of future generations.

References

Andina, Tiziana. 2022. *A Philosophy for Future Generations: The Structure and Dynamics of Transgenerationality*. New York/London: Bloomsbury Academic.

Andina, Tiziana and Petar Bojanic, eds. 2020. *Institutions in Action. The Nature and the Role of Institutions in the Real World*. Vol. 12: Cham: Springer.

Brandstedt, Eric. 2015. "The Circumstances of Intergenerational Justice." *Moral Philosophy and Politics* 2 (1): 33–55. https://doi.org/10.1515/mopp-2014-0018

Danto, Arthur C. 1965. *Analytical philosophy of history*. Cambridge: Cambridge University Press.

Danto, Arthur C. 2007. *Narration and knowledge: (including the integral text of Analytical philosophy of history)*. New York: Columbia University Press.

English, Jane. 1977. "Justice between generations." *Philosophical Studies* 31 (2): 91–104. https://doi.org/10.1007/BF01857179

Gilbert, Margaret. 2006. *A theory of political obligation: membership, commitment, and the bonds of society*. Oxford/New York: Clarendon Press.

Gosseries, Axel, and Lukas H. Meyer (eds.). 2009. *Intergenerational justice*. Oxford-New York: Oxford University Press.

Hacker, Peter, and Joseph Raz. 1977. *Law, morality, and society essays in honour of H. L. A. Hart*. Oxford: Clarendon Press.

Held, Virginia. 2006. *The ethics of care: personal, political, and global*. Oxford-New York: Oxford University Press.

Hinton, Timothy. 2015. *The original position, Classic philosophical arguments*. Cambridge (UK)-New York: Cambridge University Press.

Jamieson, Dale. 2007. "When Utilitarians Should Be Virtue Theorists." *Utilitas* 19 (2): 160–183. https://doi.org/10.1017/S0953820807002452

McTaggart, J. Ellis. 1908. "The Unreality of Time." *Mind* 17 (68): 457–474.

Paul, Laurie Ann. 2010. "Temporal Experience." *The Journal of Philosophy* 107 (7): 333–359.

Plato, and Robin Waterfield. 1993. *Republic*. Oxford: Oxford University Press.

Pulcini, Elena. 2020. *Tra cura e giustizia: le passioni come risorsa sociale*. Torino: Bollati Boringhieri.

Rawls, John. 1999. *A theory of Justice.* Belknap Press of Harvard University Press.

Rawls, John. 2005. *Political liberalism. Expanded ed*. New York: Columbia University Press.

Singer, Peter. 1972. "Famine, Affluence, and Morality." *Philosophy & Public Affairs* 1 (3): 229–243.

Singer, Peter. 2009. *The Life You Can Save: Acting Now to End World Poverty*: Random House Publishing Group.

Tronto, Joan C. 1993. *Moral boundaries: A political argument for an ethic of care*. New York: Routledge.

Zwarthoed, Danielle. 2015. "Cheap Preferences and Intergenerational Justice." *Revue de philosophie économique* 16 (1):69-101. https://doi.org/10.3917/rpec.161.0069

7

What is Emissions Egalitarianism?

OLLE TORPMAN

How should the available carbon budget, in terms of emissions permits, be distributed among people? One suggestion is implied by Emissions Egalitarianism (EE), according to which it should be divided equally between people. However, there are many ways in which EE could be understood, and many criticisms have been raised against this view. The aim of this chapter is to clarify what EE is and what it is not, and to defend this view against some common objections.

Introduction

One of the debates in climate ethics concerns the question of how the carbon budget, in terms of emissions permits, should be distributed among people. In other words, how should the atmosphere's capacity to absorb greenhouse gases, considered as a public good, be divided (Broome 2012, 69)? One answer to this question, which is implied by a principle called Emissions Egalitarianism (EE), is that emissions permits should be divided equally between people. In other words, EE implies that every person is entitled an equal share of the atmosphere's absorptive capacity (Baatz and Ott 2017; Torpman 2019).

EE has been defended by a number of authors (see, e.g., Singer 2010; Attfield 2003; Jamieson 2005; Garvey 2008; Broome 2012; and Torpman 2019). But it has been criticized by just as many others (see, e.g., Posner and Weisbach 2010; Caney 2012; Margalioth and Rudich 2013; Roser and Seidel 2017; Baatz and Ott 2017). However, many of the debates around EE stem from confusions about how this view should be understood in detail. My aim in this chapter is to clarify how EE is best understood and to defend EE from some of the objections that have been levelled against it. EE is, strictly speaking, a family of views that can be understood in many different ways (Caney 2012: 260). As this suggests, it is not clear exactly how EE should be understood. This chapter attempts to clarify what sort of a principle EE is – and what sort of a principle it is not.

1. Is EE a Mitigation, Adaptation or Compensation Principle?

There are mainly three different kinds of responsibility regarding the climate crisis: mitigation, adaptation, and compensation. The purpose of mitigation is to prevent climate change to some extent. The purpose of adaptation is to make people capable of living good lives even with a changed climate. The purpose of compensation is to make restitutions to those whom climate change has – despite mitigation and adaptation efforts – caused loss and damage. Mitigation is thus about alleviating the causes of climate change, while adaptation and compensation are both about coping with the effects of unmitigated climate change.

One can thus ask whether EE is a mitigation, adaptation or compensation principle. Most climate ethicists assume that EE is a mitigation principle since it is supposed to help prevent climate change (Vanderheiden 2008; Risse 2008; Caney 2012; Baatz and Ott 2017). This is plausible because an equal division of emissions permits is at the very least not a means of adaptation to climate change. In terms of adaptation, EE is at most supposed to help us adapt to the fact that the atmosphere's capacity to absorb greenhouse gases is limited (Gardiner 2010, 574).

Some have argued that EE is implausible as a mitigation principle because it fails to solve the climate crisis (Baatz and Ott 2017). While it is true that EE is not sufficient to solve climate change, it should be mentioned that even if EE is a mitigation principle, it is not by itself supposed to solve the climate crisis. Indeed, EE is only one of three parts of the so-called cap-and-trade solution to climate change.

According to the cap-and-trade idea, climate stability is achieved by a three-step procedure: (i) an international agreement sets an emissions cap for all countries, (ii) this emissions cap is then divided in terms of emissions permits between countries and (iii) these emissions permits are either used directly by each nation or sold to other nations (Posner and Weisbach 2010, 137–138; Broome 2012, 68–69). As this suggests, EE can only help us achieve climate stability together with the other steps in the cap-and-trade procedure. This is true of any distributive principle for emissions permits, as they are all supposed to fill the same function in a cap-and-trade system. Hence, they all fare equally well in this sense with respect to climate change mitigation.

2. Is EE a Distributive or a Corrective Principle?

Besides the division between mitigation, adaptation and compensation principles of climate justice, there is a distinction between distributive and corrective principles. These deal with two different questions of justice in relation to climate change: one distributive question concerns how the absorptive capacity of the atmosphere should be divided, and one corrective question concerns how the costs for dealing with the effects of climate change should be divided.

Once we observe that EE is only supposed to answer how the absorptive capacity of the atmosphere should be divided, it becomes clear that it is a distributive principle. It is not supposed to answer the question of how the costs for dealing with the impacts of climate change should be divided. That question is supposedly answered by corrective climate principles, such as the Polluter Pays Principle or the Beneficiary Pays Principle (see also Grasso and Corvino in this volume).

In relation to this, EE have been criticised for disregarding the fact that different people have contributed to climate change to different extents (Roser and Seidel 2017, 154–155). The idea is that those who have caused climate change should receive fewer emissions permits. However, this criticism conflates the distributive question with the corrective question. Of course, an equal per capita division of the costs for dealing with already caused climate change would imply that poor people, who have typically contributed disproportionately little to climate change, will have to pay for the climate costs imposed by rich people. And this would be implausible. But since EE is not a corrective principle, it should not be charged for disregarding issues of causal responsibility for climate change (Vanderheiden 2008, 229–230; Vanderheiden 2011, 69; Risse 2008, 38). Indeed, EE is compatible with a corrective climate justice principle according to which the costs for dealing with climate impacts (i.e., costs related to adaptation and compensation) should be divided in proportion to the extent that different countries (or individuals) have contributed to climate change.

3. Is EE an Isolationist or Integrationist Principle?

There is a question whether emissions permits should be divided in isolation from, or in integration with, other considerations of justice – such as trade, development, poverty and health (Caney 2012; Posner and Weisbach 2010). In other words, there is a distinction between isolationist and integrationist distributive principles. While isolationist principles imply that the distribution of emissions permits should be insensitive to other considerations of justice, integrationist principles imply that this distribution should be sensitive to such considerations. Since EE recommends that everyone should have an equal share of emissions permits, however, it should be understood as an isolationist principle.

Many debaters have argued that EE is implausible precisely because it is an isolationist approach. For instance, Posner and Weisbach (2010, 129) argue that 'the intuitive attractiveness of the per capita approach [i.e., EE] depends on seeing it in isolation from all of the effects of a climate treaty and from other global policies, including other policies with distributive effects. Once we take these factors into account, the per capita approach [i.e., EE] appears far less attractive'. As this objection suggests, any principle for distributing emissions permits should be integrationist rather than isolationist (Caney 2012, 285; Baatz and Ott 2017, 14).

However, it is not clear that an integrationist principle would be superior to an isolationist principle like EE for several reasons. First, integrationist principles presuppose that emissions permits must be substitutable by other goods. But some emissions are non-substitutable. This is true especially for so-called subsistence emissions – i.e., emissions caused by processes required to fulfill basic needs (Shue 1993). It is impossible to survive without emitting, since breathing and digesting, as well as the production and transportation of food, give rise to emissions. Second, the considerations of justice that are disregarded by EE could be dealt with by other principles. Indeed, EE is a local rather than global principle in the sense that it governs a specific domain of justice. Consequently, it is only supposed to recommend how emissions permits should be distributed, not to provide recommendations for all sorts of issues (Posner and Weisbach 2010, 86). For this reason, EE should be considered one principle among many that deal with issues other than those regarding the distribution of emissions permits. Third, it seems clear that isolationist principles would fare better from the perspective of political feasibility, since most political communities would likely find it easier to agree on one issue at a time than agreeing on several at once (Singer 2010; Posner and Weisbach 2010, 86; Torpman 2021). Hence, EE should not be rejected simply for being an isolationist approach.

4. Is EE a First-level or Second-level Principle?

In ethics, there are different levels of principles, most noticeably first-level and second-level principles. Fundamental normative theories of justice and morality, such as egalitarianism, sufficientarianism, utilitarianism and Kantianism, are examples of first-level principles. These are general principles in the sense that they can be used to derive recommendations for all sorts of choices and actions. At the second level, however, we find principles that are meant to deal with specific domains of practical issues, such as those related to agriculture, war, medicine, climate change and so on. These principles are justified in turn on the basis of first-level principles, and hence they are non-fundamental.

Since EE is supposed to deal with the distribution of emissions permits in particular, rather than issues of justice in general, it is important to make clear that EE should be understood as a second-level principle. That is, even if EE is a form of egalitarianism, it is not a *fundamental* – or first-level form – of egalitarianism. This means that EE is not in itself a rival to any first-level moral principle, but rather in need of justification based on such principles (Caney 2005: 2–3).

It has been argued that one problem with EE is that it cannot be justified by any first-level principle, since its recommendation – i.e., an equal distribution of emissions permits – cannot be supported by any such theory. But this critique is mistaken. If the atmosphere can be regarded as a common good that initially belongs to everyone, it could be argued from a libertarian point of view that everyone has an equal initial right to use its absorptive capacity (Moellendorf 2011; see also Moser in this volume). And since EE would push the political process towards a solution to the climate crisis and thus increase global welfare, it could be defended from a utilitarian perspective (Singer 2010, 194). If understood as a maxim for an equal distribution of emissions permits, EE could also find support from Kantian morality, since everyone could consistently will to emit an equal share of the total amount of greenhouse gases the atmosphere can absorb. Moreover, EE seems compatible with virtue ethics, since any distribution a virtuous person would recommend could plausibly be justified along the lines of EE.

Still, it might be argued that EE is redundant because we could infer recommendations for emissions distributions directly from first-level theories of justice or morality (Caney 2012, 291–300; Posner and Weisbach 2010, 143). However, it is not clear why it would be more practicable to infer a distributive scheme directly from fundamental moral theories than from second-level principles like EE. Since first-level theories are general, they will have to take into consideration all morally relevant aspects before yielding an all-things-considered recommendation regarding emissions distributions. This leads us to the next issue.

5. Is EE a Pro Tanto or All-Things-Considered Principle?

While it is clear that EE is a local principle, governing the specific domain of how to distribute emissions permits, one can ask whether EE is an all-things-considered or *pro tanto* principle governing that specific domain. Understood as an all-things-considered principle, its recommendations impose obligations – full stop. As such, nothing could override these recommendations. Considered as a *pro tanto* principle, however, EE's recommendations should be weighed against the recommendations of other *pro tanto* principles and thus could be overridden if other such principles were to carry heavier weight.

The plausibility of EE depends, at least to some extent, on whether it is understood as a *pro tanto* or an all-things-considered principle (see, e.g., Baatz and Ott 2017; Morrow 2017, 20; Knight 2013). If it is understood as a *pro tanto* principle, then other principles – concerning other considerations than just the distribution of emissions permits – must be taken into account before anything can be said in detail about how emissions permits should be distributed. In effect, this might well imply that they should not be divided equally between people.

However, defenders of EE do not think that other principles need to be considered in order to answer how emissions permits should be divided among people. Instead, they regard it as the one and only principle for distributing emissions permits, suggesting that it is an all-things-considered principle. This is, moreover, in line with the fact that EE is an isolationist principle, disregarding other considerations of justice – such as of needs, contributions, costs, benefits and so on.

Still, even if EE is an all-things-considered principle regarding the distribution of emissions permits, it should be emphasized that it is not a general distribution principle. Instead, there are additional principles for other distributive issues than those related to climate change. For instance, a health care principle of some sort should answer issues related to how prioritizations should be made in health care.

6. Is EE an History-Sensitive or History-Insensitive Principle?

Even if EE is an all-things-considered principle governing the specific domain of how to divide emissions permits, it is possible to distinguish between two versions of such a principle: one history-sensitive version that takes into account past emissions and one history-insensitive version that does not take such emissions into account. It is not obvious whether an equal distribution of emissions permits is sensitive to the emissions people have made in the past.

Interestingly, objectors to EE have claimed that it is a history-insensitive view. Roser and Seidel (2017, 156), for instance, claim that EE 'does not include historical emissions'. Similarly, Baatz and Ott (2017, 21) argue that it ignores past emissions. On the basis of such an understanding, they have argued that EE is implausible because it disregards the unjust historical emissions of different nations and hence because it does not take into account people's different needs.

However, there is nothing inconsistent with a history-sensitive version of EE. Understood as a history-sensitive view, EE would not recommend that future emissions permits are divided equally from now on. Rather, it would suggest that those with a history of higher-than-equal emissions should have fewer emissions permits in the future. Given the unequal past emissions of different countries, the *overall* equal distribution suggested by a history-sensitive version of EE implies that *future* emissions permits should be divided unequally – at least until an equilibrium is reached (see Broome 2012, 70). Since current differences in needs between people are related to their historical emissions, the history-sensitive version of EE actually takes these differences into account.

It is still true that a history-sensitive version of EE does not take needs to be relevant *per se*. For instance, one country could have emitted a lot in the past, yet in such a wasteful way that it is nevertheless as poor as if it would not have emitted so much. And EE would not take that into direct consideration. Here, however, I think the problem at issue – regarding people's different needs – should be dealt with through the use of other principles than those regarding the distribution of emissions permits. Indeed, EE is certainly not the one-and-only second-level principle at stake. Moreover, the existence of domain-specific second-level principles does not preclude the existence of an overarching rectification principle – whose purpose would be to correct for injustices that remain after all local principles have been applied.

7. Is EE an End-State or Transition-State Principle?

EE could be understood either as an end-state or a transition-state principle. Considered as an end-state principle, the recommendation for an equal per capita distribution of emissions permits would be an ideal or long-term goal. As a transition-state principle, EE's recommendations would instead be understood best as an instrumental step towards an ideal or a long-term goal.

When deciding whether EE should be understood as a transition-state or end-state principle, it should be noted that climate ethicists are not basically interested in emissions permits *per se*, but rather in the benefits that can be produced by emissions permits (e.g., capabilities or opportunities for well-being). This is in line with the fact that EE is not directly concerned with the distribution of such benefits, but rather with the distribution of one important resource (i.e., emissions permits) considered as a means to such benefits.

This indicates that EE is most plausibly understood as a transition-state principle rather than an end-state principle. This is in accordance with Singer's view (2010, 191), according to which EE should be seen as 'a fair starting point'. A similar interpretation is made by Baatz and Ott (2017, 26), discussing EE 'as a first step towards a more just world and a global redistribution of entitlements that provides the poor and voiceless with more (bargaining) power'. Similarly, Roser and Seidel (2017, 64) argue that it 'can be understood as a natural starting point, so that all deviations from equality are in need of justification'.

Thus understood, there is nothing that precludes that future emissions permits – i.e., permits that will become available thanks to the atmospheres cyclic ability to absorb greenhouse gases – are unequally distributed. Moreover, EE might perhaps only have a role to play within a cap-and-trade solution to the climate crisis, as discussed above. Once this crisis is solved and the world's economies have become carbon-free and thus climate-neutral, the role for EE and other distributive principles for emissions permits may be outlived. There would then not be any scarcity of the atmospheric resource (for greenhouse gas absorption), and hence no need for a principle to divide it fairly. Until then, however, EE may fill an important role.

8. Is EE an Individualist or a Collectivist Principle?

EE can be understood either as an individualist or collectivist principle. Considered as an individualist principle, EE proposes an equal distribution of emissions permits directly to particular individuals. Understood as a collectivist principle, EE instead proposes a distribution of emissions permits to nation-states, where every nation-state receives emissions permits based on its number of citizens. However, most defenders of EE seem to think that its recommendations should be implemented at a collective level (Broome 2012, 70). This can be seen as problematic for several reasons.

First, the collectivist version of EE appears to identify the wrong claimants on the carbon budget, since it gives to states what should be given to individuals. It is commonly assumed that justice is owed basically to particular individuals rather than to collectives of individuals. This might be taken as evidence in favor of an individualist understanding of EE. Second, EE might thus seem to give populous states an advantage over non-populous states. In effect, it would not give countries an incentive to decrease population growth – something that would be desirable from a climate change perspective.

Third, EE appears to neglect the fact that there is also inequality within states. Indeed, not all people who live in rich countries are themselves rich, just as not all people who live in poor countries are themselves poor.

In defense of the collectivist version of EE, however, it should be mentioned that international climate agreements form the context in which distributive principles for emissions permits are debated and that the purpose of such agreements is to distribute emissions permits between nation-states (Morrow 2017). Moreover, we should recall that EE is most plausibly a transition-state principle. As such, its recommendations should be regarded as providing an instrumental step towards an ideal or long-term goal. Although the long-term goal is certainly fairness among individual people, an equal per capita distribution of emissions rights to nation-states might be a means to achieve such a goal.

In relation to this, it is not clear that an allocation of emissions permits directly to each individual worldwide would be politically feasible. For one reason, since an individual citizen's net emissions consist in part of emissions made by its nation-state's institutions, an attribution of emissions rights directly to individuals would require a tremendous knowledge apparatus in order to account for this (Morrow 2017). An allocation through nation-states is more feasible and would allow different states to decide on different concrete climate policies in order to comply with such an allocation (Margalioth and Rudich 2013, 194).

Conclusion

The purpose of this chapter has been to clarify how EE should be best understood, and to defend it against some common objections. It has been argued that EE should be understood as:

1. A mitigation rather than adaptation principle in the sense that it is supposed to fill a function in the cap-and-trade solution to the climate crisis rather than to help us adapt to climate change
2. A distributive rather than corrective principle in the sense that it aims to provide a recommendation for how emissions permits should be distributed, rather than for how the costs for climate change should be divided
3. An isolationist rather than integrationist principle in the sense that it suggests that these permits are distributed in isolation from other considerations of justice
4. A second-level rather than first-level principle in the sense that it is derived from more general and fundamental principles of justice and morality

5. An all-things-considered rather than a *pro tanto* principle in the sense that it gives us all that is needed to determine how emissions permits should be distributed
6. A history-sensitive rather than history-insensitive principle in the sense that it takes historical emissions into account
7. A transition-state rather than an end-state principle in the sense that it is supposed to function as a means rather than a normative ideal
8. Finally, a collectivist rather than individualist principle in the sense that it suggests a distribution of emissions permits to nations (rather than directly to particular individuals) in proportion to their number of citizens.

When EE is understood in this way, it becomes clear that many of the objections raised against it can be dismissed.

Acknowledgment: I gratefully acknowledge the financial support from Riksbankens Jubileumsfond (grant number M17-0372:1).

References

Attfield, Robin. 2003. *Environmental Ethics: An Overview of the Twenty-First Century*, Polity Press (Cambridge).

Baatz, Christian and Ott, Konrad. 2017, "In Defense of Emissions Egalitarianism", in L. Meyer & P. Sanklecha (Eds.) *Climate Justice and Historical Emissions*, Cambridge University Press.

Bovens, Luc. 2011. "A Lockean Defense of Grandfathering Emission Rights," in Denis Arnold (ed.), *The Ethics of Global Climate Change*, Cambridge University Press.

Broome, John. 2012. *Climate Matters – Ethics in a Warming World*, W.W. Norton & Co (New York).

Caney, Simon. 2005. *Justice Beyond Borders*, Oxford University Press.

Caney, Simon, 2012. "Just Emissions", *Philosophy and Public Affairs* 40 (4): 255–300.

Chancel and Piketty. 2015. "Carbon and inequality: From Kyoto to Paris. Trends in the global inequality of carbon emissions (1998-2013) & Prospects for an equitable adaptation fund," Paris School of Economics, Paris. http://piketty.pse.ens.fr/files/ChancelPiketty2015.pdf

Gardiner, Stephen. 2010. "Ethics of Global Climate Change", in Gardiner S.M., *et al.*, 2010. eds. *Climate Ethics –Essential Readings*. Oxford University Press (USA).

Garvey, James. 2008. *The Ethics of Climate Change – Right and Wrong in a Warming World*, Continuum (London).

Hooker, Brad. 2013. "Rule-Consequentialism", *The Blackwell Guide to Ethical Theory*, Blackwell Publishing.

IPCC. 2014. *Climate Change 2014: Impacts, Adaptation, and Vulnerability. Part A: Global and Sectoral Aspects. Contribution of Working Group II to the Fifth Assessment Report of the Intergovernmental Panel on Climate Change.* Cambridge: Cambridge University Press.

Jamieson, Dale. 2005. "Adaptation, Mitigation, and Justice", pp. 221–253 in Sinnot-Armstrong and Howarth (eds.), *Perspectives on Climate Change,* Emerald Publishing Group (Amsterdam).

Margalioth, Yoram and Rudich, Yinon. 2013. "Close Examination of the Principle of Global Per-Capita Allocation of the Earth's Ability to Absorb Greenhsouse Gas", *Theoretical Inquiries in Law* 14 (1): 191–206.

Moellendorf, Darrel. 2011. "Common atmospheric ownership and equal emissions entitlements", In: Arnold DG (ed) *The ethics of global climate change*. Cambridge University Press, Cambridge and New York, pp. 104–123.

Morrow, David R. 2017. "Fairness in Allocating the Global Emissions Budget", in *Environmental Values* 26 (6): 669–691.

Posner, Eric and Weisbach, David. 2010. *Climate Change Justice*, Princeton University Press.

Risse, Mathias. 2008. "Who Should Shoulder the Burden? Global Climate Change and Common Ownership of the Earth," Faculty Research Working Paper Series, Harvard John F. Kennedy School of Government.

Roser, Dominic and Seidel, Christian. 2017. *Climate Justice: An Introduction*, Routledge.

Shue, Henry. 1993. "Subsistence Emissions and Luxury Emissions", *Law & Policy* 15 (1): 39–59.

Singer, Peter. 2010. "One Atmosphere", in Stephen M. Gardiner, Simon Caney, Dale Jamieson & Henry Shue (eds), *Climate Ethics – Essential Readings*. Oxford University Press (USA).

Steiner, Hillel. 2009. "Left Libertarianism and the Ownership of Natural Resources", *Public Reason* 1 (1): 1–8.

Torpman, Olle. 2019) "The Case for Emissions Egalitarianism", *Ethical Theory and Moral Practice* 22: 749–762.

Torpman, Olle. 2021. "Isolationism and the Equal per Capita View", *Environmental Politics* 30 (3): 357–375.

Vallentyne, Peter, and van deer Vossen, Bas. 2014. "Libertarianism", *The Stanford Encyclopedia of Philosophy* (Fall Edition), Edward N. Zalta (ed.). https://plato.stanford.edu/entries/libertarianism/

Vanderheiden, Steve. 2008. *Atmospheric justice: A Political Theory of Climate Change,* Oxford University Press.

8

The Polluter Pays Principle and the Energy Transition

FAUSTO CORVINO

One of the guiding principles of the energy transition is the polluter pays principle (PPP): getting those who engage in emission-generating activities to internalise the social cost of carbon. In this chapter, I distinguish between the backward-looking and forward-looking versions of the PPP. In the first part, I examine the two main objections that have been raised against the backward-looking version of the PPP: i) those who emitted GHGs in the past did not know that their actions would warm the climate, so these actors are not morally responsible for climate change and neither are their heirs; ii) even if past polluters were morally responsible for their emissions, their heirs could not have done anything to prevent past polluters from emitting, and so it is unfair to hold present people responsible for past emissions.

I argue that none of the answers proposed so far to these two objections allow us to hold present individuals morally responsible for all GHGs emissions that occurred in the past. I then introduce the forward-looking version of the PPP (FL-PPP). On the one hand, I explain how the implementation of the FL-PPP is indispensable for the energy transition to live up to ambitious mitigation targets. On the other hand, I argue that the FL-PPP calls for developed countries to transfer part of the resources stemming from the internalisation of the social cost of carbon to developing countries, as developing countries are more exposed to climate harms. I conclude by positing that it is possible to justify the duty of developed countries to finance the transition to low-carbon energy, adaptation policies and compensation measures in developing countries on the basis of the PPP, but without relying on its backward-looking version.

Introduction

Climate change is mainly a matter of global negative externalities prolonged for about two hundred years. From the Industrial Revolution onwards, humans have sustained modern economic growth by failing to internalise the social cost of greenhouse gasses (GHGs), and of carbon dioxide in particular. Until now, those that have emitted GHGs have only paid for the private cost of emission-generating activities (which is a function of both the market interaction between supply and demand for fossil fuels and of the cost of producing emission-generating goods and services) without bearing the cost of collective climate harms caused by GHG emissions at the moment in which GHGs accumulate in the atmosphere (see Fleurbaey et al. 2019; Metcalf 2019, 36–37; Stern and Stiglitz 2021). This of course does not mean that if the social cost of emission-generating activities had been internalized, there would have been no economic benefits for polluters, but these benefits would have been lower (Shue 2015).

Once we introduce the climate variable into the economic history of the last two hundred years, we realise that past generations have maximised their prosperity by borrowing against future generations, i.e., by leaving us with part of the costs associated with the benefits they reaped (see Meyer and Sanklecha 2017) – obviously, some of these benefits have also come to us, in terms of infrastructure, technological know-how, financial capital, etc. These costs manifest themselves today, among other things, in the need to implement a rapid climate transition aimed at limiting global warming to 1.5°C above pre-industrial levels – of course, if present generations had acted earlier, the costs would have been lower (IPCC 2018). In order to do this, it is necessary to reach net-zero CO_2 emissions by 2050 and net-zero GHG emissions no later than 2070 (IEA 2021; IPCC 2018, 12). Furthermore, since some GHG emissions cannot be abated and we may emit more than is compatible with the 1.5°C mitigation target, it will be essential to achieve so-called negative emissions: removing CO_2 from the atmosphere through anthropogenic actions, either by enhancing natural CO_2 sinks or through appropriate technologies and/or geo-engineering solutions (see Peacock 2021; Gardiner and McKinnon 2020).

One of the cornerstones of the climate transition is the energy transition: decarbonising the global energy system while meeting a growing energy demand worldwide due to population growth and the economic ambitions of developing countries (other sectors that play a central role in the climate transition are agriculture, food, waste production and management, land preservation/restoration and so on, see also Nogrady 2021). The energy transition requires two main things to be effective and one to be fair. First, the price of fossil fuels must rise so that economic agents internalise the social cost of GHGs as much as possible (Baranzini et al. 2017). In this way we can, on the one hand, incentivise the abandonment of fossil fuels, and, on the other hand, collect economic resources (e.g., carbon revenues) to be used to offset the negative effects of the energy transition on short- and medium-term net-losers (e.g., people who lose their jobs or are penalised by the rising cost of non-renewable energy). Second, the price and reliability differential between polluting technologies and clean technologies (e.g., those using renewable energy sources) must fall substantially – which can only be achieved through major investments in technological research (see Bourban 2021; Helm 2020; Gates 2021). Third, the negative short- and medium-term social impacts of the energy transition must be offset, especially when workers are the victims.

Nonetheless, no energy transition can guarantee the mitigation of all climate threats. Climate change is already occurring, and developing countries are suffering the most. Accordingly, adaptation to present and future climate threats, i.e., natural, economic and social adjustments that can minimise the negative impacts of global warming, must be provided wherever possible (Page and Heyward 2017). Where, however, both mitigation and adaptation fail – either because adaptation encounters unavoidable limits or because it was not done in time – the question of responsibility for climate-induced damage arises (Wallimann-Helmer et al. 2019). Moreover, the energy transition must be contextualised in a historical perspective. If no one can afford to base future economic growth on the exploitation of fossil fuels, this is a consequence of the past unequal appropriation of the carbon budget – i.e., the cumulative CO_2 emissions compatible with a given climate mitigation target, such as the 1.5°C target (of all GHGs, CO_2 is the main driver of climate change, and the vast majority of CO_2 emissions are caused by the combustion of fossil fuels). This raises an issue of fairness vis-à-vis developing countries, insofar as they are asked to give up a similar fossil fuel-based growth path followed by developed countries.

In brief, ambitious climate mitigation based, among other things, on an effective and (globally) fair energy transition, plus an international redistribution of resources, which aims to ensure means of adaptation and loss-and-damage compensation for the blameless (or less blameless) victims of climate change, are the main ingredients of a just climate transition. One normative principle that seems to fit perfectly with this challenge is the polluter pays principle (PPP): you broke it (and/or you cannot stop breaking it, at least in the near future), you fix it. Applied to the climate transition, the PPP would recommend making those who emit GHGs internalise the social cost of their pollution and use this money to invest in green energy technologies, transfer these technologies at affordable prices to developing countries and finance adaptation and loss-and-damage compensation.

There are, however, two possible versions of the PPP. The first version emphasises making sure that anyone emitting GHGs internalises the resulting social cost from now on. This can be done either by introducing a carbon tax on top of the market price of fossil fuels or by capping the maximum amount of CO_2, distributing emissions permits and then allowing private actors to trade these permits (whoever emits more than the permits they have been given will have to buy more permits, and this is tantamount to a tax – the so-called cap-and-trade system) – public authorities, moreover, can internalise the social cost of GHGs by including it in the assessment of public policies and investments (i.e., the higher the social cost of GHGs, the easier it will be to justify stringent climate regulations and green investments). The second version of the PPP emphasises making sure that anyone who has emitted GHGs up to now (thus contributing to the climate problem) internalise *ex post* the social cost of their pollution. This can be achieved by presenting present individuals (which most likely are also high-emitters themselves) with an additional bill for the pollution by their predecessors.

The two versions are not equivalent. The first version of the PPP implies that those who reap the benefits of past emissions now will take advantage of the fact that their predecessors have externalised the social cost of these emissions to third parties; and this means that the burden of the climate transition will be distributed in a way that is history-insensitive (let bygones be bygones, we could say) – this is the forward-looking version of the PPP (see also Mittiga 2019, 167–173; Tilton 2016). The second version, on the other hand, implies that those who have inherited wealth unfairly (based on the social costs of GHG pollution that have not been fully internalised) should bear a proportionately higher burden of the climate transition – this is the backward-looking version of PPP (see Roser and Seidel 2017, 118–120). Though different, the two versions are not mutually exclusive. Indeed, many believe the PPP applies in both temporal directions.

This chapter is structured into two parts. In the first part, I analyse the philosophical obstacles encountered by the backward-looking version of the PPP, and I maintain that it falls short of covering all past emissions. In the second part, I explain why the forward-looking version of the PPP is indispensable to the energy transition, focusing in particular on how it can reorient market choices in an efficient way; at the same time, if we adopt a global accounting of the social cost of carbon, the FL-PPP can demand an international redistribution of resources to finance mitigation, adaptation and loss-and-damage compensation, without linking this duty to the complex issue of historical climate injustice. While the PPP is therefore limited in its historical intergenerational reach, it remains relevant to global climate justice in its future-oriented component.

1. The Backward-Looking Version of the PPP

The backward-looking version of the PPP (BL-PPP) is the environmental application of a more general moral principle that many people intuitively consider appropriate in responding to a wrong that occurred in the past: you created the problem, now you solve it. Usually, when it comes to historical injustices, B's heirs (Bh) demand compensation for harms inflicted by A on B, and they address their claims to A's heirs (Ah) if A is no longer there. Take, for example, the case of artworks stolen from Jewish families by the Nazi regime during World War II and the request for compensation made by the heirs of these families to the governments in countries such as Austria or Germany that came into possession of these artworks after the defeat of the Nazis (the story is very well described in the movie 'Woman in Gold', directed by Simon Curtis, 2015). Here is a case of both an intra-generational harm (A→B) and an intra-generational compensatory claim (Bh→Ah), which is grounded in two inter-generational relations (B→Bh; A→Ah).

Historical climate injustice, by contrast, differs from classic cases of historical injustice insofar as those who emitted more than their fair share of GHGs in the past directly harm those who live today and are more vulnerable to the negative effects of climate change because GHG emissions accumulate into the atmosphere and deploy their negative effects many years later (see also Meyer and Roser 2010, 230). Accordingly, historical climate injustice consists of both inter-generational harm (A→Bh) and an intra-generational compensatory claim (Bh→Ah), and the latter is based on one single inter-generational relation (A→Ah), which in turn is called into play for the mere fact that A is no longer here.

If we look at global data on historical emissions, it is easy to see the appeal of BL-PPP. There was a first historical phase, from the beginning of the Industrial Revolution and lasting until the end of the eighteenth century, in which almost all GHG emissions were attributable to United Kingdom (UK). A second phase, from the beginning of the nineteenth century and lasting until 1880, in which emissions overwhelmingly belonged to the UK, the United States (US) and the countries that are now part of the European Union (EU). A third phase began in 1880 and lasted until to the beginning of the 1970s, in which Canada, Australia, other members of the EU and some developing Asian countries joined the global polluting club with progressively increasing shares (Ritchie and Roser 2020; Friedrich and Damassa 2014).

Basically, we are faced with a situation where, at least until the 1970s, the responsibility for emissions that caused climate change lies almost exclusively with developed countries. And while it is true that developing countries started emitting a lot in absolute terms after that date and then surpassed even developed countries in relative terms, two factors mitigate the responsibility of developing countries for their emissions. The first is that many more people live in developing countries like India and China than in Europe and the US. In 2020, more than 1.4 billion people lived in China and more than 1.3 billion people lived in India, while just over 300 million lived in the US and less than 500 million lived in the EU (see Worldometer 2022). Thus, if we look at per capita emissions, rather than total emissions, developing countries' contribution to climate change drops dramatically. Secondly, many developing countries are net exporters of CO_2, meaning they emitted to produce goods that were then exported to developed countries, especially in the second half of the twentieth century, so many argue that the responsibility for these emissions should be placed on importers rather than exporters (Duus-Otterström and Hjorthen 2019; Grasso 2016).

Given the history of climate change, the BL-PPP opens up two different normative arguments, one strictly compensatory and the other mainly distributive. The compensatory argument suggests that those who caused climate change should take on a proportional burden of coping with climate harms, e.g., by financing adaptation policies and/or compensatory measures in vulnerable countries. The distributive argument, instead, is based on the premise that the carbon budget is a global common good, and therefore those who have arbitrarily appropriated large shares of it are henceforth entitled to fewer emission permits than the heirs of historical under-appropriators (see Roser and Seidel 2017, 118–129; Meyer and Roser 2010). Since the climate transition, however, necessitates that all countries achieve net-zero emissions by the middle of the twenty-first century, the distributive argument turns into the demand that historical over-appropriators finance the transition to low carbon energy in those countries that have historically emitted the least (see also Gajevic Sayegh 2017). Yet, whether BL-PPP is intended as a distributive principle for emissions, or whether it is used only for compensatory purposes, or both, it must prove to withstand the objections that have been raised in the literature to the notion of historical responsibility for climate change. Given the limited space available in this chapter, I focus on two of the strongest objections: the excusable ignorance objection (EIO) and the powerlessness exemption (PE).

1.1 The Excusable Ignorance Objection (EIO)

The EIO challenges the moral blameworthiness of inter-generational climate harm (A→Bh). Past polluters, the EIO maintains, undeniably harmed present people, leaving them with less opportunity to cheaply industrialise through fossil fuels and exposing them to today's climate threats. Yet, past polluters were completely unaware of the effects of their emissions; hence, it is unreasonable to hold them morally responsible for inter-generational climate harms. Starting from a given historical reference point, we can assume that the scientific evidence on climate change has become incontrovertible – such as 1990, for example, when the first Intergovernmental Panel on Climate Change (IPCC) report was published (Gosseries 2004, 39–40). According to the EIO, everything that happened before this date should have no moral consequences either in terms of compensatory duties or the distribution of the remaining carbon budget.

A first, possible reaction to the EIO is to bite both the theoretical and the empirical bullet: only post-1990 GHG emissions are morally relevant. This first response would reduce the scope of the BL-PPP only in part, given that more than half of historical emissions have been produced in the last 30 years. A second possible solution is to challenge the empirical premises of EIO. Humans did not really need to wait for the first IPCC report to realise that the burning of fossil fuels, whether for production, consumption or transport, was pushing the planet into a zone of risk, however unspecified. Many people like to remember Svante Arrhenius, a Swedish chemist and Nobel laureate who lived in the second half of the nineteenth century, as the 'father of climate change'. Arrhenius was the first person to conduct a quantitative analysis on the nexus between CO_2 and water vapour in the atmosphere, on the one hand and global warming on the other (Sample 2005). Meanwhile, the first warnings about the risks of global warming (induced by increasing concentrations of CO_2 in the atmosphere) were probably given by Gilbert N. Plass (1956), a Canadian physicist working at John Hopkins University. Moreover, there is little doubt today that oil companies have known since at least the late 1970s about the risks that fossil fuels posed to the climate – and indeed many accuse them of deliberate cover-ups (McGreal 2021).

In this chapter, I do not dwell on this dispute, which is primarily the concern of hard scientists and historians – even though I discuss the moral implication on this scientific uncertainty at the end of this section. I focus, instead, on the argumentative strategies that resist the theoretical, rather than empirical, challenges to the EIO. The first theoretical rejoinder to the EIO consists in adopting a strict-liability interpretation of responsibility and maintaining that knowing about the effects of someone's actions is irrelevant for this person to acquire remedial responsibility. Applied to the climate case, the strict-liability argument tells us that developed countries appropriated a much larger share of the carbon budget than they were entitled to, thus causing climate alterations. Thus, irrespective of whether this was intended and/or known, the over-appropriation by A has left Bh with less than they might otherwise have had, and this calls for compensation from Ah to Bh (Neumayer 2000, 188).

This is quite a strong position, in my view. And it is also difficult to justify because it is based on a moral principle that many would find intuitively inadmissible in most other cases: namely, that A has a compensatory duty towards B, beyond what is established by the law, for any harm caused by action X, even including harms that have nothing to do with the reason why A performed X. Suppose, for example, that A and B are playing a game of poker. A cheats and makes B lose. B is full of debt, he loses the last money he has in the game and kills himself. Can we say that A, who certainly performed a morally reprehensible action X by cheating, but was completely unaware of B's economic situation, is responsible for B's death? Can we say further that A must also compensate B's family for any loss or damage that results from B's death? I believe that most people would consider this excessive in the sense that it goes beyond the faults that are morally attributable to A. And if this were the case, there would be no reason why the climate affair should be different from the poker example just described.

A second theoretical rejoinder to the EIO consists of a sort of backward induction: after people acquired scientific certainty about climate change (in the late twentieth century), their emissions levels did not decrease, rather they increased; hence we may infer that even if people had known about climate change in the past (e.g., in the twentieth century), their emissions trend would have remained the same. This is empirically plausible and is also based on a reasonable assumption of anthropological continuity. The problem, however, as rightly observed by Simon Caney (2010, 208-210), is that this solution passes over established arguments of procedural justice. That is, we would consider A morally responsible for something they did not commit – i.e., continuing to emit GHGs despite knowing the negative effects on the climate – just because Ah, which we assume to be anthropologically similar to A, did the same thing under similar circumstances. This does not seem like a very solid solution.

A third theoretical rejoinder to the EIO relies on the disentanglement of moral responsibility for climate change from moral blameworthiness. Alexa Zellentin (2014), for example, has argued that the people that emitted CO_2 before knowing the negative effects of climate change were acting outside their 'sphere of secure competence', and hence, the standards of care were more stringent than with other activities. According to Zellentin, A acts outside her sphere of secure competence with respect to an action X if, after taking all precautions, she is unable to foresee all the risks arising from X. And this may come, I would add, either from the fact that A uses instruments or technologies that go beyond the general knowledge acquired up to that moment – think, for example, of the use of a drug before experimentation is complete – or from the fact that the instruments and techniques in use involve risks, known and quantifiable in probabilistic terms, that are not fully controllable. It is not morally forbidden to act outside one's sphere of secure competence (at least not always), Zellentin maintains, but when someone does so and something goes wrong, the agent is responsible for compensating for any damage that occurs, even if all reasonable precautions have been taken.

Let us consider a case where a person decides to launch fireworks in an apparently isolated area. Let us also assume that this person takes all necessary and legally required precautions. A firework, however, does not follow the expected trajectory and causes damage to a house a few kilometres away. According to Zellentin's theory, although this person is not blameworthy, she is acting outside her sphere of secure competence and therefore has a moral duty of compensation to the owner of the house (in addition to the possible legal duty of compensation that falls on the person who produced the faulty firework). In other words, although this person is not morally blameworthy for launching fireworks, which are commonly sold and can be used legally if all precautions are taken, she is 'outcome responsible' for the damages caused by the faulty fireworks because she should have understood that she was dealing with something that she could not fully control once fireworks speed up and soar into the sky.

Zellentin's argument suggests that pre-1990 emissions are comparable to fireworks in my example. If correct, and the theory of 'outcome responsibility' outside someone's secure sphere of competence seems reasonable, then the EIO fails. A full discussion of outcome responsibility, which in any case seems reasonable, lies beyond the scope of this chapter. Instead, I focus on the empirical assumption, namely that when people started burning fossil fuels for industrial purposes, they were in a position to foresee that what they were doing could cause risks to the environment (and therefore also to people) that they were unable to control. It is certainly true that, since the start of the Industrial Revolution, everyone had the feeling that humans were interfering with the environment in an unprecedented way. Although numerical estimates of the effects of coal combustion on the environment and on people were not produced until the twentieth century, it is true that anyone living near the industrial towns of northern Britain in the early nineteenth century could see that the sky was blackened by factory effluents. And the effects of pollution on people's health, and even on children's growth, were common knowledge (see Hatton 2017; Hanlon 2020).

What is more difficult (and perhaps wrong) to assume is that factory owners and workers in the nineteenth century, and also in the first half of the twentieth century, could have understood the effects of CO_2 emissions on the climate. That is, while it is correct to argue that our predecessors understood very well the negative and sometimes lethal effects of coal burning on themselves and their communities – and probably also on the people who would live on that land in the future – it is very unlikely that they could have predicted that, many decades later, the GHGs released in Manchester would contribute to increasing the likelihood of floods in Bangladesh, or to desertification in some parts of central Africa. If this were the case, as I believe it is, then it would still be possible to raise the EIO against the BL-PPP as a principle of global climate justice. The EIO would instead fail with respect to local pollution cases, but this would certainly be unsatisfactory from a broader climate perspective.

1.2 The Powerlessness Exemption (PE)

The PE holds that a person cannot be considered morally responsible for another person performing an action if the first person is powerless in the face of the second person carrying out this action. In an inter-generational perspective, the PE would maintain that Ah cannot be considered responsible for A carrying out X if Ah could have done nothing to prevent X from occurring (Gosseries 2004, 41–42; see also Moss and Robyn 2019, 275-278). If this argument is sound, it would imply that present people cannot be ascribed compensatory duties for emissions caused before they were born, or also, I would add, before acquiring both individual freedom and political agency (i.e., it would not make sense to hold a 4-year-old child responsible for their emissions). While the EIO challenges the inter-generational harm claim $(A \rightarrow Bh)$, the PE remains neutral on this and questions, instead, the inter-generational relation $(A \rightarrow Ah)$ as a vehicle of accountability for historical actions, and in doing so, the PE rejects the intra-generational compensatory claim $(Bh \rightarrow Ah)$.

As Gosseries (2004, 41-42) correctly points out, there are two possible ways to resist the PE. Either the general moral argument at the basis of the PE could be challenged (showing, for example, how in many intra-generational affairs we pass over the PE and ascribe compensatory responsibility for an action to a person different from the one who performed this action), or the theoretical problem raised by the PE could possibly be circumvented by adopting a collectivist rather than an individualist methodological perspective, thus justifying the claim that compensatory responsibility does not pass from one generational cohort to the next, but remains with the community that endures over time.

The first solution does not work for an empirical reason that is clearly explained by Gosseries (2004, 41–42). In intra-generational cases where A performs an action X that harms B, and C is called upon to pay for the harm caused by A (because A is not in a position to bear the cost or because the legal system prevents A from being held liable to pay for the harm caused by *X*), C is always in a position of authority vis-à-vis A; a position which may be given, for instance, by an employment contract or by parental authority. In inter-generational dynamics, by contrast, the relationship of authority between the current generation and the previous one does not apply (at least as long as the two generations do not overlap in the political sense of the term, i.e., as long as the younger generation is born – and grows up – it is impossible to claim that the younger generation exercises authority over the older one), and so it is much easier for Ah to claim they are not responsible for A's decision to perform *X*.

The second strategy is more complex. It holds, for example, individuals are not responsible for the emissions caused by their predecessors, but rather the community to which they belong, which replicates itself over time and expresses its will through public institutions (see also de Shalit 1995). The community survives (usually) the individuals that make it up at any given time, and therefore the community can (and must) take responsibility for historical injustices. There are three basic problems, at least in my view, with this kind of argument.

First, if we give up on methodological individualism, we cannot embrace moral individualism either. And without moral individualism, we cannot uphold a cosmopolitan conception of justice. Missing the latter, we risk losing solid normative arguments to explain why a person living in Stockholm should accept that public money from her country should be invested in the mitigation of a phenomenon that, at least at the moment, affects people living in developing countries with less means to adapt to climate change (i.e., the classic obstacles we encounter when trying to globalise a theory of justice that has a communitarian basis – see also Page 2006, 120–121).

Second, we would need to hold that states are moral agents despite changes to governments and to forms of government. This would lead us to conclusions that are morally problematic (Caney 2006, 469–471; Mittiga 2019, 164–165). For example, should we accept that a people that has fought and gained independence should bear responsibility for the actions of the state that oppressed them, or that the countries that emerged from the dissolution of the Soviet Union are responsible for the emissions of the Soviet government, or that a people that have recently freed themselves from an authoritarian government should be bound by the public debt (or the emissions) incurred by the (possibly corrupt) elite that previously ran the country? (see also Pogge 2008, 118–122).

Third, even if we lighten our collectivist position and maintain that states can generate inter-generational obligations 'if they possess some specific moral properties' (e.g., being democratic), as rightly emphasised by Caney (2006, 470), this would not suffice to cover all past emissions. Some countries that emit a lot today have never been democratic. Some countries, such as Venezuela, for example, have gone through phases, marked by intense oil exploitation, that some would find difficult to describe as democratic. And, finally, many countries now considered democratic only became fully democratic after they started emitting GHGs for industrial purposes, and still others have had intermediate phases, in the past century, in which they were governed by authoritarian governments – think of Italy, Germany, Chile, Argentina, Portugal, Spain and so forth.

Finally, the fact that a country makes decisions in a democratic way does not necessarily mean that everyone is responsible for these decisions – at least, not everyone who spent time, effort and perhaps even money struggling against these decisions. Can we say, for example, that an activist who dedicates her life to convincing people that wars are useless and unjust is responsible for her country's military interventions or at least as responsible as her fellow citizens who have passively accepted these wars or even supported them? (see also Corvino and Pirni 2021). If we believe that this is not the case, then the collectivist response to the PE loses some of its force.

1.3 Can Present People Be Held Responsible for Past Climate Injustice Despite the EIO and PE?

In sum, neither of the two objections (the EIO and the PE) to the BL-PPP allow us to hold present individuals morally responsible for all GHG emissions that occurred in the past. Specifically, by applying the theory of outcome responsibility presented by Zellentin, the EIO can only be withstood to a very limited extent. Only some actors, in fact, can be considered to have acted outside their sphere of secure competence before 1990 – for example, the oil companies that understood that the use of fossil fuels could cause environmental and climate problems. However, the challenge posed by the PE remains: even if we can show that past emitters have compensatory duties towards the current victims of climate change, we should explain why the heirs of historical emitters inherit these duties despite being powerless over past polluters. As we have seen, none of the collectivist solutions analysed above is well suited to the climate problem.

Moreover, in this chapter I do not take into consideration a third, metaphysical objection to BL-PPP, the so-called 'non-identity problem', which holds that members of the present generation cannot claim to be harmed by the polluting activities of the past generations, because without these activities there would have been alternative reproductive choices and, therefore, we would have completely different individuals today, see Parfit 2000; I also discuss this issue in Corvino 2019.

It would therefore seem to follow that only those post-1990 emissions (and also the limited pre-1990 emissions for which emitters are outcome responsible) that are not subject to the PE legitimise the BL-PPP; but this would mean that the BL-PPP ends up forgiving a large part of the climate debt accumulated by developed countries over the last two centuries. Accordingly, many seek to overcome these theoretical limitations by holding that the emissions attributable to A have caused harm to Bh, and Ah have benefited from these emissions (in terms of economic growth, infrastructure, technological development and so forth) more than Bh. There may therefore be reasons of either compensatory or distributive justice that require a reallocation of resources from Ah to Bh, irrespective of the moral blame of A (EIO) and of the transferability of the moral responsibility from A to Ah for the unjust actions committed by A (PE). So, the Beneficiary Pays Principle (BPP), rather than the PPP, explains the moral relevance of past emissions. There are two possible interpretations of the BPP. A compensatory interpretation holds that if A inflicted a harm on Bh, and Ah have received benefits from this harm, then Ah cannot condemn the harm committed by A without also forfeiting the benefits they have derived (Butt 2014, 2017; Page 2012). A distributive interpretation, by contrast, suggests that if A seized more GHG emissions than is fair, and Ah received the largest share of the indirect benefits of A's emissions, Ah must contribute more to the climate transition than Bh (Roser and Seidel 2017, 133). I do not enter here into the issue whether the BPP is normatively sound. I merely point out three theoretical challenges for the supporters of the BPP. First, to prove that the mere benefit Ah derive from the harm inflicted by A on Bh is sufficient to justify either compensatory or distributive duties of justice on the part of Ah towards Bh (see also Huseby 2016). Second, to demonstrate that it is morally wrong for Ah to benefit from the harm inflicted by A on Bh, even if this harm is excused; only in this case can the BPP pass the EIO test and apply to emissions far in the past (see Barry and Kirby 2017, 296–297). Third, to identify sufficiently clearly what share of Ah's current wealth is due to A's GHG emissions.

2. The Forward-Looking Version of the PPP

The forward-looking version of the PPP (FL-PPP) is the environmental equivalent of a more general moral principle that applies to cases in which one agent (or group of agents) repeatedly causes harm to another agent (or group of agents), yet the harm is so embedded in certain social structures that until these social structures are reformed, the former party cannot stop causing harm to latter party but can nevertheless compensate the victims (see, for example, Pogge 2008, 202–221). The energy transition aims at reforming the major social structures that warm the climate. When global GHG emissions become net-zero, the remaining positive emissions cease to bear a social cost. For obvious reasons, this cannot happen overnight. What we can do in the meantime, however, is to compel those that emit CO_2 to internalise the social costs of their pollution. This is necessary both to align the market cost of emission-generating activities with their real cost (efficiency reasons) and to make GHG emitters compensate the victims of harms induced by these emissions (fairness reasons). In other words, the FL-PPP differs from the BL-PPP in that the latter aims to cover the social cost of emissions that have already occurred (and are now manifesting themselves in current climate threats), while the former aims to contain climate threats in the future.

The FL-PPP has both a practical and a theoretical advantage over the BL-PPP. The first advantage is that the internalisation of the social cost of present emissions not only serves a justice purpose but also acts as a market incentive – the latter being indispensable for the energy transition. The second advantage is that the FL-PPP does not imply intergenerational switches, and hence it is not susceptible to the two theoretical objections discussed in the previous section. The FL-PPP does, however, have one major drawback: it does not address past emissions, which might seem to lead to the conclusion that it favours the effectiveness of the energy transition at the expense of global climate justice (Mittiga 2019; see also Tilton 2016, 118).

In fact, this drawback is much less serious than it may seem if we consider that the social cost of carbon (and of GHGs more generally) has a global dimension, i.e., a dimension that encompasses the economic value of future climate damage. Climate damages do not manifest themselves homogeneously throughout the world; rather, climate damages affect developing countries more, both because these countries are geographically and economically more exposed to climate change and because they have fewer possibilities to invest in adaptation (see Tol 2019). This implies that, under the FL-PPP, at least part of the revenues accruing to developed countries from internalising the social cost of current and future emissions should be redistributed to developing countries in the form of climate finance (see also Cramton et al. 2017). Let us assume that country Q (which polluted massively in the past and continues to do so today) decides to introduce a carbon tax at a rate Y per tonne of CO_2 (tCO_2), where Y is the marginal social cost of emitting $1tCO_2$. Each person that performs an action that causes a given amount W of CO_2 emissions would then have to pay for the private cost of *W* plus the social cost of *W*, which is given by *W* x *Y*. If *Q* is proportionally less exposed to climate harms than developing countries, then the FL-PPP demands that at least part of *Y* should be passed on to more exposed countries – indeed the radical application of the FL-PPP would maintain that the largest part of *Y* should be redistributed internationally.

A potential problem with global accounting (and consequent revenue redistribution) of the social cost of carbon could be that a government of a developed country that introduces carbon pricing measures gets less revenue than is needed to offset the increased cost of energy and/or protect the net losers of the energy transition domestically (as some of revenues have to be transferred abroad). However, this would be a practical problem that could be solved by supplementing revenues with additional public money collected through other taxes or even by placing a tax on CO_2 emissions higher than their social cost for reasons of efficacy. A matter of normative concern with the global accounting of the social cost of carbon, by contrast, is that there may be developing countries that currently pollute significantly but which are not particularly exposed to climate change. This might imply that these countries could claim almost nothing from developed countries. Without skirting around it too much, two big countries that emitted relatively little in the past but emit massively today are China and India. However, both countries are vulnerable to climate damage and certainly have limited means of adaptation, so they are entitled to retain a larger percentage of the carbon revenues collected domestically than historical polluters.

Lastly, a strong objection entertained by Mittiga (2019, 186–187) with respect to the FL-PPP is that if the energy transition works, then at some point the revenues from GHGs taxation will cease, and then the PPP will not be able, on its own, to explain why industrialised countries have to bear the burden of climate damage that will inevitably occur in developing countries. Therefore, he proposes that in an advanced mitigation stage the FL-PPP should be supplemented with the ability to pay principle (APP), where everyone contributes in proportion to their capabilities. This is certainly a pressing issue; however, if green technologies were to become competitive with polluting ones (both in terms of costs and reliability), then developing countries could follow a development path similar to historical polluters without polluting. At this point, global climate justice would mainly consist of patent and technology transfer on the one hand, and climate finance for adaptation and loss-and-damage compensation on the other. Both things could more easily be financed by developed countries once they have reached the final stage of the energy transition, both because renewables can bring major savings on energy costs in the long term and because developed countries would be progressively relieved of the burden of supporting the net-losers of the energy transition domestically (primarily workers tied to fossil fuel industry and poor households). If these circumstances occur, it is therefore possible to assume that the FL-PPP will be able to fulfil its task as a principle of global climate justice at least until climate neutrality is achieved - as the latter does not presuppose the absolute zeroing of GHG emissions (and consequently of carbon revenues), but only net zeroing.

Conclusions

In this chapter, I distinguished between the backward-looking and the forward-looking versions of the Polluter Pays Principle – respectively, the BL-PPP and FL-PPP. First, I discussed the two strong objections that limit the applicability of the BL-PPP: the 'excusable ignorance objection' and the 'powerlessness exemption'. I argued that the various responses proposed in the literature to the two objections fail to substantiate a long-range duty of inter-generational compensation for past emissions. However, this should not dishearten us about the relevance of the PPP. Indeed, what the energy transition calls for is the implementation of the FL-PPP, not the BL-PPP. If all incumbent polluters compensated developing countries for past climate injustice tomorrow morning, this would have no effect in terms of incentives to abandon fossil fuels in favour of clean energy – except indirectly, by making poorer countries richer, but there would be no guarantee that these resources would be used by developing countries for climate mitigation rather than for increasing emission-generating activities.

What would seem to be missing by giving up the BL part of the PPP is the global redistribution of resources, which is a key part of the just climate transition. However, it should be noted that without the BL part, only the historical responsibility of the PPP is lost, not the global one. The FL-PPP, in fact, is based on the internalisation of the social cost of carbon and developing countries are more exposed to climate damages: this implies that the FL-PPP justifies a global redistribution of resources to finance mitigation, adaptation and loss-and-damage compensation for developing countries. The global redistribution dictated by the FL-PPP is probably lower than the one that we would have with the BL-PPP, but it has the advantage of being linked to present and future damages caused by people that have the possibility to compensate the major victims of a warmer climate.

References

Azad, Rohit and Shouvik Chakraborty. 2019. "Balancing climate injustice: a proposal for global carbon tax". In *Handbook of Green Economics*, edited by Sevil Acar and Erinç Yeldan, 117–134. London. Academic Press.

Baranzini, Andrea, Jeroen C. J. M. van den Bergh, Stefano Carattini, Richard B. Howarth, Emilio Padilla, and Jordi Roca (2017). "Carbon pricing in climate policy: seven reasons, complementary instruments, and political economy considerations". *WIREs Climate Change* 8 (e462): 1–17. https://doi. org/10.1002/wcc.462

Barry, Christian, and Robert Kirby. 2017. "Scepticism about Beneficiary Pays". *Journal of Applied Philosophy*, 34 (3): 285–300. https://doi.org/10.1111/ japp.12160

Bourban, Michel. 2020. "Ethics, Energy Transition, and Ecological Citizenship". In *Volume 9: Renewable Energy and the Environment*, edited by Trevor Letcher. In *Encyclopedia of Comprehensive Renewable Energy, 2nd edition,* edited by Ali Sayigh. Amsterdam and San Diego: Elsevier. https://doi. org/10.1016/B978-0-12-819727-1.00030-3

Butt, Daniel. 2014. "'A Doctrine Quite New and Altogether Untenable': Defending the Beneficiary Pays Principle". *Journal of Applied Philosophy* 31 (4): 336–348. https://doi.org/10.1111/japp.12073

Butt, Daniel. 2017. "Historical Emissions: Does Ignorance Matter?". In *Climate Justice and Historical Emissions*, edited by Lukas H. Meyer and Pranay Sanklecha, 61–79. Cambridge: Cambridge University Press. https://doi. org/10.1017/9781107706835.004

Caney, Simon. 2006. "Environmental degradation, reparations, and the moral significance of history". *Journal of Social Philosophy* 37 (3): 464–482. https:// doi.org/10.1111/j.1467-9833.2006.00348.x

Caney, Simon. 2010. "Climate change and the duties of the advantaged". *Critical Review of International Social and Political Philosophy* 13 (1): 203–228. http://dx.doi.org/10.1080/13698230903326331

Corvino, Fausto and Alberto Pirni. 2021. "Discharging the moral responsibility for collective unjust enrichment in the global economy". *THEORIA. An International Journal for Theory, History and Foundations of Science* 36 (1): 139–158. https://doi.org/10.1387/theoria.21237

Corvino, Fausto. 2019. "The Non-identity Objection to Intergenerational Harm: A Critical Re-examination". *International Journal of Applied Philosophy* 33 (2): 165–185. https://doi.org/10.5840/ijap2020228126

Cramton Peter, Axel Ockenfels, and Steven Stoft. 2017. "Global Carbon Pricing. In *Global Carbon Pricing: The Path to Climate Cooperation*, edited by Peter Cramton, David JC MacKay, Axel Ockenfels, and Steven Stoft, 31–89. Cambridge (MA): MIT Press.

De-Shalit, Avner. 1995. *Why Posterity Matters: Environmental Policies and Future Generations*. London & New York Routledge.

Duus-Otterström, Göran and Fredrik D. Hjorthen. 2019. "Consumption-based emissions accounting: the normative debate". *Environmental Politics* 28 (5): 866–885, https://doi.org/10.1080/09644016.2018.1507467

Fleurbaey, Marc, Maddalena Ferranna, Mark Budolfson, Francis Dennig, Kian Mintz-Woo, Robert Socolow, Dean Spears, Stéphane Zuber. 2019. "The Social Cost of Carbon: Valuing Inequality, Risk, and Population for Climate Policy". *The Monist* 102 (1): 84–109, https://doi.org/10.1093/monist/ony023

Friedrich, Johannes and Thomas Damassa. 2014. "The History of Carbon Dioxide Emissions". *World Resources Institute*. May 21, 2014. https://www.wri.org/insights/history-carbon-dioxide-emissions

Gajevic Sayegh, Alexander. 2017. Climate justice after Paris: a normative framework, Journal of Global Ethics 13 (3): 344–365. https://doi.org/10.1080/17449626.2018.1425217

Gardiner, Stephen and and Catriona MacKinnon, eds. 2020. "The Justice and Legitimacy of Geoengineering" (special issue). *Critical Review of International Social and Political Philosophy* 23 (5).

Gates, Bill. 2020. *How to Avoid a Climate Disaster: The Solutions We Have and the Breakthroughs We Need*. London: Allen Lane.

Gosseries, Axel. 2004. "Historical Emissions and Free-Riding". *Ethical Perspectives* 11 (1): 36–60. http://dx.doi.org/10.2143/EP.11.1.504779

Grasso, Marco. 2016. "The Political Feasibility of Consumption-Based Carbon Accounting", *New Political Economy* 21 (4): 401–413. https://doi.org/10.1080/13563467.2016.1115828

Hanlon, W. Walker. 2020. "Coal Smoke, City Growth, and the Costs of the Industrial Revolution". *The Economic Journal* 130 (626): 462–488. https://doi.org/10.1093/ej/uez055

Hatton, Tim. 2017. "Air pollution in Victorian-era Britain – its effects on health now revealed". *The Conversation*, November 14, 2017. https://theconversation.com/air-pollution-in-victorian-era-britain-its-effects-on-health-now-revealed-87208

Helm, Dieter. 2020. *Net Zero: How We Stop Causing Climate Change*. London: William Collins.

Huseby, Robert. 2016. "The Beneficiary Pays Principle and Luck Egalitarianism". *Journal of Social Philosophy* 47 (3): 332–349. https://doi.org/10.1111/josp.12154

IEA. 2021- *Net Zero by 2050*. Paris: IEA. https://www.iea.org/reports/net-zero-by-2050

IPCC. 2018. 'Global Warming of 1.5°C. An IPCC Special Report on the impacts of global warming of 1.5°C above pre-industrial levels and related global greenhouse gas emission pathways, in the context of strengthening the global response to the threat of climate change, sustainable development, and efforts to eradicate poverty'. https://www.ipcc.ch/sr15/

McGreal, Chris. 2021. "Big oil and gas kept a dirty secret for decades. Now they may pay the price". *The Guardian*, June 30, 2021. https://www.theguardian.com/environment/2021/jun/30/climate-crimes-oil-and-gas-environment

Metcalf, Gilbert. 2019. *Paying for Pollution: Why a Carbon Tax is Good for America*. Oxford: Oxford University Press.

Meyer, Lukas H. and Dominic Roser. 2010. "Climate justice and historical emissions". *Critical Review of International Social and Political Philosophy* 13 (1): 229–253. http://dx.doi.org/10.1080/13698230903326349

Meyer, Lukas H. and Pranay Sanklecha, eds. 2017. *Climate Justice and Historical Emissions*. Cambridge: Cambridge University Press.

Mittiga, Ross. 2019. "Allocating the Burdens of Climate Action: Consumption-Based Carbon Accounting and the Polluter-Pays Principle". In *Transformative Climates and Accountable Governance*, edited by Beth Edmondson and Stuart Levy, 157–194. Cham, Switzerland: Palgrave Macmillan

Moss, Jeremy and Kath, Robyn. 2019. "Historical Emissions and the Carbon Budget". *Journal of Applied Philosophy* 36 (2): 268–289. https://doi.org/10.1111/japp.12307

Neumayer, Eric. 2000. "In defence of historical accountability for greenhouse gas emissions". *Ecological Economics* 33 (2): 185–192. https://doi.org/10.1016/S0921-8009(00)00135-X

Nogrady, B. 2021. *Climate Change: How We Can Get to Carbon Zero*. WIRED guides – Penguin Randon House.

Page, Edward A. 2012. "Give it up for climate change: a defence of the beneficiary pays principle". *International Theory* 4 (2): 300–330. https://doi.org/10.1017/S175297191200005X

Page, Edward A. and Clare Heyward. 2017. "Compensating for Climate Change Loss and Damage". *Political Studies* 65 (2): 356–372. https://doi.org/10.1177/0032321716647401

Page, Edward. 2006. *Climate Change, Justice and Future Generations*. Cheltenham, UK – Northampton, MA, USA: Edward Elgar.

Parfit, Derek. 2000. "Energy Policy and the Further Future: The Identity Problem". In *Climate Ethics: Essential Readings*, edited by Stephen M. Gardiner, Simon Caney, Dale Jamieson, and Henry Shue, 112–121. Oxford: Oxford University Press.

Peacock, Kent A. 2021. "As Much as Possible, as Soon As Possible: Getting Negative About Emissions". *Ethics, Policy & Environment*, online first. https://doi.org/10.1080/21550085.2021.1904497

Pogge, Thomas W. 2008. *World Poverty and Human Rights – Second Edition*. Cambridge: Polity Press.

Ritchie, Hannah and Max Roser. 2020. "CO_2 and Greenhouse Gas Emissions". *Our World In Data*. https://ourworldindata.org/co2-and-other-greenhouse-gas-emissions#citation

Roser, Dominic and Christian Seidel. 2017. *Climate Justice: An Introduction*. Translated by Ciaran Cronin. London & New York: Routledge.

Shue, Henry. 2015. "Historical Responsibility, Harm Prohibition, and Preservation Requirement: Core Practical Convergence on Climate Change". *Moral Philosophy and Politics* 2 (1): 7–31. https://doi.org/10.1515/mopp-2013-0009

Stern, Nicholas and Joseph E. Stiglitz. 2021. "Getting the Social Cost of Carbon Right". *Project Syndicate*. February 15, 2021. https://www.project-syndicate.org/commentary/biden-administration-climate-change-higher-carbon-price-by-nicholas-stern-and-joseph-e-stiglitz-2021-02

Tilton, John E. 2016. "Global climate policy and the polluter pays principle: A different perspective". *Resources Policy* 50: 117–118.

Tol, Richard S. J. (2019). "A social cost of carbon for (almost) every country". *Energy Economics* 83: 555–566. https://doi.org/10.1016/j.eneco.2019.07.006

Wallimann-Helmer, Ivo, Lukas Meyer, Kian Mintz-Woo K., Thomas Schinko, Olivia Serdeczny. 2019. In: *The Ethical Challenges in the Context of Climate Loss and Damage*, edited by R. Mechler, L. Bouwer, T. Schinko, S. Surminski, J. Linnerooth-Bayer, 39-62. Cham: Springer.

Worldometer. 2020. Countries in the world by population (2022). https://www.worldometers.info/world-population/population-by-country/

Zellentin, Alexa 2015. "Compensation for Historical Emissions and Excusable Ignorance". *Journal of Applied Philosophy* 32 (3): 258–274. https://doi.org/10.1111/japp.12092

9

Taking Climate Change Seriously: The 'Values Approach'

SUE SPAID

In contrast to ideal arguments whose 'just targets' reflect definable/ measurable metrics and/or impersonal corporate/national policies destined to mitigate climate change, this chapter proposes a nonideal approach for thwarting global warming. This chapter begins by demonstrating the limitations of ideal principles meant to mitigate climate change, whether distributive justice schemes (such as the polluter pays principle, beneficiary pays principle, emissions egalitarianism or ability to pay principle) or Simon Caney's forward-looking scheme that commits parties to 'just burdens' in order to achieve 'just targets'. Alternatively, the Values Approach, which requires us to outline identity-defining values, commits us to act in ways that cohere with our values, lest we be deemed hypocrites. Unlike virtues, the values that we 'make explicit' play normative roles since values are a subset of substantial self-knowledge. We thus adjust our habits to accord with how we see ourselves. Moreover, voters who value their environments are apt to pressure democratically elected leaders to implement appropriate policies.

1. Moving from Backward-Looking to Forward-Looking Schemes

To be congruent, a philosophical approach to climate change mitigation must assiduously persuade the greatest number of climate change contributors (both producers and consumers) to modify offending actions to achieve the public's environmental goals. As we shall see, citizens routinely articulate their desire for governments to implement strategies to mitigate climate change, yet too many consumers have failed to adopt the requisite lifestyle changes needed to reduce their carbon footprints, thus hindering governmental efforts to advance the public's environmental values. Consider that the continued rise in sport utility vehicle (SUV) purchases worldwide has caused gasoline consumption to far exceed carbon footprint savings earned by newly constructed energy efficient power plants and industrial factories (IEA 2021).

More worrisome still, transnational corporations tend to treat economic sanctions meant to deter them from degrading environments as the 'cost of doing business', rather than as prohibitions against undesirable actions or penalties befitting such actions. This is the well-known failure of backward-looking schemes such as retributive justice. Even if victims consider payouts fair (possibly enormous in terms of their economic standards), no amount of money can truly compensate inhabitants whose environments have been degraded and/or properties destroyed (Spaid 2020b, 146). This may sound outdated, but the Danish engineering firm FLSmidth and its funders, the Danish export credit agency and Danish Pension Fund, were blamed for developing an opencast mine in 2014 that they knew would cause severe environmental and agricultural losses in Teghut, Armenia (Malling 2019). Home to 32 metal mines, profits readily flow to foreign investors and Armenian oligarchs despite concerted efforts by the United Nations Human Rights Council and numerous non-governmental organisations (NGOs) to protect the townspeople from mining disasters.

Having learned that retributive justice schemes rarely provide adequate recompense, the UN implemented a different tack, which David Caron and Brian Morris (2002) call 'practical justice'. Following Iraq's seven-month occupation of Kuwait in 1990, UN staff members worked extra hard to implement programs that delivered aid directly to victims, rather than enacting economic sanctions that are distinctly retributive (Caron and Morris 2002). Caron and Morris (2002, 188) define the UN Compensation Commission's notion of 'practical justice' as being the sort of justice whose delivery is 'swift and efficient, not rough'. On this level 'practical' refers to how justice is carried out, rather than a particular type of justice. The UN avoids retributive schemes, because they fail to persuade bad actors to reform and tend to encourage 'business as usual' practices so long as profits outweigh extant fines (Spaid 2020b, 149–150). In contrast to 'practical values', I use 'environmental values' to indicate inhabitants' environmental preferences and 'ameliorative values' to reflect tools available to boost human and nonhuman well-being alike, whereas 'explicit values' are those that value-holders regularly express. Not all environmental values are explicit and not all explicit values concern the environment.

Unlike the earlier Kyoto Protocol (1997/2005), whose signatories were primarily developed nations, the Paris Agreement (2015) and Katowice

Climate Package (2018) gather the world's nations together in entirely voluntary agreements. Treating climate change as a shared problem is meant to motivate signatories to establish their own targets for reducing emissions in order to limit the rise in temperature by 2100 to 1.5°C above pre-industrial levels. 'Every five years, countries are supposed to assess their progress toward implementing the agreement through a process known as the global stocktake; the first is planned for 2023. Countries set their own targets, and there are no enforcement mechanisms to ensure they meet them' (Maizland 2021). Given each nation's varying per capita carbon footprints, the belt-tightening required to achieve this consensus-driven global target varies enormously. Moreover, noncompliant countries such as Australia, Brazil, China and the United States are sometimes called 'cheaters', but they have thus far escaped penalties.

No doubt, 'practical values' underlie the move to ensure that each nation's pledge remains voluntary. Unlike retributive and even most distributive justice schemes, practical justice focuses on redressing the offended parties, rather than offsetting offenses. Given the number of nations that have contributed nothing to climate change yet are regularly forced to weather its dramatic effects, it is no wonder their delegates regularly pressure the United Nations Framework Convention Climate Change (UNFCCC) to ratify loss and damage mechanisms meant to deliver practical justice. According to Dáithí Stone, '[Paris included] a very explicit statement that [loss and damage] is not liability or compensation. [But] it is still unclear, I think, coming out of Warsaw, Cancun and Paris exactly what loss and damage is under the UNFCCC process' (CarbonBrief 2017).

To compensate nations for costs accrued by mitigation and adaption strategies meant to 'ward off' climate change's impact on coastal cities, river cities, flooded farms, islands and rural villages, several philosophers have proposed distributive justice schemes. When framed such that polluters owe an 'ecological debt' (Hayward 2005, 193), wealthy nations have a 'responsibility' to help those less fortunate (Beitz 2005) or nations whose economies have directly benefited from greenhouse gas emissions owe a 'negative duty' (Pogge 2002, 178), such ideal principles as the 'polluter pays principle', the 'ability to pay principle' and the 'beneficiary pays principle' prove no less backward-looking than retributive justice schemes. Whether funds are paid out directly or pooled and distributed on a case-by-case basis, citizens inhabiting payer nations are rarely burdened by said sanctions and thus fail to alter their climate-warming lifestyles. Changing our bad habits requires constant reflection on the costs and burdens paid by others for our 'poor' choices. I imagine the act of empathising with animals inhabiting 'factory farms' prompts eaters to switch protein sources. Similarly, reflecting upon poorly paid miners slaving for hours in diamond pits or each gold band's having generated 20 tonnes of mining waste could lead people to purchase secondhand rings.

Such ideal principles, formulated by philosophers to compensate potential victims, extend John Rawls's notion of distributive justice, a socio-political and/or economic template available to shared language-users inhabiting a political community to select and implement their conception of fairness (Rawls 1971, 274–284). Not only are values implicit in each member's conception of fairness, but they play a central role in his original conception. 'All social values—liberty and opportunity, income and wealth, and the social bases of self-respect—are to be distributed equally unless an unequal distribution of any, or all, of these values is to everyone's advantage. Injustice, then, is simply inequalities that are not to the benefit of all' (Rawls 1971, 55). Additionally, 'the utilities represent the worth of the alternatives for this person as estimated by his scheme of values' (Rawls 1971, 150).

Environmental justice, which addresses inequitable conditions befalling environments that sustain both human and nonhuman inhabitants, is neither redistributive nor compensatory, since no amount of money can reverse environmental degradation. Distributive justice may remedy global injustices resulting from financial burdens, property losses or unequal resource allocations, but it makes a farce of climate justice since it fails to prevent consumers from actions that exacerbate global temperature rise. So long as backward-looking schemes fail to thwart bad actors, they fail to achieve global environmental justice, let alone climate justice. In 2018, Simon Caney introduced a forward-looking notion of distributive justice, yet as the next section demonstrates, the fact that climate change targets are global while burdens (and values) are local engenders asymmetries between targets and burdens, making his ideal principle unworkable. As we shall see, even the most disciplined actors reject burdens deemed onerous or irrelevant.

To diminish the asymmetry between local burdens and global targets, this chapter proposes the Values Approach, which builds upon the significance of 'nonhuman rights' whose 'articulation' empowers value-holders to acknowledge, monitor and safeguard values made explicit in word and/or deed ('Making Values Explicit' plays on Robert Brandom's *Making it Explicit*, 1994). Implementing the Values Approach requires value-holders to routinely reflect upon their lifestyles' immediate and long-term impacts in order to modify present habits and preferences - in tying our personal identity to our lifestyle choices, the Values Approach goes one step further than Kant's 'Categorical Imperative', for which adherents imagine the outcome were everyone to do what they are considering doing. According to the Values Approach, mitigating climate change depends on value-holders who view their every action as expressing core beliefs that indicate their particular identity. Non-inhabitants and those who claim contrary beliefs, such as 'foreigners' and 'climate deniers', are no less culpable for causing harmful outcomes elsewhere. Since sound judgement is a universal value, harmful outcomes convey people's terrible choices and/or poor judgement. A nonideal principle, the Values Approach requires regular assessment of our ecological impact, independent of our lax attitudes. People whose actions indicate 'I don't give a damn' or 'I care little about others' effectively lack sound judgement. In motivating people to act on their environmental values, the Values Approach aims to safeguard agreed-upon commitments. Despite being voluntary, I envision people concerned by climate change finding it easier to adhere to the Values Approach, since people's choices reflect their identities, which is hardly the case for current mitigation strategies.

2. An Asymmetry: Just Targets vs. Just Burdens

Given the extraordinary income gap between wealthy and poor nations, Conference of the Parties (COP) delegates to the UNFCCC have implemented forward-looking variants of distributive justice, such that wealthier countries voluntarily incur greater burdens to offset the (opportunity) costs of burdens that prove unfair for poorer countries. What counts as a burden varies from situation to situation. Sometimes, poorer nations access the Green Climate Fund to offset expenses caused by implementing renewable energy projects or building factories that emit fewer greenhouse gases. Alternatively, wealthier nations have agreed to implement more expensive, greenhouse gas-reducing energy generators since their economies have benefited greatly from earlier, less-efficient energy-generating systems.

Needless to say, the results of the Paris Agreement, whose efficacy is largely a matter of self-policing, are disappointing. If current practices continue unabated, the average global temperature is predicted to rise from between 2.9–3.2°C above pre-industrial levels by 2100, nearly twice the internationally agreed-upon limit. Moreover, two German non-profits, Climate Analytics and the New Climate Institute, not only consider national targets insufficient, but their slow implementation rates are likely to drive average global temperatures 2.7°C above pre-industrial levels by 2100 (Maizland 2021). Since global citizens have already driven temperatures to 1.1°C above pre-industrial levels, leaving room for only a 0.4°C rise over the next 80 years, I worry that citizens either deem participation too burdensome or don't realize their options.

Therein lies the rub. How do nations with many millions of people motivate their citizens who not only can afford today's conveniences, but have worked extra hard to establish lifestyles built on convenience, to now adopt 'inconvenient lifestyles' (local burdens) such that they must limit air travel, forego online purchases, curtail electronics, pay more for locally sourced food, opt for public transportation and heat/cool less in order to achieve global targets? Only a severe belt-tightening would enable nations to 'regularly adjust' their emissions-reducing pledges to meet goals established elsewhere. After all, strong economies are built on the principle that 'affordability' entitles one to 'the right' to purchase goods and services. How do we now redirect the attitudes of millions of people whose fundamental motivation to earn more money is to acquire previously inaccessible goods? This is no less difficult a prospect than placing a nation's citizens on a voluntary diet in order to achieve internationally agreed-upon healthy weights.

Covid-19's routine lockdowns, varying curfews and enforceable controls (social distancing, mask wearing, hand washing and testing requirements) have shed light on nations' citizens varying attitudes toward similar, if not identical, local burdens aimed at achieving a global target – the eradication of a deadly virus. One could say that each nation tacitly agreed to enforce whatever controls were needed to contain the virus and its variants. Varying outcomes reflected each nation's citizens' capacities to endure local burdens. One notable outlier was Sweden, whose Covid-19 deaths per million far exceeded those of other Nordic nations and actually put its neighbors on alert given its more lax approach to preventing community spread. Finland, Norway and Denmark kept their Swedish borders closed long after opening up borders to essential travelers. Swedish researcher Therese Sefton attributes Sweden's different approach to the dismantling of the welfare state in the late 1980s and the fragmentation of government responsibilities among federal, regional and municipalities (Jakobsen 2020). She adds, 'Swedish authorities have had major problems in formulating a strategy that is clear and unambiguous and that brings people together in solidarity' (Jakobsen 2020). Adding insult to injury, the Organisation for Economic Co-operation and Development assigned Sweden the 'lowest possible grade' for its poor handling of Covid-19 (Jakobsen 2020).

Simon Caney's forward-looking approach to distributive justice identifies the potential harms (just burdens) people incur in order to achieve mutually agreed-upon outcomes (just targets) (Caney 2018, 666). The comprehensiveness of his ideal principle is admirable, especially his strong-integrationist approach that stresses gathering a plurality of burdens to identify some general principle of justice that can be used to distribute responsibilities aimed at both mitigation and adaption. He terms this approach holism, in contrast to atomism, which requires distinct principles to determine the distribution of responsibilities owing to the cost of each burden. What worries me, however, is the assumption that parties will voluntarily carry out their commitments to achieve their joint goals just because they have agreed to just targets and just burdens.

When it comes to climate change, Swedes have proven their willingness to accept costly burdens, such as taking long-distance international trains instead of flying, having Europe's smallest families (1.8 children on average) and paying fees for cars to enter cities, such as Göteborg, where Volvo's headquarters is located. Until 2022, Sweden consistently outranked every other nation on the Climate Change Performance Index (CCPI 2022). Such disciplined behavior articulates Swedes' environmental values. Moreover, their identity reflects their natural environment, so they clearly consider it imperative to defeat global warming before its effects reach Sweden. Extreme climes affecting the health of forests, lakes and agricultural lands elsewhere pose a huge threat to Swedes' fish and berry diet, if not their outdoor lifestyles.

In my view, pragmatic Swedes defied Caney's forward-looking model during the 2020 Covid-19 pandemic. Despite just targets (minimise community spread), they refused to endure just burdens (globally recommended measures) as if the opportunity costs far outweighed their advantages (saving lives, while minimising stress on healthcare workers, first responders and hospital supplies). To everyone's surprise, Swedes resisted confinement for months on end: they remained unmasked and kept childcare, primary schools, offices and restaurants and bars open, while neighboring nations enforced confinement. Numerous scholars have posed varying explanations for why Swedes initially refused practices known to minimise the spread of Covid-19. One view suggests that Swedes trust scientists and government officials more than citizens of other Nordic nations, who listen to politicians. Another view suggests that the early 1990s economic reforms that destroyed the welfare state broke the system from top to bottom and destroyed communication mechanisms. I do not pretend to know what motivated Swedes to seriously delay confinement procedures enforced elsewhere. What interests me here is their pragmatism. When it comes to climate change, this chapter's primary concern, their incredible discipline must be motivated by perceived opportunity costs (loss of berries, fishing and skiing) rather than conformity or rule-following. Sweden was not unique. Even when targets reflect consensus, there are outliers who either deem burdens ridiculous or seriously underestimate their risks. This is especially the case for global warming, since like the spread of Covid-19, CO_2 concentrations rise geometrically, thus increasing everyone's ordinary risks (SkepticalScience 2011).

In 2002, Thomas Pogge remarked that it is difficult for societies to curb consumption and pollution when opportunity costs are perceived as exceeding long-term benefits, thus anticipating the way local burdens (lost opportunity costs) make achieving just targets exceedingly difficult. To counter such problems, he proposed the global resource dividend (GRD), a tax calibrated to each nation's resource extraction rate. He claimed that the 'GRD reform can produce great ecological benefits that are hard to secure in a less concerted way because of familiar collective-action problems, each society has little incentive to restrain its consumption and pollution, because the opportunity cost of such a restraint falls on it alone while the costs of depletion and pollution are spread worldwide and into the future' (Pogge 2002, 206). As already noted, pecuniary penalties that curb consumption and pollution cannot compensate citizens for degraded land. In 2018, the Intergovernmental Science-Policy Platform on Biodiversity and Ecosystem Services described 75% of Earth's surface as already 'substantially degraded' (IPBES 2018).

One solution under consideration is to identify 'sector goals', rather than global goals, such that members of either geographic or economic groups (such as bilateral treaties, city or state initiatives and the Group of 20) or industrial sectors (such as energy, aviation, steel, forestry and technology) police their members' incapacity to achieve just targets. Such a view accommodates Carol Gould's 'common activity' principle, whereby 'people with shared aims and goals… have a valid claim to the participation in the decisions necessary to achieve those aims or goals' (Dahbour 2005, 610). While it may be easier for sectors to establish and achieve just targets than global initiatives, members can still abandon 'just burdens' when perceived opportunity costs outweigh the just target's advantages.

Moreover, behind every industrial sector are individual human actors who make decisions on behalf of their enterprises. Decision makers who value their transnational corporation's quarterly profits over protecting environments from degradation make a mockery of just burdens, while paying lip service to just targets. Fortunately, UNFCCC meetings regularly assemble thousands of delegates committed to protecting their own environment. These thousands of value-holders exemplify the view: 'How can you destroy a world if you're *hooked* into it?' which Jeremiah Day uttered during his improvisatory video *What Would You Do for Love and Country? (*The Sivens Example, 2020-21) on view during "Citoyenne Reprise' at NETWERK in Aalst, BE.

3. Articulating 'Nonhuman Rights'

Elsewhere, I have argued that one explanation for the limited success of the Paris Agreement and Katowice Climate Package is that both ignore the role of water in cooling the planet. Waterway and groundwater restoration remain crucial tools for cooling our planet and reducing sea-level rise, making hydrological justice an indispensable component of climate justice (Spaid 2020b, 144). My focus here, though, concerns what I consider the greatest strength of the pluralist COP process in generating international agreements. Since each COP delegate hails from a unique environment, attendees feel duty-bound to defend their particular territory in order to ensure their community's well-being and prosperity, as well as that of future generations and inhabitants (Andina 2020). By expanding the number of delegates well beyond that of the typical UN member-states' representatives, the COP process enables many more voices to be heard, which makes this process invaluable. As a result, many more stories are voiced, recorded and evaluated than would be otherwise. Moreover, such a diverse range of global witnesses increases conference attendees' capacities to empathise with and understand situations that differ from most delegates' personal experiences.

One of the implicit advantages of the COP process is the thunderous chorus of prevailing values, which has led the world's citizens to recognise increasingly and understand the importance of attributing rights such as 'liberty, autonomy, equality and fairness' to nonhuman actors such as waterways, plants and animals. Nonhuman rights are considered on par with the UN's 1948 Universal Declaration of Human Rights, for which only 48 out of 58 UN members originally voted. Yet over 70 years later, it is widely accepted. Since 1948, additional conventions have been drawn up and signed to protect rights related to children, refugees and disabled persons, among others. No doubt, the global community will continue adding more rights.

Nonhuman rights currently in play across the planet recognise and protect the rights of local environments such as forests, rivers, glaciers, mountains, ancient territories and even wild rice. The move to recognise nonhuman rights differs from the Nonhuman Rights Project, an NGO focused on animal rights. Meanwhile, communities in nations such as Australia, Bangladesh, Columbia, Ecuador, India, New Zealand and the United States have adopted laws, and in some cases constitutions, to protect nonhuman rights (Bresler 2020). Switzerland's constitution recognises the dignity of individual plants, while Colombia's supreme court ruled in 2018 that the Amazon River is a 'rights-bearing entity'. In adopting such documents, citizens have made their environmental values, which include nonhuman rights, explicit.

Consider the proposed World Water Law, which articulates a plan to protect waterways:

> We, citizens of Earth, call for and commit to working together
> to ensure that a binding international law is put in place for the
> immediate and universal protection of all Water, as the first

vital step towards global cooperation for effective, worldwide social and ecological healing.

The World Water Law requires:
1. the uncompromising protection and restoration of all natural water sources, watersheds, aquifers, rivers, lakes, wetlands, estuaries and oceans
2. the rewilding of ecosystems, necessary for the restoration of the planetary watercycle
3. the guaranteed free access of all humans and animals to natural, uncontaminated Water

The *World Water Law* holds all governments, corporations, communities, and individuals, fully accountable for their impact on all waters everywhere. This one Law serves as a unifying foundation for all governments and citizens to work together with community-led wisdom and stewardship councils in ways that effectively serve the health and vitality of the whole (WaterCodes 2021).

No doubt, enacting nonhuman rights across the globe will be insufficient to curtail bad actors from profiting off of other people's cherished environments. Just as the citizens of nearly 90 nations have pushed their political representatives to draft constitutions that address human rights, locals will elect politicians whose platforms aim to defend popularly held environmental values, including nonhuman rights. Such rights publicise people's special appreciation for beings that are ordinarily overlooked and thus risk going unacknowledged. When environmental values are made explicit, local inhabitants' defense of nonhuman rights is warranted. Bound together in a joint venture, value-holders feel authorised to govern and protect invaluable beings in their shared care, granting special attention to particular places' nonhuman inhabitants. Even if such values are merely local, the value-holders' authority is rather universal, since outsiders will be forced to recognise such rights, however *outré*. Such values are likely to reflect obscure populations' peculiar proclivities, leaving transnational corporations no other option than to deploy corporate lobbyists to fight the popularisation of nonhuman rights, precisely because values made explicit empower value-holders to take legal action against harmful outsiders.

4. From Passions to Shared Values

One argument against granting non-human beings rights on par with human beings is that what drives us to value nonhuman rights is rather 'emotionally-infused', making such actions morally unjustifiable since we lack valid reasons. So long as our reasons remain out of reach, critics can claim that COP delegates who find others' testimonies persuasive or citizens who elect politicians because they promise to extend nonhuman rights are simply swayed by their emotions. As such, nonhuman rights advocates are merely responding to happy pictures of cute animals, sad stories regarding species loss or anger-prompting images of industrial wastelands. Politicians may even use such tactics to lure votes. As we shall see, we are right to worry that basic emotions such as sadness, happiness, fear, disgust, anger and surprise might influence our thoughts since they sometimes engender beliefs that inadvertently alter or conflict with our values. Moreover, images that catch our attention and trigger emotional responses rarely prompt us to take action. A picture or story may alert us to problems caused by nonhuman beings' lack of rights, but taking action requires firsthand knowledge of the environment's condition, the belief that what we value demands protection plus an awareness of our options.

Even if 'emotional responses signal values, they are not *sine qua non*. What people go to bat for is what they truly value, and thus appreciate' (Spaid 2020a, 130). In *Valuing Emotions*, Michael Stocker (1996, 27) argues that emotions are extremely important for the good life, but they are not reducible to either desires or reasons. He does not consider awareness of our emotions key, since a person could feel shame but not be able to explain why (Stocker 1996, 22). Moreover, emotions can be misleading. Even if we fear bees, we are unlikely to lead a campaign to eradicate them. Despite our fears, we are just as likely to be passionate about preventing colony-collapse disorder. We do not simply pay lip service to our preferences and dislikes.

According to Bruun et al. (2009), 'It might be the case that although emotions do not themselves represent or present values, they are nevertheless associated with some sort of intentional relation to values… Emotions should therefore be described as reactions to felt danger, injustice or other values or to what seems to be the felt danger of objects, the felt injustice of situations, etc.'. This view seems right since emotions seem to signal values (Spaid 2020a). The problem is that some philosophers do not consider acting on feelings on par with acting for a reason. To sidestep this quagmire, Julie Tannenbaum (2002, 321) distinguishes between the manner of doing something (how) and one's motive (why) for doing something. In so doing, she demonstrates that emotionally infused actions such as 'compassionately comforting one's own child' can be considered acting from duty and are thus no less morally motivated.

Since our emotional responses can be misleading, it is imperative that we double-check them as we do our ordinary perceptions. For example, many people get nervous and agitated when a bee flies by. They obviously fear getting stung, but when asked what they think of bees, they typically recognise bees' importance as pollinators. Our fears are real, but our emotions are not veridical in the sense that they are not fool proof (reliable indicators of our thoughts). Similarly, people who find mouldy food disgusting tend to discard it straight away. Others find it no less disgusting, but they have little qualms about scraping mould off jam or rinsing it off peels prior to preparing marmalade. As we age, we learn when to distrust disgust (Spaid 2021) and how to substantiate our emotional responses.

One widespread complaint regarding websites that request 'likes' is that clicking 'like' merely conveys positive feelings. In fact, the term 'slacktavist' was invented to account for the way people click 'like', yet contribute nothing, or what Slavoj Žižek argues confuses 'interpassivity' with 'interactivity' (Spaid 2019, 674). It is no wonder that our emotional responses to the surfeit of images depicting catastrophic fires, hurricanes and desertification have yet to prompt changes to daily routines that admittedly exacerbate global warming. Consider that 70% of American voters (up from 48% in 2017) claim that the United States (US) government must 'take action to address climate change' (Milman 2020), yet SUVs still outsell cars at a rate of two to one. Despite this disconnect, the US's energy-related CO_2 emissions fell in 10 of the past 15 years thanks to a greater reliance on renewable energies and the use of natural gas instead of coal-fired electric plants (EIA 2021b). Following a record 10% drop in 2020, greenhouse gas emissions rose by 6.2% in 2021 due to truck traffic and a rise in coal power, as higher natural gas prices caused a '17% spike in coal generation' (NPR 2022). Initially, this seems exemplary of citizens not accepting burdens required to meet targets, but in fact energy producers likely made this decision on behalf of their consumers in order to avoid profit losses that affect investors more than consumers. Were consumers given the chance to choose between paying 17% more for gas to avoid burning coal, as they do when they pay on average 42% more for organic produce, they could make their values explicit. So long as consumers lack opportunities to act on their values in order to make them explicit, fearing climate change's effects or grieving over species loss remain temporary, albeit spontaneous emotional responses that bear little weight on people's conscience.

5. Actions as Values

When we instigate harmful actions, we are either bad actors, whose actions align with our perverse values; or our terrible judgements facilitated events that have effectively betrayed our values. Destructive choices reveal far more about our values than rational pleas, let alone emotional responses, which are easily primed if not misleading. And primed emotions, especially fear, tends to pervert our judgements. Those of us who self-identify as doing our part to mitigate climate change regularly engage in value-juggling, even if we do not literally weigh the impact of every choice versus our articulated values. In prioritising family safety over carbon emissions, most consumers deem SUVs safer than cars. Fortunately, our identities play selective roles, helping us to modify our actions, but only if we have made our values (backed by knowledge) explicit. Some consider climate change to pose a far greater danger to their descendants than the risk of head-on collisions, where cars are increasingly at risk of getting hit by larger vehicles. SUVs perceived safety mostly reflects their predominance.

Let us return to nonhuman beings' rights. Those of us who get wrapped up in advocating for particular environmental causes, such as nonhuman rights, are driven by our passions, which are what Stocker (1996, 153) calls 'enacted emotions'. In other words, even if we do not know the content of our beliefs, we soon discover the things for which we are willing to go to bat. No doubt, 'The fact of caring affects our judgements by making us recognise certain facts as reasons to act' (Bruun et al.). Similarly, feelings of love, admiration, disappointment or resentment effectively prompt direct action even though we may not be able to grasp, let alone explain, the actual causes or underlying reasons for said feelings and any resultant actions. On this level, our actions serve as evidence for our beliefs, which eventually guide us to articulate our values.

Emotions that are felt but remain unarticulated or unnoticed are not beliefs. By contrast, when we act on our passions, they become dispositional like beliefs. We, or perhaps our friends, glean our passions from our actions, enabling our passions to become belief-like (articulatable). Similarly, values that remain unarticulated become belief-like when we act on them. If we come for a less resource-exploitative and energy-intensive car but leave with an SUV since the salesperson stressed family safety, we are articulating different values even if the salesperson tricked us. When we do not exercise good judgement, we risk acting on beliefs coloured more by emotions than values. By contrast, those who are passionate about watersheds because they connect waterways, cool the planet and nourish living beings are acting on far more than mere emotions, such as surprise or delight.

In *Mind and World*, John McDowell (1996, 140) challenges Donald Davidson's view that only a belief can count as a reason for beliefs. McDowell offers perceptual experience as an alternative reason for beliefs. What gets lost in this debate is the way perceptual experiences alter our beliefs, making McDowell's insight pertinent, yet Davidson's assessment prevails. Our passions are typically the result of revelatory perceptual experiences that alter prior beliefs, making us even more committed to new beliefs than earlier ones. As a result, people who advocate for nonhuman beings' rights articulate beliefs that reflect shared values regarding living beings deemed worthy of protection, such as their very own children, which not only reflects their identities but commits them in ways that laws rarely do.

Conclusion: Virtue Ethics vs. Explicit Values

Virtue ethics posits ethics as a kind of long-winded checklist of virtues (Rachels 1999, 176) for which we relate to a handful, though certainly not all. For example, tolerance 'for the sake of tolerance' is rather indifference. Moreover, civility, dependability, industriousness and tactfulness can easily become vices. By contrast, values are 'personally held', idiosyncratic beliefs that reflect our character because they are infused with our personal identities. Values are a subset of what Quassim Cassam (2014, 29) terms substantial self-knowledge, which includes knowledge of our character, values, abilities, attitudes, reasons, emotions and what makes us happy. When framed as idiosyncratic beliefs, values appear to lack normativity, but in fact, their distinct relationship to our identity grants them normativity, because they reflect our commitment to bear witness to our identities. We (and our observers) discover our values through routine daily actions. When our actions do not cohere with the values we claim, we feel pressured to change one or the other, lest we be deemed hypocrites. The Values Approach also applies to politicians and corporations whose inconsistent policies and poor judgements harm environments and citizens alike.

While virtues belong to someone else (whichever philosopher assembled the list), our values are our own, which means that we can pass them to friends, fellow inhabitants and future generations to become shared values. Psychologists Rozin *et al.* note that values, rather than mere preferences, drive the social practices that get passed from generation to generation.

> Moralization converts preferences into values and in doing so influences cross-generational transmission (because values are passed more effectively in families than are preferences), increases the likelihood of internalization, invokes greater emotional response and mobilizes the support of governmental and other cultural institutions (Rozin *et al.* 1997, 67).

Similarly, Alexandra Plakias (2018, 198) argues that food ethics 'protect us from foods embodying toxic values – values that threaten, not our physical health, but our ideological integrity and therefore our very identity'. This notion coheres with research that shows that vegetarians who are vegetarians for moral, rather than health reasons, register stronger pangs of disgust towards meat (Rozin *et al.* 1997, 71). Being a 'moral' vegetarian reflects values that inform people's self-identities.

When we live according to our values, we feel pride precisely because our values are ours and not someone else's. Rather than worry what people will think if we do not own a car, public transport advocates take pride in their transportation choices. Activists working to make the World Water Law as common as the rights of the disabled or children are acting on their own values. Our original concerns regarding climate change may have been prompted by pictures or stories, but modifying our habits such that we actually reduce our carbon footprint requires that we grasp the situation at hand and commit to making our values explicit so that we are prepared to uphold ameliorative values, environmental values, nonhuman beings' rights or some combination thereof.

References

Andina, T. (2020), 'Climate Issue: The Principle of Transgenerational Responsibility', *Rivista di Estetica* 75: 17–32.

Beitz, C. (2005), 'Cosmopolitanism and Global Justice', *The Journal of Ethics* 9: 11–27.

Bresler, A. (2020), 'Seven Countries that have Legally Recognized Rights of Nature', *Matadorn Network*, https://matadornetwork.com/read/countries-legally-recognized-rights-nature/

Brock, G. (2017), 'Global Justice', *The Stanford Encyclopedia of Philosophy* (Summer 2022 Edition), Edward N. Zalta (ed.). https://plato.stanford.edu/archives/sum2022/entries/justice-global/

Bruun et al. (2009), 'Emotions, Values and Norms - The Emotion-value connections'. https://www.unige.ch/lettres/philo/recherche/research-groups/thumos/projects/emotions-values-and-norms

Caney, S. (2018), 'Climate Change', in Serena Olsaretti (ed.), *Oxford Handbook of Distributive Justice* (pp. 664–688). Oxford: Oxford University.

CarbonBrief. 2017. 'Explainer: Dealing with the "loss and damage" caused by climate change', May 9. https://www.carbonbrief.org/explainer-dealing-with-the-loss-and-damage-caused-by-climate-change/

Caron, D. D. and Morris, B. (2002), 'The UN Compensation Commission: Practical Justice, not Retribution', *European Journal of International Law* 13 (1): 183–199.

Cassim, Q. (2014), *Self-Knowledge for Human Beings*. Cambridge: Cambridge University Press.

CCPI, (2022). https://www.climate-change-performance-index.org/

Dahbour, O. (2005), 'Reviewed Work: *Globalizing Democracy and Human Rights* by Carol C. Gould', *Social Theory and Practice* 31 (4): 607–612.

Day, J. (2021), 'What Would You Do for Love of Country? (The Sivens Example)'. Berlin: Jeremiah Day.

EIA. (2021a). 'After 2020 decline, EIA expects energy-related CO2 emissions to increase in 2021 and 2022', 26 January. https://www.eia.gov/todayinenergy/detail.php?id=46537

EIA. (2021b). 'U.S. Energy Facts Explained', 14 May. https://www.eia.gov/energyexplained/us-energy-facts/

Ginsborg, H. (2006), 'Reasons for Belief', *Philosophy and Phenomenological Research* 72 (2): 286–318.

Hall, S. (2015), 'Exxon Knew about Climate Change almost 40 years ago', *Scientific American*, October 26. https://www.scientificamerican.com/article/exxon-knew-about-climate-change-almost-40-years-ago/ . Accessed 1 February 2022.

Hayward, T. (2005), *Constitutional Environmental Rights*. Oxford: Oxford University Press.

IEA. (2021). 'Carbon emissions fell across all sectors in 2020 except for one – SUVs', 15 January. https://www.iea.org/commentaries/carbon-emissions-fell-across-all-sectors-in-2020-except-for-one-suvs

IPBES. (2018). 'Media Release: Worsening Worldwide Land Degradation Now 'Critical', Undermining Well-Being of 3.2 Billion People', March 23, https://ipbes.net/news/media-release-worsening-worldwide-land-degradation-now-%E2%80%98critical%E2%80%99-undermining-well-being-32

Jakobson, S. E. (2020), 'The border between Norway and Sweden is closed for the first time since 1954. Will the pandemic ruin their special friendship?' *Science Norway,* https://sciencenorway.no/borders-covid19-nordics/the-border-between-norway-and-sweden-is-closed-for-the-first-time-since-1954-will-the-pandemic-ruin-their-special-friendship/1780394

Maizland, L. (2021), 'Global Climate Agreements: Success and Failures, *Council on Foreign Relations*. https://www.cfr.org/backgrounder/paris-global-climate-change-agreements

Malling, J. (2019), Armenians Fight for Their Environment, *Le Monde Diplomatique*, February.

McDowell, J. (1996), *Mind and World*, Cambridge: Harvard University Press.

Milman, O. (2020), 'Guardian/Vice Poll Finds Most 2020 Voters Strongly Favor Climate Action', *The Guardian*, https://www.ecowatch.com/2020-election-climate-poll-2647822815.html

NPR, 2022. 'U.S. Greenhouse Gas Emissions Jumped in 2021, a Threat to Climate Goals', January 10. https://www.npr.org/2022/01/10/1071835575/u-s-greenhouse-gas-emissions-2021-climate

Plakias, A. (2019), *Thinking Through Food: A Philosophical Introduction,* Peterborough: Broadview Press.

Pogge, T. (2002), *World Poverty and Human Rights: Cosmopolitan Responsibilities and Reforms*, Cambridge: Polity Press.

Prinz, J. (2011), 'Emotion and Aesthetic Value', In Elisabeth Schellekens and Peter Goldie (eds), *The Aesthetic Mind* (pp. 71–78). Oxford: Oxford University Press.

Rachels, J. (1999), *The Elements of Moral Philosophy*. New York City: McGraw-Hill College.

Rozin, P., Markwith, M., and Stoess, C. (1997), 'Moralization and becoming a vegetarian: The transformation of preferences into values and the recruitment of disgust', *Psychological Science* 8 (2): 67–73.

SkepticalScience. (2011). 'Monckton Myth #3: Linear Warming', January 18. https://skepticalscience.com/monckton-myth-3-linear-warming.html

Spaid, S. (2019), 'Surfing the Public Square: On Worldlessness, Social Media, and the Dissolution of the Polis', *Open Philosophy* 2 (1): 668–678.

Spaid, S. (2020a), 'Emotions and Empirical Aesthetics', *Aesthetic Investigations* 4 (1): 122–132.

Spaid, S. (2020b), 'Tying Climate Justice to Hydrological Justice', *Rivista di Estetica* 75: 143–163.

Spaid, S. (2022), 'Value Disgust: Appreciating Stench's Role in Attention, Retention and Deception, *Rivista di Estetica* 78: 74–94.

Stocker, M. (1996), *Valuing Emotions*, Cambridge: Cambridge University Press.

Tannenbaum, J. (2002), 'Acting with Feeling from Duty'. *Ethical Theory and Moral Practice* 5 (3): 321–337.

Voelk, T. (2020), 'Rise of S.U.V.s: Leaving Cars in Their Dust, With No Signs of Slowing', *New York Times*, May 20. https://www.nytimes.com/2020/05/21/business/suv-sales-best-sellers.html

WaterCodes. 2021. 'World Water Law', https://www.codes.earth/waterlaw

10

Forward-Looking Transitional Climate Justice

KIRK LOUGHEED

Transitional justice usually refers to approaches to justice that focus on addressing past injustices in a way that allows societies to move forward peacefully. Sonja Klinsky and Jasmina Brankovic (2018) argue that transitional justice also applies to issues of climate justice. With respect to climate change, certain actors have been wronged by other actors, and the former are owed transitional justice. This type of justice is most often backward-looking in that it focuses on past wrongs. In this chapter, I develop a form of transitional justice that is forward-looking. This type of justice focuses on anticipating and addressing harms that will occur in the future. I argue that if forward-looking transitional justice is legitimate, then certain individuals in wealthier countries who will be negatively impacted by more effective climate policy are owed transitional justice.

Introduction

For the purposes of simplicity, I assume that climate change poses a significant existential threat to the human species. Likewise, I assume that all adult humans have *some* responsibility to help resolve the climate crisis, though this responsibility varies by degree (of course, this also means assuming it would be a bad thing if the humans species went extinct, though this is not uncontroversial – see, for example, Benatar 2008). The planet is rapidly warming, and we need to figure out how to slow this down drastically or stop it altogether.

Transitional justice usually refers to approaches to justice that focus on addressing past injustices in a way that allows societies to move forward peacefully. Sonja Klinsky and Jasmina Brankovic (2018) argue that transitional justice also applies to issues of climate justice. With respect to climate change, certain actors have been wronged by other actors, and the former are owed transitional justice. This type of justice is most often backward-looking in that it focuses on past wrongs. After explaining Klinksy and Brankovic's position, I develop a form of transitional justice that is forward-looking. This type of justice focuses on anticipating and addressing harms that will occur in the future. I apply forward-looking transitional justice to climate change in wealthier nations and demonstrate how doing so can help make it easier to enact better climate policies. I use the example of the Alberta Tar Sands in Canada to help illustrate these points. Furthermore, my theory supplements the current discourse on inter-generational justice in the climate justice literature. It is consistent with integrationist approaches to the climate problem in general, and sufficientarism and egalitarianism (among others) about inter-generational climate justice in particular.

1. What Is Transitional Climate Justice?

Transitional justice is discussed most commonly in situations where serious injustices have occurred, including those surrounding racist and colonial procedures and policies, in addition to war crimes and genocide. A key feature of transitional justice is that it emphasises strategies needed to move forward peacefully even if these strategies sometimes involve sacrificing certain aspects of justice in order to achieve peace:

> The goal of transitional justice processes is to recognize and at least partially remedy injustices while also building a sense of unity or solidarity. Such processes have included a range of mechanisms, most particularly accountability measures such as amnesties, prosecutions, and truth commissions; reparations for those harmed; and institutional reforms that aspire to prevent future harms (Klinsky and Brankovic 2018, 3).

Finally, in an encyclopedia entry on transitional justice, Nir Eiskovits (2017) writes,

> The term 'Transitional Justice' has come, in recent years, to designate a field of academic inquiry, as well as political practice, concerned with the aftermath of conflict and large-scale human rights abuses. Theorists and practitioners of transitional justice focus on the most effective and legitimate ways of addressing past wrongs and moving towards the (re) establishment of a decent civil order.

Notice that these definitions of transitional justice focus almost exclusively on the aftermath of wars and/or human rights abuses (for more examples, see Elster 2004, Futamura 2008 and Graybill 2020; for a more detailed account of transitional justice, see Teitel 2000). In *The Global Climate Regime and Transitional Justice*, Sonja Klinsky and Jasmina Brankovic (2018) seek to challenge these definitions by applying concepts from transitional justice to issues of climate justice. They explain that many of the people most adversely impacted by climate change are in developing countries. This is yet another way that developing countries suffer at the hands of wealthier nations. The result of such suffering indicates that many developing nations are entitled to some sort of compensation as a matter of basic fairness and justice. Moreover, if we cannot stop or significantly slow down climate change, then the entitlements of developing countries will only continue to increase. In this context, transitional justice approaches not only help determine appropriate compensation for the developing nations in question, but they also suggest ways to improve things going forward, including by getting wealthier nations to reduce their greenhouse gas (GHG) emissions drastically.

An important connection between transitional justice and climate justice is that for both, the question of justice occurs across space and time. Klinsy and Brankovic (2018, 21) explain that '[t]ransitional justice is... both a conceptual framework and a set of mechanisms that have come to be used around the world to redress harms that occurred in the past, as well as to address their repercussions in the present and for the future'. Similarly, some of the tensions in climate justice 'revolve around the distribution of causal responsibility for GHG emissions across time and space; the scope of obligations toward those most affected by climate change; and the relationship between climate action and broader inequalities in human well-being and access to sustainable development' (Klinsky and Brankovic 2018, 16). The global contributions of GHG production are uneven along with the purported benefits of such emissions (Klinsky and Brankovic 2018, 16). Likewise, the negative effects of climate change are not the same throughout the world, including with respect to the times in which they occur. Insofar as transitional justice offers guidance about how to address injustices that are spatially or temporally distributed, it can provide useful practical strategies for addressing climate injustice.

Klinsky and Brankovic (2018) outline a number of strategies for helping solve climate dilemmas and suggest that there are parallels between them and transitional justice. Such dilemmas are most often the result of tensions generated from wanting to hold various actors responsible for their contributions to climate change while at the same time wanting them to participate in future solutions. For example, one such solution is *fair burden sharing*. Klinsky and Brankovic (2018, 40) explain that '[i]t is hard to overstate the extent to which "fair burden sharing" has been used as a core strategy to address climate justice dilemmas... For instance, the Convention explicitly lays out that developing countries' abilities to develop must not be impeded by climate action and that developed countries are intended to take the lead on addressing change'. Another strategy is green growth and, as we will see, it is the most applicable to the strategy I develop later. Green growth rests on '[t]he idea [that] sustainable economic development is foundational in international climate policy' (Klinsky and Brankovic 2018, 44). Furthermore, '[a] recent review identified over eighty reports and policy documents published since 2009 by international and national organizations promoting "green growth" or "green economy"' (UNDESA 2012 quoted in Klinsky and Brankovic 2018, 44).

2. Forward-Looking Transitional Climate Justice

Notice that much of the above discussion of transitional justice is backward-looking. Certain parties have *already* experienced injustices as a result of climate change, so transitional justice is about what such parties are owed by various perpetrators. While these can be useful approaches to justice, I do not analyse them in this chapter. Instead, I develop what I call forward-looking transitional climate justice, which examines the future harms or injustices that will be experienced as a result of climate change. After summarising my account, I situate it within the current climate justice literature, with a particular focus on inter-generational climate justice.

2.1 What Is Forward-Looking Transitional Justice?

Forward-looking transitional climate justice has four criteria including identifying who will be harmed from climate change, accurately predict those future harms, compensating victims of climate-friendly policies, and doing so in a timely manner.

2.1.a Individuals and/or institutions who will be harmed by climate change in the future

Forward-looking transitional justice applies to countries in the southern hemisphere that will continue to be harmed by climate change in the future. Sadly, it is obvious that we have not done enough to slow drastically or stop the warming of the planet, so such individuals and institutions will continue to be harmed in the future (indeed, that harm is only likely to increase without swift and drastic changes to global environmental policy). However, notice that there is another set of individuals and institutions also entitled to forward-looking transitional climate justice. These are people in wealthier nations that will be adversely impacted by climate policies (and remember, such policies are needed as a result of climate change). None of this is to deny that such actors are partly (sometimes quite significantly) responsible for the current climate crisis. But it nevertheless remains the case that they will be harmed by climate change in the future.

2.1.b Accurately predicting the future harms of climate change

Forward-looking transitional climate justice involves relying on climate scientists to accurately predict the effects of climate change in the future. Furthermore, it also relies on social scientists (e.g., economists) to predict the outcome of environmental policies that need to be created and enforced to combat climate change. In other words, we need to know the myriad effects of climate change in order to accurately identify the parties harmed by it. A key component of forward-looking transitional climate justice is to be able to predict the harms of climate change in order to prevent them (or at least compensate victims in a timely manner).

2.1.c Compensating victims of policy changes that harm individuals and/or institutions even though such changes are necessary

This is a key criterion and likely the most controversial in that its scope is intended to apply quite broadly. The actors most negatively impacted by climate change thus far tend to be in developing nations in the southern hemisphere. However, such nations would also benefit from more effective environmental policy and enforcement in the northern hemisphere. Still, this criterion requires compensating all parties that would be harmed by better climate policy. This means, for example, possibly compensating those in wealthier nations in the northern hemisphere who will lose employment because of policy changes. Later, I discuss in more detail what effective compensation might look like in certain cases.

2.1.d Compensating victims before or as soon as these future harms become a reality

Related to (B) is the idea that victims need to be compensated immediately (or prior to) the harms becoming a reality because part of what might constitute the harm itself is the time between the initial harm and compensation. In this sense, harm comes in degrees. And the harm will be greater the longer a victim goes without receiving just compensation. For example, if a worker in fossil fuel industry loses her job because of changes in climate policy, she is harmed less if she is offered alternative employment sooner rather than later.

2.2 Forward-Looking Transitional Justice and Intergenerational Climate Justice

Now, I situate my account within the broader climate justice literature. Ethicists have argued for two broadly competing methods of addressing climate justice. The first is isolationism, which suggests that issues of climate justice ought to be addressed on their own, not in conjunction with other issues (e.g., global trade, individual incomes, etc.). One motivation for this view is that it is simpler to address issues about GHG emissions, for example, in isolation from a number of possibly competing issues. Likewise, it also allows ethicists to side-step deep disagreements over which is the correct moral theory that ought to be applied to climate justice (Caney 2020; see also Meyer and Roser 2006). Integrationists, on the other hand, counter that 'it is best to treat the ethical issues posed by climate change in light of a general theory of justice and in conjunction with other issues (such as poverty, development and so on)' (Caney 2005; 2020). While I do not deny that isolationism is simpler, I reject the idea that these issues can actually be treated in silos. Policies that impact GHG emissions inevitably affect businesses and therefore also affect employment opportunities. This is true for those working in the fossil fuels industry in wealthier nations and also for those working in tourism in small island developing states. Theorists do not need to agree on an underlying ethical theory to see how these various issues are interconnected. Forward-looking transitional justice, then, clearly favours an integrationist approach to climate justice. The prudential reasons for doing so are strong.

Much of the climate justice literature is concerned with inter-generational justice. This concerns the degree to which the current generation has a moral obligation to future generations if it has any obligation at all. Though there is widespread disagreement about the details, most ethicists agree that future generations ought to be subjects of genuine moral concern. Of course, one longstanding problem regarding duties to future persons is Derek Parfit's famous non-identity problem. Since future persons do not yet exist, how could they have interests? Likewise, our decisions today shape which future beings will come into existence. If things had gone differently, then those beings would not exist. So, if a future being wishes, for example, that the environment had been left by previous generations in better shape, they are in effect wishing themselves out of existence. Much ink has been spilled by philosophers in many different contexts on the non-identity problem. I do not wade into this technical debate here. I only note that many think the problem is solvable and that we do indeed owe future persons moral consideration, even though the solutions themselves vary widely (see Parfit 1994). It is impossible to survey this ever-growing body of literature here, but I note the ways in which forward-looking transitional justice is consistent with many different approaches to inter-generational justice.

Sufficientarianism is one view of inter-generational justice that holds that all persons must have a quality of life above a certain threshold (Caney 2020; Meyer and Roser 2009). Of course, some have observed that identifying the precise threshold would be difficult or that, in certain cases, such a threshold would not be sufficient for justice. Egalitarians claim that this theoretical threshold is fairly high, especially in light of the gross inequalities in our world. Some argue that egalitarianism not only applies to the future as a whole, but it also entails that we not increase inequalities in future generations (Caney 2018; 2020; Hoegh-Guldberg, Jacob and Taylor 2019). Another interesting proposal defends a kind of growth sustainability, which says that the current generation can only maximise their own welfare inasmuch as it means the welfare of all future generations (not only the next few ones) will grow by 25% (Lavador, Roemer and Silvestre 2015).

One benefit of forward-looking transitional justice is that it is consistent with affirming the importance of inter-generational justice. My account focuses on the future of currently existing persons. Not only is this consistent with the claim that future generations are owed moral consideration, but it also has practical implications for such generations. If transitional justice is enacted regarding workers in wealthy nations who will be adversely affected by climate policies that we need to enact, those policies will be more likely to take hold. Not only does this benefit the particular workers in question, but also future generations who need us to stop the warming of the planet. There is thus a sense in which enacting the forward-looking transitional justice framework I describe would make it easier for inter-generational justice to obtain in the future. In order for this to be true, I need not endorse any particular view of inter-generational justice. The details of how my own account is enacted may well change with the details of inter-generational justice that one has in mind. My account is not only not at odds with inter-generational justice but is actually a way of securing it. This is a feature of my view, not my bug.

Finally, my account of forward-looking transitional justice also fits well with numerous views on social discounting and climate change. Social discounting involves questions about the extent to which future persons should be considered in our current policies if at all. Time discounting is the view that we should discount the value of future persons. However, some argue that this wrongly discriminates against people based on the time at which they are alive, which is morally arbitrary (Caney 2020; Caney 2014; Parfit 1984, 480–486). Regardless of what one thinks about this type of discounting, my account does not force one to take a particular stance. Claiming that certain people that currently exist, including those in wealthy countries, are owed transitional justice, does not mean that future persons are not also owed things related to climate justice. Again, when it comes to the practical matter of enacting policies that will help improve the climate for persons that will exist in the future, forward-looking transitional justice is likely helpful.

As the reader can likely tell by now, my primary defense of forward-looking transitional justice is prudential. However, I hope I have shown that this prudential defense can also be used to bolster various underlying theories of inter-generational justice that occupy much of the literature on climate justice. Instead of providing a further technical argument in support of forward-looking transitional justice, next I show how it can be applied in a current real-life case. In the next section, I examine transitional justice for workers in the test case of the Alberta Tar Sands in Canada. Much of what I say there generalises to other relevant cases. An advantage of applying forward-looking transitional justice is that it can make enacting effective climate change policy easier than it would be otherwise. Finally, I agree with Klinsky and Brankovic when they say that, ultimately, multiple approaches to solving the climate crisis are likely necessary. I do not pretend that this is the only approach or that forward-looking transitional justice is the only type of justice at stake with respect to the climate crisis. However, this is an approach that has yet to receive sufficient attention, and it is noteworthy how well it can support other, better-established theories in the climate justice literature.

3. Applying Forward-Looking Transitional Climate Justice: The Alberta Tar Sands

In this section, I apply some of my ideas about forward-looking transitional justice to the case of the oil industry in Alberta, Canada. This example helps illustrate the importance and efficacy of forward-looking transitional justice.

3.1 The Economic Impact of the Tar Sands

The Alberta Tar Sands were discovered by settlers well over a century ago, and some form of commercial oil production began prior to World War II. Due in large part to the exploitation of the tar sands, , Alberta became one of the wealthier provinces in Canada beginning the 1970s and accelerating during the oil boom of the 2000s. Cities in Alberta, such as Fort McMurray, Edmonton and Calgary, have received significant economic benefits from the sands (Barnetson and Foster 2014, 352; Heyes et al 2018, 253). Right now, '[i]n terms of corporate taxes, oil sands producers pay federal and provincial corporate income taxes at a current combined rate of 27 percent; they also benefit from special tax provisions available to all Canadian oil and gas production' (Heyes et al 2018, 245; for more details about oil taxation in Canada see KPMG 2015).

Moreover, tax revenue in Alberta is tied to the price of oil. Between 2005–2015, the price of oil was, on average, $100 per barrel, reaching a high of over $160 per barrel in June 2008 (see Alberta Government 2022, Macrotrends n.d.). Over that period, the Albertan economy was booming and the government of Alberta took in larger tax revenues. Moreover, Canadian provinces provide a portion of their tax revenue to the federal government, and the federal government collects higher revenue from wealthy provinces and redistributes them to poorer provinces. During the period of the economic boom, therefore, Alberta was a major contributor of money that was redistributed to other Canadian provinces and territories. For example, '[a]ccording to a 2005 report by the Canadian Energy Research Institute, Ottawa will rake in $51 billion in corporate taxes from the tar sands between 2000 and 2020, while Alberta will take home only $44 billion' (Nikiforuk 2010, 158). This is not to mention the fact that at the provincial level, the Albertan government has consistently failed to collect the amount in royalties it ought to from oil companies - for the landmark public inquiry that made this painfully clear, see the Our Fair Share report (2007). For example, 'the province makes much less from its dirty oil than do Norway, Alaska, New Mexico, or even Louisiana' (Nikiforuk 2010, 158). While the federal government cannot be blamed for mismanagement at the provincial level, this problem serves to disenfranchise the average Albertan resident. However, in 2020, the price of crude oil hovered between $50–60 per barrel. As a result, Alberta's economy is struggling as royalties from oil rise and fall with the price of a barrel. Alberta is now in the unfamiliar situation of asking for financial assistance from the federal government.

In order to get a better sense of just how reliant on oil Alberta is, consider that some have suggested that Alberta suffers what is known as the 'resource curse' and 'Dutch disease'. The former occurs when a region is overly reliant on one resource such that it negatively impacts governance (perhaps in the form of lobbies and special interest groups) and also hurts educational outcomes (i.e., individuals forgo educational opportunities in favour of high-paying jobs in the booming resource sector) (Heyes et al. 2018, 253). The latter occurs when a boom in one sector pulls resources away from other sectors, sometimes to the point of causing entire industries to collapse. Many have suggested that the decline in manufacturing in Canada was at least in part sped up because of the oil boom, which in turn exacerbated the financial dependence on oil (Heyes et al. 2018, 254; see also Saches and Warner 1996 and Beine et al. 20212; for a dissenting view, see Krzepkowski and Mintz 2013). For better or worse, then, the Albertan economy is intimately connected to profits from the tar sands.

3.2 Forward-Looking Transitional Climate Justice in Alberta

The extraction and use of oil from the Alberta Tar Sands does, without a doubt, negatively impact the climate. Canada is not on track to meet its commitments under the Paris Agreement. In order to do so, it would need to enact new stricter climate policies to limit the use of fossil fuels significantly, including limiting the extraction of oil in the tar sands. However, if forward-looking transitional justice is legitimate, then it applies to those in Canada (and particularly those in Alberta) that will be negatively impacted by such policies. This is because the policies are the result of the need to address climate change. And, as such, climate change is the ultimate cause of the policies that will inevitably impact the lives of individuals in significant ways. More effective climate policy will (i) limit extraction in the tar sands and thereby limit the royalties received at both the provincial and federal levels; and (ii) in light of (i) there will be fewer direct (and indirect) jobs in the oil industry in Alberta.

There are many different ways that forward-looking transitional justice could be applied to the case of Alberta, but here are some specific recommendations:

- The creation of any 'green jobs' (both private and public) should include a mandate to hire a certain number of workers from jobs contributing to climate change (e.g., workers from the tar sands).
- The government should pay for full-time retraining if and when required. The pay should be the same (or close to the same) as what the worker received for their previous salary. This is expensive, but offering small stipends to someone with a mortgage and family will not work.
- Inasmuch as possible, workers should be given jobs similar to the previous ones they had. In many cases this may seem impossible, but many jobs have similar levels of physical and intellectual difficulty. Workers should also be given the same degree of responsibility if and when possible. Do not turn a senior manager in the oil industry into a day labourer in green energy to only wonder why they drop out of the programme.
- Do not just attract those who explicitly care about climate change to green jobs. Make the compensation and jobs themselves attractive enough to draw the brightest and most creative minds. We need as many innovators as possible. Motivating innovation with attractive jobs is not in itself wrong.

There are significant temporal and financial components required to enact these proposals. Indeed, some will hold that the relative costs are so high that the proposals are unrealistic and cannot reasonably be implemented. This is where Klinksy's and Brankovic's explanation of 'green growth' is helpful. They explain that green growth affirms that 'sustainable economic development is foundational in international climate policy' (Klinsky and Brankovic 2018, 44). Furthermore, they observe that '[a] recent review identified over eighty reports and policy documents published since 2009 by international and national organizations promoting "green growth" or "green economy"' (UNDESA 2012 quoted in Klinsky and Brankovic 2018, 44). Going green does not have to be viewed as making an economic sacrifice. This is because:

> Green development is based on three key concepts: economic growth can be decoupled form rising greenhouse gas emissions and environmental degradation; the process of "going green" can itself be a source of growth; and "going green" is part of a virtuous circle that is mutually reinforcing with growth (World Bank and People's Republic of China 2021, 217, quoted in Klinsky and Brankovic 2018, 44).

Green development includes ecological costs when calculating total costs. This allows for long-term development that will not have undue ecological costs. Indeed, '[t]he rapid increase of green growth raises hopes that it could address the very negotiating impasses highlighted as problematic in the strategy for fair burden sharing' (Klinsky and Brankovic 2018, 46; see also Zhang and Shi 2014). Finally, 'green growth provides an avenue for the private sector to be more involved in climate change action. If economic growth and ecological protection could be combined, it would present profit incentives for the private sector at all scales' (Klinsky and Brankovic 2018, 46).

This is not to say that there are no challenges when introducing green growth strategies. For example, there are numerous definitions of 'green growth', some of which conflict with each other. But if the definition of 'green growth' is too vague, it is unlikely to be sufficiently motivating (Klinsky and Brankovic 2018, 47). The most significant challenge would be if green growth turns out to be fundamentally incompatible with the current economic paradigm. In this case, there would need to be a significant change in thinking before we could properly consider green growth an option. Such a shift would no doubt be quite difficult to make in Alberta (Klinsky and Brankovic 2018, 48-57). So, I do not deny that there are challenges to implementing green growth strategies. But it seems to me that they are well worth tackling since green growth could be a way to ensure transitional justice for workers who would otherwise lose employment opportunities. We need to stop thinking of going green in terms of economic loss. This need not be the case. While my recommendations will sometimes lead to high start-up costs, these could potentially pale in comparison to future economic gains.

This is not intended to be a comprehensive guide to forward-looking transitional justice in Alberta. But it does show that there are tangible strategies that could be put in place to enact forward-looking transitional justice. Furthermore, this type of transitional justice is likely to make such policy changes easier to enact and enforce. In Alberta, environmental policies that do not come with this sort of transitional justice are viewed as a threat to the livelihood of many individuals in the province (this is a psychological claim about individuals, but my above explanation of the economics of the sands should show that it is quite plausible). Thus, the development of forward-looking transitional justice ensures that individuals are given what they are owed and can make better environmental policy significantly more feasible.

My development of forward-looking transitional justice can help make environmental policy changes in Alberta more feasible because it shows a way such policies could garner public support. Albertans have already been hurt by climate change. I realise this sounds counterintuitive given that the tar sands clearly have a negative impact on the climate. But workers in the tar sands have already lost jobs because of the decline in the price of oil, in addition to a pipeline not getting built (I do not claim the price of oil is only impacted by climate concerns - i.e., the desire to burn less fossil fuel - this clearly is not the case; however, it need only be partly responsible for the decline of oil for my point to stand). Inasmuch as these losses are the result (in part) of climate change (and policy), then workers have indeed been adversely impacted by climate change. Furthermore, as we continue to move to 'green solutions' and anti-oil and anti-pipeline sentiments continue to rise in Canada, there will be fewer opportunities for employment in the Albertan Tar Sands. Indeed, by all accounts the future is quite bleak for anyone hoping to have secure employment in the tar sands. This is not to mention the billions of dollars lost in royalties by less extraction in the tar sands and lower oil prices in general.

By providing Albertans with forward-looking transitional justice (particularly for those who currently are or have been working in the oil industry), environmental policies will be easier to implement. By offering transitional justice, Albertans would be more likely to perceive that they are being treated fairly. They are thus more likely to cooperate with the environmental policies needed to help slow down and/or stop the planet from warming.

Conclusion

Forward-looking transitional climate justice anticipates potential harms to currently existing persons caused by climate change. If forward-looking transitional justice is legitimate, then those in wealthier countries working in the fossil fuel industry have certain entitlements inasmuch as they will be harmed by more effective climate policy. I discussed the Albertan Tar Sands as an example of a region heavily dependent on oil revenues and one where individuals will be harmed by more effective climate policy. I further showed that one advantage of enacting forward-looking transitional justice in a case like Alberta is that more effective environmental policies are likely to be accepted. Yet another advantage of my view is that it is consistent with various accounts of inter-generational justice that emphasise moral obligations to future generations that do not yet exist.

References

Alberta Government. 2022. "Oil Prices". September 23. https://economicdashboard.alberta.ca/OilPrice

Barnetson, Bob and Jason Foster. 2014. "The Political Justification of Migrant Workers in Alberta, Canada. *Int. Migration & Integration* 15: 349–370.

Benatar, David. 2008. *Better Never to Have Been: The Harm of Coming into Existence.* New York: Oxford University Press.

Beine, M., C.S. Bos and Coulombe. 2012. "Does the Canadian economy suffer from Dutch disease?". *Resource and Energy Economics* 34: 468–492.

Caney, Simon. 2005. "Cosmopolitan Justice, Responsibility, and Global Climate Change." *Leiden Journal of International Law* 18 (4): 747–775.

Caney, Simon. 2014. "Climate Change, Intergenerational Equity and the Social Discount Rate." *Politics, Philosophy & Economics* 13 (4): 320–342.

Caney, Simon. 2018. "Justice and Posterity." In *Climate Justice: Integrating Economics and Philosophy*, Ravi Kanbur and Henry Shue (eds). Oxford: Oxford University Press, pp. 157–174.

Caney, Simon. 2020. "Climate Justice." In *The Stanford Encyclopedia of Philosophy*, Edward N. Zalta (ed.). https://plato.stanford.edu/entries/justice-climate/

Eisikovits, Nir. 2017. "Transitional Justice." In *The Stanford Encyclopedia of Philosophy*, Edward N. Zalta (ed.). https://plato.stanford.edu/entries/justice-transitional/

Elster, Jon. 2004. *Closing the Books: Transitional Justice in Historical Perspective*. Cambridge: The Free Press Syndicate of the University of Cambridge.

Freeman, Mark. 2010. *Necessary Evils: Amnesties and the Search for Justice*. Cambridge: Cambridge University Press.

Futamura, Madoka. 2008. *War Crimes Tribunals and Transitional Justice: The Tokyo Trial and the Nuremberg Legacy*. New York: Routledge.

Graybill, Lyn. S. 2002, *Truth and Reconciliation in South Africa: Miracle or Model?* Boulder: Lynne Rienner.

Heyes, Anthony, Andrew Leach, and Charles F. Mason. 2014. "The Economics of Canadian Oil Sands." *Review of Environmental Economics and Policy* 12 (2): 242–263.

Hoegh-Guldberg, Ove, Daniela Jacob, and Michael Taylor. 2019. "Chapter 3: Impacts of 1.5°C of Global Warming on Natural and Human Systems." In IPCC: 175–311.

Klinsky, Sonja and Jasmina Brankovic. 2018. *The Global Climate Regime and Transitional Justice*. New York: Routledge.

KPMG. 2015. *A Guide to Oil and Gas Taxation in Canada*. http://www.kpmg.com/Ca/en/IssuesAndInsights/ArticlesPublications/Documents/A-Guide-to-Oil-and-Gas-Taxationin-Canada-web.pdf

Krzepkowski, M., and J. Mintz. 2013. "Canadian manufacturing malaise: three hypotheses. *SPP Research Papers, The School of Public Policy, University of Calgary* 6 (12): 1–11.

Llavador, Humberto, John E. Roemer, and Joaquim Silvestre. 2015. *Sustainability for a Warming Planet*. Cambridge, MA: Harvard University Press.

Macrotrends. n.d. "Crude Oil Prices – 70 Year Historical Chart". https://www.macrotrends.net/1369/crude-oil-price-history-chart

Meyer, Lukas H. and Dominic Roser. 2006. "Distributive Justice and Climate Change. The Allocation of Emission Rights." *Analyse & Kritik*, 28(2): 223–249

Meyer, Lukas H. and Dominic Roser. 2009. "Enough for the Future." In *Intergenerational Justice*, Axel Gosseries and   Lukas H. Meyer (eds.). New York: Oxford University Press. pp. 219–248.

Mihai, Mihaela. 2010. "Transitional Justice and the Quest for Democracy: A Contribution to a Political Theory of Democratic Transformations", *Ratio Juris* 23 (2): 183–204.

Nikiforuk, Andrew. 2010. *Tar Sands.* Vancouver, Greystone Books.

Our fair share: report of the Alberta Royalty Review Panel to the Hon. Lyle Oberg, Minister of Finance. (18 September 2007). https://open.alberta.ca/publications/our-fair-share-report-of-the-alberta-royalty-review-panel-to-the-hon-lyle-oberg-minister-of-finance-

Parfit, Derek. 1984. *Reasons and Persons*. Oxford: Clarendon Press.

Parlee, Brenda. 2015. "Avoiding the resource curse: indigenous communities and Canada's oil sands. *World Development* 74: 425–236.

Rabson, Mia. 2018. "No G20 member has climate plan strong enough to meet Paris targets: reports." *National Observer.* https://www.nationalobserver.com/2018/11/15/news/nog20-member-has-climate-plan-strong-enough-meet-paris-targets-report

Sachs, Jeffrey D., and Andrew M. Warner. 1995. "Natural resource abundance and economic growth." *Working paper* 5398, National Bureau of Economic Research Cambridge, MA. https://www.nber.org/papers/w5398

Teitel, Ruti. 2000. *Transitional Justice*. New York: Oxford University Press.

UNDESA. 2012. "A Guidebook to the Green Economy." www.undsd2012.org/content/documents/528Green%20Economy%20Guidebook_100912_FINAL.pdf

United Nations. 2004. "The Rule of Law and Transitional Justice in Conflict and Post-Conflict Societies: Report of the Secretary- General." http://www.un.org/ruleoflaw/files/2004%20report.pdf

Zhang, Yong-Sheng and He-Ling Shi. 2014. "From Burden-Sharing to Opportunity-Sharing: Unlocking the Climate Negotiations." *Climate Policy* 14 (1): 63–81.

11

Let's Be Rational: A 'Fair Share' Approach to Carbon Emissions

DANIEL BURKETT

The climate crisis represents a serious threat to our way of life. Without drastic changes to our carbon emissions, we will see considerable and irreversible harm to the environment. There is a growing literature regarding the moral obligations of states, sub-state actors (such as local governments) and industries to curb their carbon emissions (see, for example, Moss 2015). There are also arguments that we, as individuals, have similar obligations to curb our own emissions. Usually, such arguments are based on the claim that our emissions cause or increase the likelihood of harm to others (see, for example, Lawford-Smith 2016 and Broome 2019). But detractors raise questions about how much – if at all – our individual actions truly make a difference (see, for example, Sinnott-Armstrong 2005). This chapter focuses on a relatively novel approach to individual carbon emissions; an approach that sees these emissions as a scarce communal resource, that, like other scarce resources such as a food, water or medical supplies, needs to be rationed. Under this approach, the wrongness of our individual carbon emissions does not depend on harm being caused. Instead, it requires that an individual has consumed more than her 'fair share'.

1. The Problem with Harm

The simplest way to point out the moral wrongness of an action is, perhaps, to demonstrate that it causes harm. Such a move relies upon the harm principle, which – at its simplest – is the presumption that (all other things being equal) we have a moral obligation not to perform an act that causes harm to others. Put another way, actions are *prima facie* morally impermissible when they make a negative difference in the welfare of other people. But while the collective emissions of a country's coal-fired power industry, for example, may clearly affect people's welfare in this way, it is much more difficult to establish that our individual emissions are capable of making such a difference. This is due to the fact that, in most cases, our personal emissions do not directly cause harm in any robust sense. Suppose I engage in what Kingston and Sinnott-Armstrong (2018) refer to as 'joyguzzling' – that is, going for a wasteful Sunday drive in my gas-guzzler. Even if I do so, climate change – and thus harms related to climate change – will not occur unless many other people engage in the same kind of behaviour. Given this, my high-emitting choice is neither necessary nor sufficient for climate change, and therefore makes no difference to the welfare of others. Joyguzzling – like any high-emitting individual activity – is, to use Sinnott-Armstrong's (2005, 290–291) analogy, like pouring a quart of water into a river that is going to flood downstream due to excessive rains. My individual action makes little-to-no difference.

One way around this problem might be to adopt a more general harm contribution principle, according to which we have a moral obligation not to make problems worse. But, as Sinnott-Armstrong (2005, 293) notes, this principle does not get us much further. Climate change only becomes worse if the harm it causes becomes worse – in other words, if more humans or animals are harmed or harmed to a greater extent. But there is no reason to think that my single joyride will make any particular human or animal worse off, so once again the problem of difference-making arises.

Perhaps, however, our moral responsibilities regarding our individual carbon emissions need not rely on them making a demonstrable difference in others' welfare. While an individual high-emitting action (like joyguzzling) might not actually be harmful, the expectation of harm associated with this action might be enough to make it wrong. As John Broome (2019, 114) puts it, '[W]hen you consider whether or not to joyguzzle that Sunday afternoon, you cannot know what good or harm will actually result from what you do. The result may be a typhoon or a child's death, or it may be good'.

Holly Lawford-Smith (2016, 68) explains how this uncertainty of outcomes arises from the highly probabilistic connection between greenhouse gas (GHG) concentrations and global temperatures. Further, the existence of micro-level thresholds (the GHG concentrations required to cause an extreme weather event like a flood, drought, bush fire or typhoon) and macro-level thresholds (the GHG concentrations required to cause irreversible climate-related catastrophes like massive tree death in the Amazon or the thawing of Arctic ice and the Serbian permafrost) makes it even harder to pinpoint which specific individual's actions made the difference (Lawford-Smith 2016, 68-70). But this does not entail that nobody's action makes a difference. Lawford-Smith (2016, 76, emphasis added) notes that duty must instead be understood probabilistically in such cases, so that individuals 'have duties not to perform the actions that can be reasonably expected to cause certain kinds of harms'. Emitting carbon makes the planet warmer, and the warmer the planet gets, the higher the likelihood of extreme (harm-causing) weather events becomes. Indeed, while methodologies differ, one calculation estimates that the average American causes – through her GHG emissions – the serious suffering and/or death of one or two future people (Nolt 2011, 9). Suppose, to use entirely artificial numbers, that my Sunday joyride is shown to have a 0.0002% chance of triggering a catastrophic weather event that will cause one million deaths. If this is so, then the expected cost of my joyguzzling is two human lives. When couched in these terms – and when the paltry benefit I gain from my joyride is truly considered – it becomes difficult to see how there could be any argument for the moral permissibility of such behaviour.

However, applying this argument to individual contributions to climate change in the real world is much more difficult. For one, it is worth considering whether we really wish to accept this kind of argument at all. Garrett Cullity (2015, 152–153) makes a parity argument, noting that there is a correlation between ambient noise levels and the incidence of aggressive behaviour. Given this observation, it is therefore possible for me to reduce the probability of a violent incident by being quieter. Despite this, it seems hard to accept the claim that I act wrongly if I do not do all I can to minimise my noise levels. Of course, as Cullity (2015, 153) notes, there are important factors that distinguish between these two cases: specifically, the expectation of harm (which is much greater in the case of carbon emissions), the associated of cost of offsetting (which is much less in the case of carbon emissions) and the remoteness of the causal contribution (which is less in the case of carbon emissions, since the ambient noise example relies on an additional third party to actually *engage* in the violent behaviour being causally contributed to).

Nevertheless, these factors complicate matters. While there may be a better case for a moral obligation to reduce individual contributions to carbon emissions than to reduce individual contributions to ambient noise, is it good enough? In order to answer this question, we would need to compare the relative size of the expected harm with the cost of offsetting and arrive at some kind of conclusion regarding when the scales are tipped in favour of a moral obligation to offset. At what level of expected harm are we expected to act? Precisely what kind of cost can we be fairly expected to pay for this offsetting? Further, we need to consider what effect the strength (or weakness) of our causal connection to the expected harm will have on our moral obligations. Does it discount the amount of offsetting expected from us? Given these complexities with a harm-based approach, it seems prudent to consider whether there might be another way to provide us with a moral reason to avoid excessive personal carbon emissions. That is what I examine in the remainder of this chapter.

2. Zeno's Punch Bowl

The fair share approach is perhaps best illustrated with an analogy. Suppose there existed a magical item called Zeno's Punch Bowl – so named for its ability to complete an infinite series of philosophical social events. The exact mechanism by which the punch bowl operates remains a mystery, but its characteristics are universally known among philosophers. The bowl is an ornate copper receptacle, etched with the faces of academic titans. It is capable of holding 101 250-millilitre cups of the most delicious beverage ever enjoyed by man. Herein lies the magic: if, at the end of any social event, at least one cup of punch remains in the bowl, the bowl will magically replenish its entire reserve of punch in time for the next social event.

The bowl is dutifully transferred from one philosophical event to the next: from the Annual Convocation of Free Will Compatibilists to the Central Meeting of the Rawlsian Veilers. At each event, the attendees make sure to leave at least one cup of punch remaining in the bowl, so that future assemblies of philosophers can continue to enjoy this wonderful resource. Demand is so high for the bowl, however, that no academic can expect to come across it more than once in their lifetime.

The bowl has now found its way to my local regional philosophy conference – its magical properties known to all in attendance. There are precisely 100 individuals (including myself) present at the conference. During the lunchtime frivolities, I pour myself a single cup of punch from the bowl. I taste it and am overwhelmed with joy at the flavour. I quaff my cup, then look guiltily at the still half-filled bowl. What harm could one more taste do? I quickly refill my cup, then hurry from the scene.

What, precisely, is wrong with my action? We might claim that by taking a second cup for myself, I am causing harm to other individuals by causing a reduction in their welfare. The punch bowl holds precisely 101 cups of punch, one of which must remain in order for the punch bowl to retain its magical abilities. There are, then, 100 'consumable' cups of punch in the bowl. Since there are one hundred attendees, there is only sufficient punch for one cup per attendee. By taking a second cup, I harm the other attendees by limiting the punch that they might enjoy.

But does this explanation really capture what is wrong about my actions? The punch is a continuous, not discrete, resource, so by enjoying a second cup I am not necessarily removing the ability of any other individual to enjoy the punch. Where the remaining 99 attendees could once share 99 cups between them, they can now only share 98. Where each attendee would have received 250 mL of punch, now they can only receive 247.5 mL. Does it even make sense to label such a miniscule reduction as harmful? Would there even be any appreciable difference in each attendee's experience of their cup of punch and therefore their welfare? It would seem not. In fact, given the discrepancies in different individuals' cup-pouring techniques, this small reduction most likely falls well within the margin of error of a standard cup pour, meaning that many individuals may still drink exactly the same-sized cup of punch they would have drunk had I not even acted. For this reason, claiming that my action is wrong for reasons of causing harm is flawed. We must look for an alternative analysis.

Maybe we could claim my behaviour is wrong because it amounts to something called free-riding. This occurs when an individual fails to contribute to a public good that she herself benefits from, like riding the bus without paying for your fare. The public good here being contributed to is the sustained magic of Zeno's Punch Bowl so that the bowl can be used by future assemblies (and, indeed, future generations) of philosophers. Since the bowl holds 101 cups, and one cup must remain, each individual's contribution should amount to restraining themselves to drinking one cup of punch at most. I, on the other hand, show a blatant disregard for the required restraint. But while it is clear that I fail to contribute, it is less clear whether this amounts to a case of free-riding. By stipulation, we have made it clear that no individual will be able to enjoy the wonders of Zeno's Punch Bowl twice in their lifetime. As such, the benefit to which I am expected to contribute – the sustained magic of the punch bowl – is one that I will never enjoy. My action, then, is not wrong by virtue of free-riding.

3. The Wrongness of Taking More than Your Fair Share

Why is it wrong for me to take a second cup of punch? One answer might be that it is wrong because it involves taking more than my fair share of the punch. If there are 101 cups of punch in the bowl and one of these cups must remain, then there remain 100 'available' cups for attendees to drink. If there are only 100 attendees, then all else considered, it would seem that a fair division of the resource would be an *equal* division of the resource. Any individual who appropriates more than an equal division of the resource would be acting *unfairly* towards their colleagues. They would be breaching a duty to take on the burden of restraint. This is the very thing that makes my action wrong.

We can apply the same reasoning to climate change and carbon emissions. In order to avoid the most catastrophic effects of climate change, the average global temperature increase must be kept below a certain level. Doing this requires that our total carbon emissions are capped. Once the volume of previous emissions is taken into account, this leaves us with a remaining, finite carbon budget representing the total carbon emissions we can create while avoiding catastrophic climate change. Once we recognise carbon emissions as a scarce communal resource we must, by necessity, engage in rationing. Presumably, there is some equitable way of rationing between states, sub-state actors, industries and individuals. Once this is done, it is merely a case of looking to whether or not any individual is emitting more than their 'fair share' of those emissions.

This approach differs from most other approaches to individual carbon emissions as it moves away from focusing on whether or not individual emissions actually make a tangible difference to climate change, and instead focuses purely on whether individuals are overconsuming a scarce resource – a considerably easier question to answer. This approach is, in many ways, the same one we might take in considering any kind of scarcity-of-resources scenario. When food is rationed, the immorality of taking more than your fair share is not found in the fact that you might be causing harm to others (though you very well may be), but merely in the fact that you are exceeding your own fair allocation.

A further advantage of this approach is that it places moral restraints on our behaviours regardless of the actions of others. If several other conference attendees also decide to drink multiple cups of punch, this does not suddenly excuse my actions. Likewise, any other attendees who have up until this point shown restraint are not suddenly licensed to take all they want. Every one of us who decides to drink a second cup of punch is in the wrong for breaching our duty to take on the burden of restraint. Non-compliance by others is immaterial to our own obligation to take no more than our fair share.

4. Fair Shares and Carbon Emissions

How do our actual carbon emissions match with our fair share? If the former exceed the latter, then we are acting unfairly and have a moral obligation to curb our emissions in any way possible. Unfortunately, performing such a calculation is notoriously difficult. While discovering our actual individual carbon emissions per year is relatively straightforward, ascertaining precisely what amounts to our 'fair share' is much more difficult. This is largely due to the arbitrariness of many of the lines we must draw, such as, for example, the amount of harm from climate change we are willing to tolerate (some harm is already occurring and therefore inevitable) and the likelihood we are willing to accept of this harm occurring. However, even under the most generous approach, our moral obligations regarding individual carbon emissions are plainly clear.

In 2011, nearly all countries agreed to limit the global average temperature rise to no more than 2°C above pre-industrial levels – the maximum global temperature rise we can tolerate while avoiding the most catastrophic effect of climate changes (2011 United Nations Cancun Agreement). According to the Intergovernmental Panel on Climate Change, having a >66% chance of achieving this would require us to keep our global carbon emissions below 2,900 gigatonnes of carbon dioxide ($GtCO_2$) (IPCC 2014, 10). As of 2021, only 607 $GtCO_2$ remain in the global carbon budget (Evershed 2017). Divided equally amongst the 7.9 billion people on Earth, this comes out at lifetime carbon allowance of 76.8 tonnes of CO_2 per person – or around 0.895 tonnes per year over an 85-year lifespan.

It may, of course, be the case that an equal division of a resource is not necessarily a *fair* division of that resource. Suppose a university departmental meeting is catered with 20 cupcakes and that 10 professors are in attendance. Absent any other relevant factors, it seems that the fairest division of these cupcakes would be an equal division – that is, two cupcakes per professor. The introduction of other factors may modify this, however. Suppose that there were, in fact, 25 cupcakes, and that one professor arrived earlier and has already eaten five of the cupcakes. Such evidence may affect the fair division of the remaining 20 cupcakes. Suppose, alternatively, that one professor has forgotten his lunch that day and has nothing else to eat and that two other professors brought ample lunch and have no need for the catering. These may also factor into what would count as a fair division.

Calculating precisely what amounts to a fair share of a scarce communal resource is no easy feat. As seen above, factors such as (1) an individual's responsibility for the shortage of a resource, (2) an individual's need for the resource, and (3) an individual's ability to forego a resource will all be relevant. The first of these factors is particularly relevant in the context of carbon emissions. Fairness might demand that individuals living in historically high-emitting countries receive a smaller share of carbon emissions going forward, while those in historically low-emitting countries receive a larger share. Further criteria can also be found in the literature on the distribution of scarce medical resources, where the fairness of an allocation might be influenced by contribution, ability to pay and merit (Maddox 1998), as well as temporal priority (i.e., first come, first served), chance, utility (i.e., who can make best use of the resource), social worth and effectiveness (Bringedal

1992). In fact, philosophical discussions of climate change are already familiar with many of these factors, as they are the very same kinds of criteria that are considered when establishing how to divide the global carbon budget between countries.

For those living in traditionally high-emitting countries, the most generous budget can be arrived at by mapping current carbon usage against the proportional reduction required of the 2°C target. Put simply, the attainment of this target requires that emissions peak around 2020, drop 50% by 2045 and fall below zero by 2075. Carbon Brief provides an analysis of individual carbon budgets based on this metric – that is, past emissions reduced according to proportional targets. On this analysis, a child born in the United States (US) in 2017 will have a lifetime carbon budget of approximately 197 tonnes of CO_2 – or 2.31 tonnes per year over an 85-year lifespan – in order to keep to the 2°C target (Hausfather 2019).

So, how does this compare to our actual individual emissions? In 2017 (the most recent year for which data are fully available) those in the US emitted 16.16 tonnes of CO_2 on average – almost seven times the fair share proposed above (Ritchie and Roser 2020). This means that an individual born in 2017 will have used up their lifetime share of carbon emissions before their thirteenth birthday.

There is one important wrinkle in this data, however. Specifically, these emissions figures are based on per capita, not necessarily individual, emissions. In other words, theis number represents the sum of emissions in the US divided by population. It pays no attention to whether these emissions are produced by individuals personally (say, by going on a leisurely Sunday drive), by infrastructure (e.g., bitumen production for roads) or by private industry. There is a good chance that a large portion of that per capita budget is contributed to by activities over which individuals have little-to-no direct control (how electricity is produced, how public transport is fuelled, what kind of industries our economic structure incentivises, etc.). While excessive systemic emissions might provide us with moral impetus to engage in political action to change those systems, we are here concerned specifically with our moral responsibility for those emissions over which we have direct control.

However, even if the emissions over which we have direct control only make up some small part of per capita emissions, it is significant enough to lead to some strong moral conclusions. A first point of focus is often those emissions deemed 'luxury emissions' (Shue 1993). A one-way flight from Los Angeles to Auckland will contribute 1.48 tonnes of carbon – more than a US individual's annual budget – in the space of a single day (figure calculated via Carbon

Footprint). But we need not even look to luxuries to raise serious concerns. The average person living in the US releases 2.83 tonnes of CO_2 per year through residential energy use alone (Goldstein et al. 2020). This puts most Americans over budget before even considering other necessities like food, transport, clothing and pharmaceuticals. If you opt to have a child, then any hope of remaining under-budget is entirely forfeit. Ignoring any other carbon emissions created in their lifetime, a US individual who chooses to have a single child will exhaust their entire lifetime's carbon budget in less than three years (Murtaugh and Schlax 2009).

Does this mean that we should disconnect our homes from the power grid, reduce our food consumption and give up on having children? Not necessarily. For one, it is important to note that while the above sources of emissions are listed as those over which we most likely exercise direct control, this is not necessarily the case for all individuals. Consider the example provided by Christian Baatz (2014, 10) of,

> an (elderly) person living in a rural area in the US who depends on her car to buy food and to participate in social and cultural activities because no public transport system is available or she is not able to use it. Let us further assume that she lives in a poorly insulated house, lacks the means to invest in improved insulation and there are no governmental programs subsidizing credits etc.

In such a case, Baatz (2014, 10) argues, expecting this woman to lower her emissions would be asking too much, given the burdensome hardship resulting from such a reduction. Put simply, 'her life would not be decent anymore'.

This much is certainly true, and expecting this woman to reduce her emissions by foregoing food or participation in social and cultural activities would no doubt be entirely unreasonable. It is hoped, however, that whatever metric we adopt for calculating a 'fair share' of emittable carbon would take into account these very needs. Indeed, we might argue that any purportedly 'fair share' that did *not* accommodate needs such as these would not in fact be fair at all. Such a case may, for example, be a strong counterexample to the presumption that a fair share is an equal share (since an equal division of emittable carbon is unlikely to give her enough to fulfil these necessities of life).

All of that aside, just because *some* emissions of *some* individuals cannot be reduced under any that just theory, that does not mean that many if not most individual emissions are open to scrutiny. While there are many things we cannot do without, there are many things that we can. In this way, the fair share approach requires us to seriously reconsider how we live, how we consume, how we travel and even how we reproduce. We can no longer avoid blame by claiming that we are already doing our part by merely recycling (reducing our emissions by 0.21 tonnes of CO_2 per year on average), washing our clothes in cold water (a reduction of 0.25 tonnes of CO_2 per year on average) or upgrading our lightbulbs (a reduction of 0.10 tonnes of CO_2 per year on average) (Wynes and Nicholas 2017). While such measures are important, they must be coupled with far greater behavioural changes.

It is also unacceptable to say that in the grand scheme of things our individual carbon emissions 'make no difference'. Writing in *The Guardian*, Leah Stokes (2019) argues:

> The next time you feel that you are to blame for climate change – because you forgot to hit the light switch, or you took that flight to see your ailing mother – remember the Ohio electric utilities and their coal subsidies. Remember the politicians who gave them... [a] billion-dollar bailout after receiving personal favors, like a flight to Trump's inauguration on a corporate jet. And know that one day after signing this bill, Ohio Governor DeWine attended a Trump fundraiser hosted by coal baron Bob Murray. Fighting the climate crisis is not about purifying yourself. It's about dismantling corporate power. Electric utilities are a great place to start.

But this is the very mindset against which the fair share approach is directed. While effective action on climate change will require the dismantling of corporate power, we cannot point exclusively to this solution and excuse ourselves of moral blame. Even if our emissions cause a miniscule amount of harm in comparison to the emissions of large industries, we are still morally blameworthy for the lifestyle choices we make that lead to excessive carbon emissions. The upshot is simple: even on the most generous account of what an individual carbon budget might be, those of us in developed nations like the US are – simply by virtue of the individual emissions over which we have direct control – currently producing far more than our fair share of carbon emissions. In this way, we are treating other members of our global community unfairly and, for this reason, are in breach of our moral obligations towards others. While the realities of infrastructural and political systems might make it impossible for us to completely avoid this breach for now, the onus on us is to do all we can to minimise it by reducing our individual emissions in any way possible.

5. Three Potential Objections to the Fair Share Approach

So far, we have examined the argument that excessive individual carbon emissions are wrong because they use more than one's fair share of a scarce communal resource. This means, however, that the entire argument rests on the very important first premise that emittable carbon is a scarce communal resource in the first place. One objection to the fair share argument might therefore be that this is simply untrue. 'Emittable carbon' is not a consumable resource. It is not finite let alone scarce in the same way that food, water or medicine might be. Every individual could, in theory, emit an infinite amount of carbon without removing from any other individual his or her right to also emit an infinite amount of carbon.

When we discuss the scarcity of the resource of emittable carbon, we are instead discussing something like a finite 'right to emit'. Emittable carbon is not scarce for factual reasons (as say, water is scarce when the town supply tanks are close to depletion), but rather for *moral* reasons. Given this, an independent moral argument is required for why this is the case – that is, *why* we must keep our global carbon emissions below a certain level.

This, however, is not a problem that is unique to the fair share approach. In fact, any argument that argues for a reduction in carbon emissions – be it by individuals or, more likely, by states and industries – will have this same first premise. Even those who argue against any kind of moral obligation to reduce individual carbon emissions usually grant the premise that there are independent moral reasons to establish an international carbon budget (see, for example, Sinnott-Armstrong 2005). They merely argue that this does not lead to any corollary obligation for individuals, since individuals do not make a difference.

The fair share approach argues from this same widely accepted first premise (that we have a moral reason to keep our total global carbon emissions below a certain level) that we each have a moral obligation to minimise our individual carbon emissions not because we make a difference, but because it is what fairness demands when dealing with a scarce communal resource. I contend that the existence of an exceedingly small (and rapidly diminishing) carbon budget automatically places an obligation on individuals to not exceed their 'fair share' of emissions, however miniscule that fair share might be and however little effect those emissions might have on the welfare of others.

A second potential objection has been raised by Ewan Kingston and Walter Sinnott-Armstrong (2018, 182), who claim that the fair share approach must be based on either the principle that:

A) When an existing political or social structure encourages cooperation and discourages defection in a collective action problem, individuals have a moral requirement to obey the rule of that existing institution.

Or the principle that:

B) When no existing political or social institution adequately encourages cooperation and discourages defection in a collective action problem, individuals have a moral requirement to obey the rules of a hypothetical institution that would be adequate or ideal.

Principle A, they note, seems true – but no such institution currently exists that has formally apportioned fair shares of emittable carbon. All that remains then, is Principle B, and it is this principle to which Kingston and Sinnott-Armstrong object. They note that this principle fails in certain cases, citing one specific example: 'Imagine a country with an inequitable health-care system, that places the lives of many vulnerable people in jeopardy, increasing inequality. The inequities would be removed if everyone paid 3% more for taxes, and that money funded equitable health care for all' (Kingston and Sinnott-Armstrong 2018, 182–183). In the absence of such a system, they argue, there would be no moral requirement for an individual to contribute the same amount of money to the government in the hope that the government would provide such services. In this way, Principle B is refuted.

This, however, may be too hasty. The example provided by Kingston and Sinnott-Armstrong shows that Principle B sometimes fails to generate specifically positive duties (that is, duties to actively do something). It seems, however, that it successfully generates negative duties (that is, duties to refrain from doing something). Suppose, to modify their example slightly, the inequity of this particular healthcare system was due to the active drug-seeking behaviour of a small portion of the population. These individuals exploit a loophole in the medical system that allows them to receive duplicate (and unnecessary) prescriptions, leading to a shortfall in the system's thinly stretched medical resources and subsequently creating inequities for other citizens. The legislature is currently attempting to close this loophole, but in the meantime it would seem there is a moral obligation on drug-seeking individuals to adhere to a negative duty to refrain from seeking extra medication.

In light of this observation, we might modify Principle B as follows:

B*: When no existing political or social institution adequately encourages cooperation and discourages defection in a collective action problem, individuals have a moral requirement to assume the negative duties of a hypothetical institution that would be adequate or ideal.

Principle B* is narrower than Principle B, but is all that we require in order to motivate the fair share approach. Better yet, it is not vulnerable to the types of problem cases raised by Kingston and Sinnott-Armstrong.

A final potential objection surrounds what we are to say about those individuals who already emit less than their fair share of carbon. Would fairness not dictate that those individuals be permitted to emit more carbon than they currently do? A potential response to this objection is two-fold. First, we must carefully consider whether or not such individuals even *exist*. There are, to be sure, countries whose citizens emit carbon at a very low rate. Pakistan, for example, emitted 1.12 tonnes of CO_2 per capita in 2017 – a mere fraction of what the US emitted (Ritchie and Roser 2020). However, if the carbon budget of individuals living within Pakistan is arrived at via the same process as the carbon budget for individuals living in the US (that is, by mapping their current carbon usage against the necessary proportional reduction required of the 2°C target), individuals within Pakistan are still coming in over budget, with those born in 2017 receiving an annual carbon budget of only 0.11 tonnes (Hausfather 2019).

But suppose that under-budget individuals *do* exist. On the fair share approach, would it be morally permissible for them to emit more? The simple answer is 'yes'. If an individual with a carbon budget of 1.5 tonnes were only emitting 1 tonne of carbon per year and decided to increase their emissions up to 1.1 tonnes, then they would still be taking less than their entitlement and, as such, there could be no fair share-based moral objection to their behaviour. Nevertheless, this would not be the end of the discussion. Their decision to increase their emissions purely on the basis that they are entitled to more may be morally blameworthy for a raft of other reasons external to the assessment we have outlined here. Perhaps their actions would be morally blameworthy because they were wasteful: consuming resources and emitting carbon when they did not need to. Alternatively, there may be a compelling case for some kind of 'slack-taking' principle, according to which any individual who *can* consume less than their fair share should consume less than their fair share, in order to offset those individuals who inevitably consume more than their fair share. In any case, the fair share approach will not and does not purport to exhaust all grounds for the moral impermissibility of intentionally increasing one's individual carbon emissions.

Conclusion

The fair share approach to individual carbon emissions is based on considerations of what is fair when rationing a scarce communal resource. It does not require us to point to actual or potential harm caused to current or future persons. Instead, it derives the moral impermissibility of excessive individual carbon emissions from the fact that it is unfair to take more than one's fair share of a scarce communal resource. Given even the most generous allowances for an individual carbon budget, it appears that most if not all of us are already exceeding our 'fair share' of carbon emissions. Given this, there is an onus on us to do all we can to reduce our future emissions by any way possible.

References

Baatz, Christian. 2014. "Climate Change and Individual Duties to Reduce GHG Emissions". *Ethics, Policy & Environment* 17(1): 1–19.

Bringedal, Berit. 1992. "Distributional Principles in Health Care". Institute for Research on Labour and Employment Working Paper # 40-92.

Broome, John. 2019. "Against Denialism". *The Monist* 102: 110–129.

Carbon Footprint. 2021. "Carbon Calculator". *Carbon Footprint*. Accessed August 28, 2021. https://www.carbonfootprint.com/calculator.aspx

Cullity, Garrett. 2015. "Acts, omissions, emissions". In *Climate Change and Justice* edited by Jeremy Moss. 148-164. Cambridge: Cambridge University Press.

Evershed, Nick. 2017. "Carbon countdown clock: how much of the world's carbon budget have we spent?". *The Guardian*. Last modified January 19, 2017. https://www.theguardian.com/environment/datablog/2017/jan/19/carbon-countdown-clock-how-much-of-the-worlds-carbon-budget-have-we-spent

Goldstein, Benjamin, Dimitrios Gounaridis, and Joshua P. Newall. 2020. "The carbon footprint of household energy use in the United States". *Proceedings of the National Academy of Sciences of the United States of America* 117 (32): 19122–19130.

Hausfather, Zeke. 2019. "Analysis: Why children must emit eight times less CO2 than their grandparents". *Carbon Brief*. Last modified April 10, 2019. https://www.carbonbrief.org/analysis-why-children-must-emit-eight-times-less-co2-than-their-grandparents

IPCC. 2014. *Climate Change 2014: Synthesis Report*. Geneva: IPCC.

Kingston, Ewan, and Walter Sinnott-Armstrong. 2018. "What's Wrong with Joyguzzling?" *Ethical Theory and Moral Practice* 21: 169–186.

Lawford-Smith, Holly. 2016. "Difference-Making and Individuals' Climate-Related Obligations". In *Climate Justice in a Non-Ideal World*, edited by Clare Hayward & Dominic Roser, 64–82. Oxford: Oxford University Press.

Maddox, P.J. 1998. "Administrative Ethics and the Allocation of Scarce Resources". *The Online Journal of Issues in Nursing* 3 (3).

Murtaugh, Paul A., and Michael G. Schlax. 2009. "Reproduction and the carbon legacies of individuals". *Global Environmental Change* 19: 14–20.

Moss, Jeremy (ed.). 2015. *Climate Change and Justice*. Cambridge: Cambridge University Press.

Nolt, John. 2011. "How Harmful Are the Average American's Greenhouse Gas Emissions?". *Ethics, Policy & Environment* 14 (1): 3–10.

Ritchie, Hannah, and Max Roser. 2020. "CO$_2$ and Greenhouse Gas Emissions". *Our World in Data*. Last modified August 2020. https://ourworldindata.org/co2-and-other-greenhouse-gas-emissions

Shue, Henry. 1993. "Subsistence Emissions and Luxury Emissions". *Law & Policy* 15 (1): 39–59.

Sinnott-Armstrong, Walter. 2005. "It's Not *My* Fault: Global Warming and Individual Moral Obligations". In *Perspectives on Climate Change: Science, Economics, Politics, Ethics. Advances in the Economics of Environmental Resources*, Vol. V, edited by Sinnott-Armstrong, Walter, and R. B. Howarth. 285–307. Amsterdam: Elsevier.

Stokes, Leah C. 2019 "While the planet overheats, Ohio's coal industry gets a bailout'. *The Guardian*. Last modified July 29, 2019. https://www.theguardian.com/commentisfree/2019/jul/28/planet-overheats-ohios-coal-industry-gets-a-bailout

Wynes, Seth, and Kimberley A. Nicholas. 2017. "The climate mitigation gap: education and government recommendations miss the most effective individual actions". *Environmental Research Letters* 12: 074024.

SECTION THREE

NORMATIVE PERSPECTIVES ON THE CLIMATE NEUTRALITY AMBITION

12

Climate Justice from Theory to Practice: The Responsibility and Duties of the Oil Industry

MARCO GRASSO

Thirty percent of the global industrial greenhouse gas emissions between 1965–2018 can be traced to the activities of 15 companies in the oil industry. Based on this evidence and on a number of morally relevant facts, this chapter proposes a normative framework for establishing the positive responsibility that oil companies have in relation to climate change. Then, the analysis articulates this responsibility in the form of two duties: a duty of reparation and a duty of decarbonisation. The duty of reparation implies rectification through disgorgement of funds for the wrongful actions of oil companies, which resulted in negative climate impacts, starting from the most vulnerable groups affected by climate change. The duty of decarbonisation entails a large-scale transformation that oil companies ought to undergo in order to reduce and eventually eliminate carbon emissions from their business model. Finally, the chapter indicates possible practical implications of these duties.

Introduction

Climate change is essentially a matter of justice. Philosophers and other scholars, as well as politicians, activists, religious leaders and many others have long highlighted and explored the numerous ethical considerations and challenges that are inseparable from discussions of the causes, consequences and potential human responses to anthropogenic climate change (Grasso and Markowitz 2015). A prominent and long-lasting concern of climate justice is the question of 'who counts' – that is, which agents (individuals and/or groups) should be at the centre of moral debates about climate change.

Beyond the current state-centric perspective of the international system, which considers states the primary agents of climate justice, there is a spirited debate about other possible agents. For example, some environmentalist rhetoric focuses on the role of individuals, both in terms of reducing ones' own emissions and for advocating for large-scale change. Although this perspective has gained attention in recent years, it should be complemented by forms of collective responsibility that do not exclude individual responsibility, but which rather integrate the two perspectives, paying particular attention to novel or neglected collective agents of justice.

Among these, given their unique and distinctive role, responsibility and duties in the context of climate change, oil and gas companies – for sake of simplicity hereafter referred to also as 'oil companies' or the 'oil industry' – are possibly the most significant overlooked group of agents. The oil industry, through the emissions generated by the fossil fuels it processes, has significantly increased atmospheric concentrations of greenhouse gases (GHG) (IEA 2021). Therefore, this industry has contributed directly to anthropogenic climate change.

It is worth stressing that this argument does not imply that the oil industry should become the only agent responsible for addressing climate change, or even that oil companies are the most important players. Consumers, civil society, businesses and other stakeholders all play a role in causing climate change and have consequent responsibilities in addressing climate change. The goal of the chapter is to draw attention to oil companies' responsibility for causing climate change, the duties this responsibility creates and the consequent implications for climate justice.

Oil companies should play their part in global climate governance, along with states, individuals and other agents. That part is significant, since they have played a crucial role in causing, shaping, advancing and defending the current, unsustainable fossil fuel-dependent global economy. By continuing to produce fossil fuels and feed consumer demand, they have been dictating the rules of the game to the global economic system. Based on these considerations, this chapter first outlines the direct contribution the oil industry has made to climate change in terms of its cumulative emissions and their impacts. Then, after specifying the unique agency of oil companies, the chapter argues that their activities have violated the negative responsibility of 'doing no harm'. Therefore, these companies have a positive moral responsibility in the context of climate change to 'clean up the mess' they caused. Subsequently, the chapter articulates oil companies' positive responsibility in the form of two duties, which include different actions to rectify the harm done: a duty of reparation and a duty of decarbonisation. The duty of reparation encapsulates the requirement that oil companies rectify the injustices resulting from the harm the industry has generated, while the duty of decarbonisation entails an obligation by the industry to eliminate carbon emissions from their activities to prevent future harm. Finally, the chapter briefly indicates some possible practical implications of the duties of reparation and decarbonisation.

1. The Oil Industry's Direct Contribution to Climate Change

Recent studies by Richard Heede and colleagues focus on the contributions of large carbon producers to global GHG emissions (Heede 2014; Frumhoff et al. 2015; Heede and Oreskes 2016). 'Carbon majors', as these studies term large carbon businesses, are the world's largest public and private investor-owned, state-owned and government-run oil, gas, coal and cement producers. The primary finding of Heede and colleagues is that just 100 currently operating carbon majors have produced 71% of global industrial emissions since 1988 (according to Heede's figures, the top emitters and the large majority of producers are fossil fuel corporations, whereas cement producers are a small minority among carbon majors; the original 2014 database, for instance, included only 7 cement producers whose emissions amounted to 1.45% of carbon majors' cumulative total, see Heede 2013, table 4, 17). Further ground-breaking work in attribution science – the burgeoning science of attributing weather events to specific emitters and of assessing loss and damage associated with climate impacts – has made it possible to trace specific harm-generating climate impacts to carbon majors. Ekwurzel et al. (2017) showed that carbon majors' fossil fuel-related activities substantially contributed to relevant climate impacts, namely increased global mean surface temperature (GMST) and increased global sea level (GSL). For instance, the emissions of just 90 major carbon producers are responsible for ~29–35% of the rise in GMST and ~11–14% of the rise in GSL since 1980; three of them – BP, Chevron and ExxonMobil – have caused more than 6% of the rise in GSL. By the same token, Licker et al. (2019) demonstrate that 88 of the carbon majors were responsible for 55% of observed ocean acidification 1880–2015, with as yet inestimable damage to ecosystems and marine life, not to mention the fishing industry so vital to myriad coastal communities.

Oil companies are the largest and most numerous carbon majors. Generally, oil and gas are owned by states, or, in weak and failed states, by the subjects who exert irregular coercive control over them (Wenar 2015). Yet the oil industry is the conveyor that moves oil and gas from below the ground, irrespective of its ownership and localisation, and into the global economy. This industry comprises international oil companies (IOCs) and national oil companies (NOCs) – this analysis excludes two other typologies of oil and gas companies, given their irrelevance in terms of global GHG emissions, the so-called 'independents' (smaller companies that operate only in the upstream segment of the oil industry's operations) and 'oilfield service companies' that provide services and outsourcing needs to the oil industry. IOCs are private entities whose business operations traditionally cover the full cycle from exploration, through production and refinement, to distribution of petroleum products. NOCs are largely similarly structured, being either fully or majority-owned by a national government. The activities of the oil industry are divided into upstream operations of exploration and production, and downstream operations of refining and distribution. Given the high entry costs, the world's largest oil companies are typically highly vertically integrated, i.e., they carry out both upstream and downstream activities. Exploration includes prospecting, seismic and drilling activities that take place before the development of a proper oil field; production involves the extraction of oil from below the ground through onshore and offshore drilling; refining concerns the separation of unwanted components in order to obtain clean hydrocarbons marketable into different usable products; and finally, in the distribution phase, such products are transferred to consumers through pipeline networks, tankers, railway tanks and trucks.

The oil industry's contribution to cumulative emissions of GHGs is impressive. Between 1965–2018, only 10 oil and gas companies have accounted for almost 25% of all emissions and only 15 have accounted for almost 30%, as shown in Table 12.1. Additionally, the oil industry holds fossil fuel reserves that, if burned, would increase the earth's average surface temperature well above 1.5°C above pre-industrial levels. Welsby et al. (2021) claim that by 2050 nearly 60% of oil and gas and 90% of coal reserves should remain unburned in order to meet the 1.5°C target.

2. The Unique Agency of Oil Companies

Since the dawn of climate policy, states have been the primary, 'direct' agents involved in addressing climate change. Other stakeholders, such as civil society, private-sector actors, local authorities and communities, international institutions and individuals were mostly considered secondary, 'indirect' agents. In the last decade, however, the lines between actors have been blurring, paving the way for a new framework of hybrid multilateralism – an 'intensified interplay between state and non-state actors in the new landscape of international climate cooperation' (Backstrand et al. 2017, 562).

There is an agreement that all stakeholders share common but differentiated responsibilities in the context of climate change, as stated by the United Nations Framework Convention on Climate Change (UNFCCC 1992). This means that all stakeholders must do their part – proportional to their contribution in terms of emissions and their capacity –to combat climate change. Relative to their prominence and contribution to the problem, the oil industry appears to be truly neglected in the current global climate discourse. Oil companies have contributed greatly to causing climate change and have perpetuated the climate crisis by supporting the status quo. They are causing, shaping, advancing and defending the current, unsustainable fossil fuel-dependent global economy. Through their informed and self-advantageous choice to continue the exploration, production, refinement and distribution of fossil fuels after the risks of doing so became public, carbon majors have essentially imposed on the global socio-economic system a carbon-intensive model of development. Rather than engaging in a large-scale search for alternatives and phasing out fossil fuels, as warranted by the urgency of the climate crisis, oil companies have continued their fossil fuel-dependent business models for decades. In light of this, it is morally unacceptable to equate oil companies' position and responsibility to those of other stakeholders or to those of the private sector in general. Global climate governance should reflect the unique agency of the oil industry, as it has played a very particular and significant role in causing the climate crisis and should contribute to addressing it accordingly.

In fact, oil companies currently have no special responsibilities or duties in global climate governance, despite their substantial contribution to the problem, the wealth and benefits they have obtained through fossil fuel-related activities and their political influence and technical expertise that would have granted them a relatively smooth transition to less carbon-intensive products (Frumhoff et al. 2015; CIEL 2017). As with other corporate agents, oil companies are only subject to the binding emissions limits imposed by national and sub-national political authorities. At best, similar to other corporations, oil companies assume voluntary obligations to disclose their carbon emissions and integrate abatement strategies into their business models. Given the nature of their core business, though, this is not enough.

To be clear, oil companies have a truly unique role in the current global socio-economic system: these companies have been dictating the rules of the game to other businesses in terms of their reliance on fossil fuels. Through their informed choice to continue the extraction, refinement and distribution of fossil fuels in the 1990s, oil companies have perpetuated the dependency of other industries on their products – industries that had to shape their business models around fossil fuels. Therefore, oil companies should have more stringent responsibilities than other industries in combatting climate change. Other industries that depend on supply from oil companies should be attributed fossil fuel-related duties only after the rule-of-the-game shapers (i.e., oil companies) have met theirs. Identifying oil companies as a stand-alone group, with very precise and unique responsibilities, is crucial to advancing efforts to combat climate change.

Given the scientific knowledge and consensus about climate change, fossil fuels may be considered a harmful product, the use of which affects the health, lives and well-being of present and future generations of humans and non-humans. Attribution science goes even further in trying to identify and ascribe climate impacts to specific sources; a source could be a particular agent (e.g., an oil company), a sector or an activity (Burger et al. 2020). Hence, source attribution would make it possible to identify a specific amount of anthropogenic climate harm that was caused by individual oil companies. This attribution is based on the proportional contribution of the company's fossil fuels to changes in the chemical composition of the atmosphere, the extrapolation of the proportional contribution to localised events and the identification of actual harms caused by those impacts (Burger and Wenz 2018). In other words, it seems that a sound causal chain linking anthropogenic climate change to harm and the consequent monetary costs and then to emitters – oil companies, for instance – is now possible.

Cases in which the harmfulness of a product was confirmed by scientific evidence have occurred in the past and reshaped whole industries. Like companies previously working with tobacco, asbestos or lead-based paint, oil companies should assume some responsibility for their involvement in producing a harmful product and for the harm produced.

Not all oil companies operate in wealthy states, which indicates the complex structure of the current global socio-economic system. According to Heede (2014), substantial emissions have originated in somewhat less-developed countries, such as Brazil, China, India, Iran, Mexico, Saudi Arabia and South Africa. Recognising oil companies as important players in global climate change and holding them responsible for their fossil fuel-related activities would, among other things, help bridge a simplistic divide between 'the rich' and 'the poor' worlds. It could lead also to a fairer distribution of the burden of fighting climate change among state and non-state actors around the world.

Introducing oil companies as moral agents in the context of climate change opens up a new avenue for normative inquiries in climate ethics, which may have major implications for global climate governance. For example, an alternative mode of assigning responsibility to different agents in the global system could alter approaches to rectification for harm and the related distribution of burdens and benefits, influence the patterns of well-being among agents and change the flows of financial and other resources between peoples and generations.

Recognition of the prominent role of oil companies in causing and perpetuating climate change does not mean that they should become the only or primary subjects of climate justice. States, consumers, civil society, businesses and other stakeholders all have responsibilities to do their fair share in resolving climate change. Crucially, states are the main agents responsible for providing appropriate legislative and political frameworks for ensuring that carbon majors act based on their duties. And indeed, consumers have responsibility too. However, there are ethical questions about how much responsibility they actually have for the harm caused by their emissions, which are, in the grand scheme of things, minuscule. According to the International Energy Agency (2021) individual behavioural changes would only account for about 4% of the reductions in GHGs needed to achieve a net-zero target by 2050. Additionally, there are positive moral questions regarding individual responsibility, given the political and economic constraints on action as well as the oil industry's entrenched mindset of deflecting blame by framing the question of climate change as one of individual, consumption-based responsibility and thus preventing the general public from understanding the climate crisis as a structural problem largely driven by the oil industry's denial, misinformation, lobbying and disablement of climate policy.

At any rate, this chapter does not intend to obscure the role of other stakeholders. Rather, the goal is to draw attention to a significant and utterly neglected group of agents, whose unique and distinctive role and responsibility in causing climate change should be translated into much-needed policies to support current climate efforts. Oil companies should play their part in global climate governance, which is adequate and appropriate to their role in causing climate change, along with states, individuals and other agents.

3. Oil Companies' Responsibility

One of the clearest and strongest imperatives of all forms of morality is the 'no-harm' principle (Shue 2015; Mayer 2016). This principle states that agents have a negative responsibility to refrain from acting in certain ways to prevent and/or avoid causing harm to others. The moral imperative to do no harm is central to mainstream notions of justice, and it has shaped and guided societies for generations. Considering empirical evidence of the harm that comes from oil companies' activities, these entities are clearly in violation of their negative responsibility to do no harm. In light of this, it is a societal judgement to individuate the most appropriate forms of positive responsibility as shaped by morally relevant facts associated with the violation of the no-harm principle.

The concept of responsibility raises serious concerns in relation to climate change that should be addressed pluralistically (Caney 2010; Jamieson 2010;

Jamieson 2015) and requires a contextual investigation in order to ground and develop duties applicable and appropriate to the oil industry. It is also worth noting that most authors use 'responsibility' and 'duty' interchangeably (e.g., Shue 2017). This chapter, however, distinguishes between the two concepts and adopts the view that responsibility is a condition that implies an ability to act at one's own will, whereas a duty involves a moral commitment that denotes an active willingness to do or not do something.

This analysis relies on a few conceptual distinctions related to the scope and objectives of the notions of responsibility (Miller 2008; Jamieson 2010; Jamieson 2015; Shue 2015). Responsibility can be 'negative' and require agents to refrain from action (as the responsibility that requires agents to do no harm) or 'positive' and require agents to act in specific ways (the kind of responsibility discussed in this chapter in relation to the actions required of oil companies). Additionally, responsibility can be 'special' and pertain only to some agents (the affected agents; here, the harmed agents) or 'general' and be owed to all humanity and possibly to the earth. Another distinction is between 'backward-looking' responsibility (that demands that agents act based on something that has occurred in the past) and 'forward-looking' responsibility (that implies that agents act because they are in the position to do something to improve the situation for the future). This chapter also distinguishes between 'causal' and 'moral' responsibility. Causal responsibility can be understood as 'causal contribution', while a more stringent notion of moral responsibility is based on the appraisal of agents' intentions and assesses their voluntariness, control and knowledge. These conceptual distinctions are important but should not be overstated since they are often blurred when applied to specific issues.

Oil companies' positive responsibility ought to be established in a pluralistic and non-arbitrary way to justify and outline their consequent duties. To this end, it is first necessary to point out the morally relevant facts related to oil companies' activity, which determine their positive responsibility and shape their consequent duties. Presenting the facts helps clarify the conduct of oil companies and the moral context within which they operate. The morally relevant facts listed below provide a normative foundation for oil companies' positive responsibility for causing climate change and the consequent duties they have for addressing the climate crisis (for the full specification of facts 2 through 6, see Grasso 2022, chapter 2):

- Fact 1: The largest 60 oil companies contributed to more than 40% of all global industrial emissions between 1988–2015 (Carbon Majors Database – 2017 Dataset Release). According to the 2018 Carbon Majors Database, just 10 oil and gas companies accounted for almost

25% of all global industrial emissions between and just 15 for almost 30% between1965–2018.

- Fact 2: Some oil companies have had knowledge about the harmful effects of burning fossil fuels in causing climate change (CIEL 2017). For instance, at the celebration of the one-hundredth anniversary of the world's first commercial oil well in 1959, organised by the American Petroleum Institute in New York, the renowned physicist Edward Teller warned oil executives, government officials and scientists with startling prescience about the correlation between carbon dioxide and global warming.
- Fact 3: Most of Big Oil's emissions were released between 1988–2015 (Carbon Majors Database – CDP Carbon Majors Report 2017). Additionally, the five largest IOCs – BP, Chevron, ExxonMobil, Shell and TotalEnergies – plan to invest around $3.5 billion (only 3 percent of their 2019 capital expenditures) in low-carbon technologies, while roughly $110.5 billion will be put into oil and gas exploration and production (InfluenceMap 2019).
- Fact 4: Big Oil had the possibility to reduce the harmful effects of its business and to adjust its business model to become less carbon-intensive; some investor-owned oil corporations had this opportunity over 40 years ago (CIEL 2017). At the end of the 1980s, the US oil industry owned or controlled the largest share of solar panel production in its homeland, maintaining its prominence in this technology well into the 2000s. If these technologies had been developed and deployed, the oil industry could have had a major impact on reducing carbon emissions and accelerating the shift toward a low-carbon future. But the prospects of the higher costs of carbon-saving technologies, at least initially, slashing the oil industry's profits meant that it chose to not go down this path.
- Fact 5: Leading investor-owned oil companies actively opposed and, in many cases, successfully prevented policies to reduce GHG emissions and, in some countries, funded climate denial efforts (Oreskes and Conway 2011; Frumhoff et al. 2015). The evidence of the oil industry's denial is overwhelming.
- Fact 6: oil companies have made substantial profits that have greatly increased the wealth of their shareholders through their activities related to fossil fuels (Frumhoff et al. 2015).

Fact 1 suggests that Big Oil has propelled climate change by exploring, producing, refining, distributing and burning fossil fuels. This fact establishes causal responsibility, which is a necessary but not sufficient condition for the more stringent notion of moral responsibility. Moral responsibility requires that agents are aware of the consequences of their actions, can form intentions about their actions and can carry them out (Miller 2004). Since at least since the first Intergovernmental Panel on Climate Change (IPCC) report in 1990,

Big Oil has known about the harmful consequences of its business model (Fact 2). Despite this knowledge, oil companies have released most of their emissions within the past three decades (Fact 3) when they were able to limit their harmful actions (Fact 4). In addition, some oil companies intentionally blocked initiatives to address climate change and funded climate denialism (Fact 5). And all oil companies have accumulated substantial wealth through their fossil fuel-related activities (Fact 6). This latter fact is not in itself morally wrong; however, it is still morally relevant since it strengthens and better clarifies oil companies' responsibilities and duties related to climate change. Fact 6 corresponds to a moral logic that distributes the burden of rectificatory actions in proportion to the benefits derived and also to the ability to pay.

In sum, these facts provide a justification for assigning oil companies moral responsibility for climate change (Grasso, 2020). In particular, it is possible to assign oil companies 'collective' moral responsibility. They are, in fact, conglomerate collectivities, whose 'identity is not exhausted by the conjunction of the identities of the persons in the organization' (French 1984, 13). Conglomerate collectivities have the following features: (a) an identity larger than the sum of the identities of their members; (b) decision-making structures that enable the inputs of members' judgements to be translated into collective judgements as outputs; (c) consistency over time; and (d) self-conception as a unit. Accordingly, oil companies are indeed conglomerate collectivities, which can qualify as moral agents and, therefore, can have different forms of responsibility.

Based on these morally relevant facts, and in line with the notion of moral responsibility of collective entities enunciated above, the oil industry must be held morally responsible for their contributions to causing climate change. Specifically, these facts justify assigning oil companies with positive, special, backward- and forward-looking moral responsibility for climate change.

4. Articulation of oil companies' duties of reparation and decarbonisation

Such a composite notion of oil companies' positive responsibility is a normative construct focused on their conduct and intentions in the context of the violation of the no-harm principle. It provides the moral basis for duties compelling oil companies to act in certain ways: the duties of reparation and of decarbonisation. These duties should be understood as informal 'sanctions' imposed by the moral nature of the oil industry's responsibility for climate change (Jamieson 2015) and are grounded in corrective justice which, originating from wrongful harm-doing, helps focus on the past and present harm caused by oil companies and elaborate on the resulting actions required to rectify injustices produced by such harm (Meyer and Roser 2010).

The duties of reparation and decarbonisation resonate with the core claim of climate justice movements. These claims demand, by and large, that richer agents, including corporations, repay their climate debt, divided into an 'impacts debt' and an 'emissions debt'. The ultimate objectives of this request – which, to a large extent, are consistent with the UNFCCC's core ethical ambitions – are to take democratic control over the economy, govern climate change in a participatory way and lessen the injustices involved. The impacts debt – embodied by the duty of reparation – implies, by and large, a rectification of the harm brought about by climate change, while the emissions debt – inscribable in the duty of decarbonisation – requires action to reduce carbon emissions and associated future harms, possibly in conjunction with some form of historical contribution to the problem as demanded, for instance, by the Lofoten Declaration.

To articulate the corrective justice perspective in relation to oil companies' duties of reparation and of decarbonisation, it is necessary to identify:

1. The duty-bearers (i.e., the agents who should bear the financial and other burdens of rectificatory actions);
2. The moral basis of the injustice (i.e., the moral principles that justify and define rectificatory actions);
3. The structure of the duties imposed on oil companies and the forms that rectificatory actions should take (i.e., the concrete means through which rectification of harm done should be carried out); and,
4. The duty-recipients (i.e., the subjects entitled to rectification and the modality of the allocation of the rectificatory actions among them envisaged by the duties).

The rest of this section addresses point (ii) since it is common to both duties of reparation and decarbonisation. The following section addresses points (iii) and (iv) in relation to each of the two duties individuated. A thorough answer to point (i) is pleonastic, since this analysis obviously considers oil companies as duty-bearers and, more broadly, as moral agents.

4.1 The moral basis of the injustice

Point (ii) concerns the moral principles that justify the rectificatory actions included in the duties imposed on oil companies by their positive moral responsibility. The climate ethics literature (e.g., Caney 2005; Shue 2015) usually refers to two backward-looking principles – the 'polluter pays principle' (PPP) and the 'beneficiary pays principle' (BPP) – and one forward-looking principle – the 'ability to pay principle' (APP). The PPP distributes the financial and other burdens associated with rectificatory actions in proportion to past contributions agents have made to the overall level of emissions. The BPP holds instead that proportionality in such a distribution should be calculated based on the benefits that agents have derived from emission-generating activities. Finally, the APP posits that the quota of burdens should be proportional to agents' relative capacity to bear such burdens.

All of the abovementioned principles aim to establish and justify positive responsibilities for sharing the burden of rectifying the unjust situation created by the actions that have caused climate change. Instead of relying on any one principle, this moral analysis employs the hybrid version developed by Shue (2015). Shue (2015, 16) argues that 'those who contributed heavily to creating the problem of excessive emissions thereby both benefitted more than others and became better able to pay than most others'. This convergent principle appears to fit the case of oil companies perfectly and provides a moral justification for their duties of reparation and decarbonisation. This hybrid, convergent understanding of the moral bases of oil companies' duties generates different rectificatory actions included in the duties of reparation and decarbonisation.

5. Duties of reparation and decarbonisation: structure and duty recipients

To prevent harming humanity and the planet, responsibility requires oil companies to undertake actions (a) to better cope with the effects of climate change through rectification of the harm already done and prevention of future harm, and (b) to stop causing climate change through the reduction and eventual termination of their harmful activities. These actions can be articulated respectively in the form of the duties of adaptation and mitigation, as usually discussed in the relevant literature (e.g., Caney 2010; Vanderheiden 2011). The duty of adaptation requires moral agents to support efforts aimed at preventing climate change, adapting to its impacts and compensating for non-adapted or mitigated impacts. The duty of mitigation requires moral agents to curb anthropogenic GHG emissions and/or enhance their sinks to avert dangerous interference with the climate system.

A distinction between duties of adaptation and mitigation is undoubtedly a helpful one in the general context of climate ethics. However, in the specific case of oil companies, these duties require a more contextualised and nuanced interpretation. The current analysis articulates the duties of reparation and decarbonisation as specific manifestations of the duties of adaptation and mitigation, respectively. These names reflect and emphasise the kinds of actions required of oil companies in light of their moral responsibility and unique agency related to climate change.

5.1 The Duty of Reparation

5.1.a Structure of the duty and forms of rectificatory actions

The duty of reparation captures the need to ensure that oil companies rectify injustices faced by those who undeservedly suffer harm from climate change caused by burning fossil fuels (Vanderheiden 2011; Shue 2015). This duty posits that oil companies should 'disgorge' part of the funds they have accumulated from their harmful activities to help those affected by climate change to prevent and/or adapt to climate impacts and to compensate for non-adapted or non-mitigated impacts.

To frame and better understand the duty of reparation, as well as the form it should take – point (iii) – it is useful to consider oil companies as moral agents that, through their harmful fossil fuel-related activities, have benefitted from the suffering of others. According to Pasternak's (2014) categorisation of wrongful beneficiaries, oil companies would be 'voluntary beneficiaries', since they know of the wrongdoing and could have avoided it without incurring unreasonable costs, but which instead have sought and welcomed it (Facts 2, 4 and 6). As 'voluntary beneficiaries', oil companies must rectify the harm done by supporting the affected parties in relation to the harm they caused. There are different ways to support them: from immaterial approaches, like public acknowledgment and apologies, 'naming and shaming' or establishment of the truth, to material rectification of historical wrongdoing. In the context of climate change, much remains to be done in practical terms to reduce the harmful impacts of fossil fuel production. Rectification, therefore, must be primarily material and ought to aim at minimising climate impacts through practical actions.

There are different forms of material rectification, too. For example, restitution implies returning misappropriated things to the rightful owners or their successors, and compensation means compensating the rightful owners or their successors for the harm done. Unfortunately, applying the duties of restitution and compensation is highly problematic considering the complex nature of climate change since both require identification of the recipient of such duty (Goodin 2013). Given the substantial temporal and spatial lags between carbon emissions and their impacts, it is virtually impossible to identify the rightful duty-recipient or a legitimate successor with certainty. Moreover, in the case of restitution, the context of climate change makes it close to impossible to identify the 'misappropriated thing' apart from a rather abstract notion of atmospheric absorptive capacity, which was wrongfully overconsumed by carbon majors' emissions.

Whereas restitution and compensation approaches fail, disgorgement appears to be more appropriate. Disgorgement requires only the relinquishment of the fruits of historical wrongdoing: in the case of carbon majors, their tainted benefits. Unlike restitution and compensation, the disgorgement form of rectification focuses on the duty-bearer, not the duty-recipients and their welfare (Goodin 2013). A remarkable example of implementing the moral provisions of disgorgement has already occurred in the case of art stolen by the Nazis from heirless Jews during World War II. After the war, the art was sold, and the proceedings were put into a fund providing support to Holocaust survivors (O'Donnell 2011). Disgorgement does not require the identification of a particular duty-recipient or speculation over how she would have been today had the past wrong not occurred. The potential and the advantage of disgorgement lies in its informational parsimony that makes it much more feasible, especially in the complex situations created by climate change.

It is worth noting that not all benefits that are attributable to oil companies' historical wrongdoing should be viewed as 'tainted'. For example, tainted benefits would not include charity donations or benefits to communities that emerged as a result of oil-related operations. A satisfactory theoretical proxy and a sound pragmatic measure for oil companies' tainted benefits can be their profits; yet, not even all profits would count as such. In case of the oil industry, the notion of wrongdoing reasonably applies to their emissions since 1992 (the presentation of the first IPCC assessment report at the Rio Conference). After this point in time, ignorance about the consequences of carbon emissions and alleged impotence of oil companies to reduce them became inexcusable. The profits of oil companies since 1992 offer a practical measure of the tainted benefits they should disgorge.

5.1.b Duty recipients

Finally, to articulate the duty of reparation, it is necessary to identify who should be entitled to the disgorged funds. The agents most vulnerable to the harmful impacts of climate change should be the rightful duty-recipients. Vulnerability to climate change impacts is not simply about the risks of harmful events occurring. Rather, it is about the preparedness and capacity of different groups to cope with these effects. In this light, it is useful to clarify the notion of vulnerability, which, applied to social systems, is also termed social vulnerability (Brooks et al. 2005). Social vulnerability could be broadly understood as a state of well-being pertaining directly to individuals and social groups. Its causes are related not only to climate impacts but also to social, institutional and economic factors, such as poverty, class, race, ethnicity, gender, etc. (Paavola and Adger 2006). Social vulnerability produced by climate impacts endangers a number of critical aspects of well-being, such as life, health, livelihood, etc.

The degree of social vulnerability can be used for defining duty recipients' level of entitlement to the disgorged funds: the greater their social vulnerability, the larger the rectification through disgorged funds. Shue's (1999) third general principle of equity clearly endorses a stringent normative imperative of putting the most socially vulnerable first.

At the same time, there is another group of vulnerable agents, perhaps not subject to actual climate harm, but who could suffer a different kind of loss deriving from the shrinking financial capacity that the duty of reparation imposes to the oil industry (and, indeed, from commitments to the low-carbon transition required by decarbonisation). These agents are the displaced workers of the industries – fossil fuel and other industries, such as chemicals, transport and shipping – damaged in terms of job loss/reduction of opportunities by this transition, as well as frontline communities along the fossil fuel supply chain: they can be defined as direct victims of a low-carbon transition (Sovacool 2021). It should be emphasised that the inclusion among duty-recipients of displaced workers and impacted communities enlarges the scope of the duty of reparation beyond the strict moral boundaries of the financial rectification of the harm generated by fossil fuel-related activities. The rationale for this choice is eminently pragmatic; on the one hand, a wider scope greatly increases the acceptability and feasibility of the duty of reparation; on the other hand, the establishment of a separate fund for displaced workers and impacted communities would probably be too cumbersome for the already overburdened international governance of climate change.

In practical terms, reparation can take the form of a fund similar in its objectives to the Earth Atmospheric Trust envisaged by Barnes et al. (2008) aimed at helping people most vulnerable to climate change impacts.

5.2 The duty of decarbonisation

5.2.a Structure of the duty and the form of rectificatory actions

To address the harm produced by its fossil fuel-related activities, the duty of decarbonisation requires the oil industry to eliminate carbon emissions from its business model (Shue 2017). Decarbonisation means adopting non-carbon intensive business models to eliminate carbon emissions from companies' operations and products. To decarbonise its products, an oil company would have to either cease its operations completely or transition to distributing low- or zero-carbon-intensive products, such as renewable energy. Such efforts would be consistent with the mounting pressure for phasing out fossil fuels (Grasso 2022).

A broad understanding of decarbonisation should not be confused with two narrower interpretations. One would only compel oil companies to comply with binding emissions limits set by some legitimate political and regulatory bodies (e.g., states, environmental agencies, local, national, regional, international authorities with enforcement power, etc.). This narrow commitment to decarbonisation depends on the willingness of legitimate authorities to set and enforce binding emissions limits, while a broader notion of decarbonisation entails much thornier governance-related behavioural and institutional issues. The second narrow interpretation implies only decarbonisation of oil companies' operations, like reducing the carbon footprint of their offices around the world. Some companies have already engaged into such actions, which in essence have served the purpose of 'greenwashing' their image. The famous case of BP rebranding itself from 'British Petroleum' to 'Beyond Petroleum' is one such example (Pearce 2008). Decarbonising operations (and not products) of oil companies is clearly insufficient, considering that these companies distribute fossil fuels to the global economy.

Carbon emissions largely result from the use of the oil industry's products by various agents, which contributes to the atmospheric greenhouse effect (see Buizza in this volume). Considering that carbon emissions are the commonly accepted 'currency' of climate justice, framing and accounting for the burden of decarbonisation imposed on oil companies in terms of emissions is the logical course of action. In this light, decarbonisation implies extensive and systematic reductions in the carbon emissions generated by the products and activities of oil companies.

Such burdens should be distributed among oil companies proportionally to their cumulative emissions, which represent a measure of their harm-generating activity over time (Grasso 2012). For example, the oil companies that contributed the most to global emissions should curb their fossil fuel-related activities at a higher rate, with a speeder pace and with larger reductions than other oil companies. Any 'carbon allowances' that may be assigned to oil companies according to this logic should be gradually reduced to zero over time.

5.2.b Duty recipients

Given the global nature and spatial unpredictability of harm-reduction gene-rated by oil companies' decarbonisation, all of humanity is the duty-recipient.

6. Possible developments

If oil companies act on their responsibilities and duties, there may be different possibilities for their actions to unfold. The harshest (and least likely) possibility would involve an abrupt dissolution of oil companies as a result of immediate termination of their fossil fuel-related activities. Let us call this option 'Sudden End'. From the perspective of justice, this abrupt termination would help prevent harm from any future fossil fuel-related activities. However, at the same time, it would rob victims of climate change from fair reparations for their suffering and for adapting to non-mitigated consequences of climate change. The 'Sudden End' scenario would also put in jeopardy some of the more vulnerable shareholders of the oil industry, such as pension funds and their individual account-holders. Thus, though attractive from the perspective of preventing future harm, this scenario is not functional from the point of view of disgorgement. In fact, there appears to be no ideal scenario from a justice perspective – all possible courses of action imply some degree of compromise among different justice concerns.

Another possibility would imply phasing out fossil fuels from oil companies' operations and products more gradually. Let us call this scenario the 'Just Transition'. Compensation and obligations towards more vulnerable shareholders make a strong case in favour of 'keeping oil companies alive' to ensure they do the maximum of what justice requires of them. This scenario would be less disruptive than the 'Sudden End' to the fossil fuel-dependent global socio-economic system, including the interests of some states (especially in the case of NOCs) and other businesses (which rely on fossil fuels, such as chemical or automotive industries). This does not change, though, the ultimate goal of the 'Just Transition', which is complete phasing out of fossil fuels from oil companies' operations and products, over the period of several decades.

The 'Just Transition' can take various shapes in terms of length and a combination of decarbonisation, compensation, business-as-usual (BAU), offsetting emissions, etc. The range of possible transition scenarios could vary from slow and ineffective BAU coupled with 'greenwashing' efforts, to BAU coupled with enhanced compensation efforts, or more rapid phasing out fossil fuels and switching to other, non-carbon-intensive business models. Notably, trade-offs between the duties of reparation and decarbonisation are inevitable. In practical terms, oil companies have finite budgets and will need to prioritise the most appropriate course of action. Yet, it is difficult to argue in favour of one strategy over another in abstract terms: both reparation and 'full' decarbonisation are critical from the justice perspective. Future research could address this conundrum in a contextualised way and offer a more nuanced exploration of the relative weight of each duty. More in-depth theoretical discussion of various perspectives on 'Just Transition', as well as its practical policy implications, is also necessary.

Conclusion

Oil companies' activities of exploration, extraction, refining, use and distribution of fossil fuels generate emissions of GHGs that are harmful for the planet and for humanity. This chapter maintains that oil companies have a positive responsibility to reduce and eventually stop their harmful activities and to rectify the harm they have caused. Such responsibility originates from oil companies' violation of the no-harm principle, which compels moral agents to refrain from acting in certain ways in order to prevent and/or avoid causing harm to others. This analysis articulates oil companies' responsibility in the form of two duties: a duty of reparation and a duty of decarbonisation.

The duty of reparation requires that oil companies disgorge their tainted benefits as an appropriate form of rectification of their historical wrongdoing. From a practical standpoint, this can be achieved by putting their post-1992 profits, a satisfactory theoretical proxy of the tainted benefits, in a fund aimed at helping the people most socially vulnerable to climate change to cope with its impacts. The duty of decarbonisation requires oil companies to engage in a large-scale transformation to radically alter their business model and progressively eliminate all carbon emissions from their operations and products.

By specifying and vindicating the duties of reparation and decarbonisation, this analysis aims to contribute to the creation of a normative basis needed to justify the inadequacy of the prevalent socio-economic practices of the oil industry in the broader context of the moral progress of humanity (Jamieson 2017). Condemning these practices as morally unacceptable could lead to the emergence of a social norm, which would delegitimise the current fossil fuel-centred behaviour of the oil industry, as happened for other, once deeply entrenched and influential socio-economic practices, such as slavery (Finnemore and Sikkink 1998).

In the absence of a moral analysis of the role of the oil industry in climate change, a normative perspective that justifies and outlines the responsibility and consequent duties of oil companies could provide a helpful initial normative framework for a reasoned dialogue with civil society and amongst political representatives belonging to different political traditions and subject to different political constraints. Despite their alleged abstractness, the duties of reparation and decarbonisation are moral provisions with immediate relevance to international climate governance.

Table

Table 12.1: Top 15 (10) oil companies' contributions to global GHG emissions between 1965–2018, measured in megatonnes of carbon dioxide equivalent ($MtCO_2e$) and percent of global industrial GHG emissions (in the same period, the latter amount to 1,410,737 $MtCO_2e$). The largest share (roughly 90%) of oil companies' global GHG emissions originated from downstream combustion (for energy and non-energy purposes) of oil and gas that Big Oil distributed within the global economic system. These emissions are defined by the Greenhouse Gas Protocol of the World Resources Institute (WRI) as scope 3 emissions. Source: Carbon Majors 2018 Data Set (released December 2020). https://climateaccountability.org/carbonmajors_dataset2020.html

Oil Company	Emissions ($MtCO_2e$)	Percent	Typology
Saudi Aramco	61,143	4.33%	NOC
Gazprom (Russia)	44,757	3.17%	NOC
Chevron (USA)	43,787	3.10%	IOC
ExxonMobil (USA)	42,484	3.01%	IOC
National Iranian Oil	36,924	2.62%	NOC
BP (UK)	34,564	2.45%	IOC
Royal Dutch Shell (UK/Netherlands)	32,498	2.30%	IOC
Pemex (Mexico)	23,025	1.63%	NOC
Petro China / China National Petroleum	16,515	1.17%	NOC
PDVSA (Venezuela)	16,029	1.14%	NOC
ConocoPhillips (USA)	15,422	1.09%	IOC
Abu Dhabi National Oil	14,532	1.03%	NOC
Kuwait Petroleum	13,923	0.99%	NOC
Iraq National Oil	13,162	0.93%	NOC
TotalEnergies (France)	12,755	0.90%	IOC
TOTAL 15 (Top 10)	**421,520** **(351,726)**	**29.88%** **(24.93%)**	

References

Backstrand, K., Kuyper, J.W., Linner, B.O. and Lovbrand, E. 2017. "Non-state actors in global climate governance: from Copenhagen to Paris and beyond". *Environmental Politics* 26 (4): 561–579.

Barnes, P., Costanza, R., Hawken, P., Orr, D., Ostrom, E., Umaña, A. and Young, O. 2008. "Creating an earth atmospheric trust". *Science* 319 (5864): 724.

Brooks, N., Adger, W. N. and Kelly, P. M. 2005. "The determinants of vulnerability and adaptive capacity at the national level and the implications for adaptation". *Global Environmental Change* 15 (2): 151-163.

Burger, Horton, M. R. and Wentz, J. 2020. "The law and science of climate change attribution". *Columbia Journal of Environmental Law* 45, 1–185.

Burger, M., and Wentz J. 2018. "Holding fossil fuel companies accountable for their contribution to climate change: Where does the law stand?". *Bulletin of the Atomic Scientists* 74 (6): 397–403.

Caney, S. 2005. "Cosmopolitan justice, responsibility, and global climate change". *Leiden Journal of International Law* 18(4), 747–775.

Caney, S. 2006. "Environmental degradation, reparations, and the moral significance of history". *Journal of Social Philosophy* 37 (3): 464–482.

Caney, S. 2010. "Climate change and the duties of the advantaged". *Critical Review of International Social and Political Philosophy* 13 (1): 203–228.

CDP 2017. The Carbon Majors Database – CDP Carbon Majors Report 2017. https://www.cdp.net/en/articles/media/new-report-shows-just-100-companies-are-source-of-over-70-of-emissions

CIEL – Center for International Environmental Law 2017. *Smoke and Fumes. The Legal and Evidentiary Basis for Holding Big Oil Accountable for the Climate Crisis*. Washington and Geneva: CIEL.

Ekwurzel, B., Boneham, J., Dalton, M. W., Heede, R., Mera, R. J., Allen, M. R. and Frumhoff, P. C. 2017. "The rise in global atmospheric CO_2, surface temperature, and sea level from emissions traced to major carbon producers". *Climatic Change* 144 (4): 579–590.

Finnemore, M. and Sikkink, K. 1998. "International norm dynamics and political change". *International organization* 52 (4): 887–917.

French, P. 1984. *Collective and Corporate Responsibility*. New York, NY: Columbia University Press.

Frumhoff, P. C., Heede, R. and Oreskes, N. 2015. "The climate responsibilities of industrial carbon producers". *Climatic Change* 132 (2): 157–171.

Goodin, R. E. 2013. "Disgorging the fruits of historical wrongdoing". *American Political Science Review* 107 (3): 478–491.

Grasso, M. 2012. "Sharing the emission budget". *Political Studies* 60 (3): 668–686.

Grasso, M. 2020. "Towards a broader climate ethics: Confronting the oil industry with morally relevant facts". *Energy Research & Social Science* 62: 101383.

Grasso, M. 2022. *From Big Oil to Big Green. Holding the Oil Industry to account for the Climate Crisis*. Cambridge MA: MIT Press.

Grasso, M. and Markowitz, E. M. 2015. "The moral complexity of climate change and the need for a multidisciplinary perspective on climate ethics". *Climatic Change* 130 (3): 327–334.

Heede, R. 2013. *Carbon Majors: Accounting for Carbon and Methane Emissions 1854 2010. Methods and Results Report*. Available from: http://carbonmajors.org/download-the-study/

Heede, R. 2014. "Tracing anthropogenic carbon dioxide and methane emissions to fossil fuel and cement producers, 1854–2010". *Climatic Change* 122 (1-2): 229–241.

Heede, R., and Oreskes, N. 2016. Potential emissions of CO_2 and methane from proved reserves of fossil fuels: an alternative analysis". *Global Environmental Change* 36: 12–20.

IEA 2021. *Net Zero by 2050. A Roadmap for the Global Energy Sector*. Paris: IEA

InfluenceMap. 2019. "Big Oil's Real Agenda on Climate Change. How the Oil Majors Have Spent $1Bn after Paris in Narrative Capture and Lobbying". March, https://influencemap.org/report/How-Big-Oil-Continues-to-Oppose-the-Paris-Agreement-38212275958aa21196dae3b76220bddc

IPCC 1990. Climate Change: *The IPCC Scientific Assessment*. Geneva: IPCC.

IPCC 2014. *Climate Change 2014: Synthesis Report. Contribution of Working Groups I, II and III to the Fifth Assessment Report of the Intergovernmental Panel on Climate Change*. Geneva: IPCC.

Jamieson, D. 2010. "Climate change, responsibility, and justice". *Science and Engineering Ethics* 16 (3): 431–445.

Jamieson, D. 2015. "Responsibility and climate change". *Global Justice: Theory Practice Rhetoric* 8 (2): 23–42.

Jamieson, D. 2017. "Slavery, carbon, and moral progress". *Ethical Theory and Moral Practice* 20 (1): 169–183.

Licker, R., Ekwurzel, B., Doney, S. C., Cooley, S. R., Lima, I. D., Heede, R., and Frumhoff, P. C. 2019. "Attributing ocean acidification to major carbon producers". *Environmental Research Letters* 14 (12): 124060.

Mayer, B. 2016. "The relevance of the no-harm principle to climate change law and politics". *Asia Pacific Journal of Environmental Law* 19: 79–104.

Meyer, L. H. and Roser, D. 2010. "Climate justice and historical emissions". *Critical Review of International Social and Political Philosophy* 13 (1): 229–253.

Miller, D. 2004. "Holding nations responsible". *Ethics* 114 (2): 240–268.

Miller, D. 2007. *National Responsibility and Global Justice*: Oxford: Oxford University Press.

Miller, D. 2008. *Global Justice and Climate Change: How Should Responsibilities Be Distributed*. The Tanner Lecture on Human Values, Delivered at Tsinghua University, Beijing, March 24–25, 2008.

O'Donnell, T. 2011. "The restitution of Holocaust looted art and transitional justice: The perfect storm or the raft of the Medusa?". *European Journal of International Law* 22 (1): 49–80.

Oreskes, N., and Conway, E. M. (2011). *Merchants of Doubt: How a Handful of Scientists Obscured the Truth on Issues from Tobacco Smoke to Global Warming*. Bloomsbury Publishing USA.

Paavola, J. and Adger, W. N. 2006. "Fair adaptation to climate change". *Ecological Economics* 56 (4): 594–609.

Pasternak, A. 2014. "Voluntary benefits from wrongdoing". *Journal of Applied Philosophy* 31 (4): 377–391.

Pearce, F. 2008. "Greenwash: BP and the myth of a world 'Beyond Petroleum'". *The Guardian*, 20 November 2008. Available from: https://www.theguardian.com/environment/2008/nov/20/fossilfuels-energy

Shue, H. 1999. "Global environment and international inequality". *International Affairs* 75 (3): 531–545.

Shue, H. 2015. "Historical responsibility, harm prohibition, and preservation requirement: Core practical convergence on climate change". *Moral Philosophy and Politics* 2 (1): 7–31.

Shue, H. 2017. "Responsible for what? Carbon producer CO_2 contributions and the energy transition". *Climatic Change* 144 (4): 591–596.

Sovacool, B. K. 2021. "Who are the victims of low-carbon transitions? Towards a political ecology of climate change mitigation". *Energy Research & Social Science* 73: 10191.

UNFCCC – United Nations Convention on Climate Change 1992. Available from: https://unfccc.int/process/the-convention/what-is-the-united-nations-framework-convention-on-climate-change

Vanderheiden, S. 2011. "Globalizing responsibility for climate change". *Ethics & International Affairs* 25 (1), 65–84.

Welsby, D., Price, J., Pye, S. et al. 2021. "Unextractable Fossil Fuels in a 1.5°C World". *Nature* 597: 230–234.

Wenar, L. 2015. "Coercion in cross-border property rights". *Social Philosophy and Policy* 32 (1): 171–191.

13

Legitimate Expectations about Stranded Fossil Fuel Reserves: Towards a Just Transition

RUTGER LAZOU

While the energy transition is needed now more than ever, it also brings adverse consequences for some agents. The question arises of whether these transitional losers are owed any kind of transitional aid. This chapter answers this question by developing a harm-based account of legitimate expectations, according to which one should not harm others by wrongfully causing false expectations. It applies this to the transitional losses of fossil fuel owners, companies and states, who see their reserves becoming worthless if the energy transition succeeds. Without entering into the parallel issue of the past injustices committed by fossil fuel producers (which may take the form of compensatory duties towards the victims of climate change), this chapter argues that private actors should receive transitional aid from states for having been provided with false expectations about regulatory stability, while states cannot make a similar claim on the global level. Transitional aid should be given in the form of compensation that is limited to investment costs induced by false expectations.

Introduction

When circumstances and knowledge change, we sometimes need to change the world or how we behave. When the changes are large-scale and desirable given a particular set of values and principles, we can speak of a transition (Hölscher et al. 2018). One of the biggest challenges humanity has ever faced, and the focus of this chapter, is the transition to a low-carbon society and economy (Edmond 2020). At least since 1995 – when the Intergovernmental Panel on Climate Change (IPCC) published their second report (Gosseries 2004, 7; Meyer and Sanklecha 2011, 460) – the world has been confronted with the knowledge of the dangerous and irrevocable effects of global warming due to the emissions from fossil fuel use: extreme weather events, ruined habitats, higher sea levels, drought, crop failure, heatstroke, increased incidence of diseases, higher weather unpredictability, etc. Combatting climate change, therefore, would avoid immense harms. Even if we leave out these harms, moreover, a zero-carbon economy might be more prosperous in the long run than a high-carbon economy (Fay 2015, 154). Even on a national scale, in the medium to long term, this outcome is likely to be net beneficial.

In the short term, however, for many agents, the energy transition will bring significant 'transitional losses': benefits that cannot be realized because of the transition. When an industry or corporation has to stop or reduce its activity, the owners of the corporation suffer from reduced stock valuations and lower profits, workers lose their jobs, the communities and states in which these corporations are established suffer from decreased economic activity, losses in tax revenue and increased expenditures on social transfers for newly unemployed workers, consumers can no longer consume the corporations' products and suppliers' resources lose value (Green and Gambhir 2020, 4–7). In economic terms, the energy transition will affect the value of their assets. Assets are resources that have value because they will benefit its owner(s) in the future. They could refer to various inputs to production and sources of wealth, including capital, labor, and natural endowments (Colgan, Green, and Hale 2021, 586). Owning these resources has beneficial consequences of an enduring nature: it brings benefits or burdens in the long run. Unfortunately, this makes them vulnerable to being stranded, which means that these assets 'have suffered from unanticipated or premature write-downs, devaluations, or conversion to liabilities' (Caldecott et al. 2013, 7; Caldecott 2017, 2) or, in other words, they 'los[t] economic value well ahead of their anticipated useful life' (Generation Foundation 2013).

While we are mostly concerned with the winners of the energy transition, the question arises of what is owed to those who are disadvantaged by the transition, to those who see their assets being stranded. One way to respond to these losses is reformative: one determines what justice requires, implements the necessary changes and lets all losses (and gains) lie where they fall (Green 2019, 3). The opposite, conservative approach favours preserving the pre-existing situation or status quo, which is also called grandfathering (Damon et al. 2019, 1; Knight 2013, 1; Knight 2014, 571; Schuessler 2017, 141). Conservative measures can vary from providing transitional aid, exemptions from new rules or paying compensation. Whether and how to deal with the potential production that is foregone or reduced, Pye et al. (2020, 2) contend, is an important challenge for reaching a just transition. Similarly, Green and Gambhir (2020, 2) stress that the losses caused by the energy transition 'raises complex normative and political questions about which of these burdens on which kinds of agents and groups should be mitigated, and how this should be done'. Kartha et al. (2016) also consider the relevance of these transitional losses as something that must be further investigated.

In trying to justify the implementation of conservative measures, authors have been arguing that high-emitters have acquired a right to their level of emissions (Bovens 2011), that the status quo is relevant because it influences how good or bad the consequences of a change are (Knight 2014) or that some grandfathering might be required to reach a political agreement. However, these accounts have been overwhelmingly rejected (Caney 2009; Gosseries and Hungerbühler 2006; Schuessler 2017). An emerging response to the question of how to deal with transitional losses focuses on the concept of legitimate expectations (LE). LE was first mentioned in the 1970s by Buchanan (1975) and Rawls (1971) but only received significant attention approximately 10 years ago by being applied to different topics and contexts: administrative law (Brown 2011; Brown 2012; Brown 2017; Brown 2018), territorial rights (Moore 2017; Waligore 2017), immigration (Carnes 2020), punishment (Matravers 2017), hospitality and membership (Weinman 2018) and climate change (Meyer and Sanklecha 2011; Meyer and Sanklecha 2014). The concept is still under development and subject to debate. This chapter will contribute to this discussion by providing an account of LE that can indicate the normative relevance of transitional losses.

In doing so, it focuses on a particular set of climate change-related transitional losses – the losses of fossil fuel reserve owners, because these present not only a large but also a clearly defined category of potentially stranded assets. These reserves can be considered as assets: when they are extracted from the ground, they can deliver energy for domestic use or be an important source of revenue (Caney 2016). When they should be left under the ground, however, they get stranded. Given that fossil fuel reserves can be owned by states or companies, this chapter investigates both whether states can refer to their expectations to justify the protection of their reserves from being stranded or any other form of transitional aid and whether companies can do so.

An example of a state where the argument readily applies is Australia. Compared to other countries, the world's second-biggest exporter of coal has been doing little to reduce pollution. The government faces pressure to take measures, but instead of phasing out coal production, they are committed to digging for more (Mao 2021). LE might justify the continuation of their plans.

An example in which companies might be able to use the argument is found in Norway. Oil companies currently pump out over 1.6 million barrels of oil a day from their offshore operations. It has been estimated that there are still 1–3 billion barrels of oil under the seabed of the Lofoten archipelago. When the parliament decided to withdraw their support for oil exploration, oil companies referred to their expectations and interest in long-term planning to oppose the change. The head of the Norwegian Oil and Gas Association reacted as follows: 'The whole industry is surprised and disappointed. [The government] does not provide the predictability we depend on' (The Independent 2019). Before assessing these arguments, I first indicate which reserves should stay under the ground according to considerations of inter- and intragenerational justice (section 1). I then develop a harm-based account of LE (section 2). Finally, I investigate whether the argument works for fossil fuel reserve owners, both on the level of states and companies (section 3).

1. Towards a Fair Fossil Fuel Reserves Stranding

Before appealing to the normative relevance of their expectations, fossil fuel owners could consider whether justice actually requires the stranding of their reserves. If this is not the case, they do not need to refer to their expectations at all. The argument of LE is only relevant for fossil fuel owners whose reserves should stay under the ground according to more general principles of justice. In what follows, therefore, I sketch what a fair fossil fuel stranding consists of.

To begin, let us consider the kinds of agents that own fossil fuel reserves. First, there are (public or private) investor-owned companies. These companies are registered and operate in a state (or in multiple states, in case of multinational companies) but are not owned by the state itself. State-owned or government-owned companies are a second category. These companies are created by a government to undertake commercial activities on the government's behalf (Kenton 2019). Heede and Oreskes (2016) distinguish a third category of owners: nation-states, in which reserves are managed by the state directly. A ministry or an administrative branch operates instead of a company. Both in the case of state-owned companies and nation-states, states are the owners of the reserves. They own 90% of the remaining proven recoverable resources (Heede and Oreskes 2016). BP (2018) also concluded that the majority of the proven reserves of coal, oil and gas are controlled by state-owned companies like Saudi Aramco and governments directly. The normative relevance of the expectations of states, therefore, is potentially more far-reaching than the relevance of companies' expectations. Nevertheless, I investigate whether the LE argument works for both kinds of agents.

Note that I presuppose the traditional view that natural resources are subject to the sovereignty of states. Other agents like companies can own these resources if it follows from a voluntary transfer or agreement. Alternative views consider the agents that own the property under which the resources are located (private individuals, groups of individuals or companies) as the rightful owners (Caney 2016, 22) or argue that natural resources are owned by all humanity (Beitz 1979, 136–143). I stick with the traditional view, as I am concerned with transitional losses as a consequence of the energy transition and not with losses as a consequence of rethinking ownership claims. Importantly, the stranding of one's resources is compatible with this view. As Caney (2016, 23) explains, 'it is widely recognized that there are moral limits on what states may do with the natural resources in their jurisdiction'.

Since the effects of climate change already exist, there is a duty towards both presently living and future generations to avoid the dangerous consequences of climate change. At the twenty-first Conference of the Parties (COP21) in Paris in 2015, 195 countries agreed that the average global temperature rise should be kept 'well below' 2°C above pre-industrial levels and that they will 'pursue efforts' to avoid an increase of 1.5°C (UNFCCC 2015, 3, article 2). From this, we can infer the remaining carbon budget – the amount of greenhouse gas emissions that can still be emitted (Meinshausen et al. 2009). Realizing the 2°C target with a 67% chance of success would leave us with a remaining budget of 1,150 gigatonnes of carbon dioxide ($GtCO_2$) starting from 2020. To not exceed 1.5°C of warming, this would be 400 $GtCO_2$. Higher or lower reductions in non-CO_2 emissions could slightly increase or decrease these carbon budgets (IPCC 2021, 29).

Unburnable carbon, then, refers to the fossil fuels that could have been burnt if there would be no need for climate change mitigation, given the availability of fossil fuel reserves (Carbon Tracker Initiative 2011, 2013, 2017; Heede and Oreskes 2016). The amount of unburnable carbon is difficult to determine as there are huge uncertainties about future production and revenues (Pye et al. 2020, 2). Burning all resources that are recoverable over all time with both current and future technology, irrespective of current economic conditions, would emit 11,000 $GtCO_2$ – 11 times the carbon budget (McGlade and Ekins 2015, 187–188). Fully producing the world's *reserves*, i.e., resources that are proved to be recoverable under current economic conditions and a specific probability of being produced (McGlade and Ekins 2015; Society of Petroleum Engineers 2008), would lead to an estimated 2,734.2 $GtCO_2$ (Heede and Oreskes 2016, 15), which is around three times the budget.

In determining whose reserves should be stranded and for whom the LE argument, therefore, is most important, there are several criteria to consider. Distributing the remaining permissible production benefits according to countries' number of inhabitants, to begin with, follows from the basic idea that each counts as one (Singer 2010, 190). However, there are a couple of reasons to deviate from this equal per capita distribution. Consider countries' past productions. People living in countries with high levels of past productions have already enjoyed benefits earlier in their lives, which entitles them to fewer benefits in the future. Fossil fuel production in the past also yielded benefits that are still present today, including the provision of infrastructure like schools, hospitals, streets and railroads (Meyer 2013, 605–607). This implies that developing countries should benefit most from the remaining carbon budget and that reserves should be stranded most in developed countries – even though some developing countries have also produced significant amounts (Heede 2014, 328). Considering needs leads to similar conclusions in favour of developing countries if one focuses on countries' level of development, as measured by, for instance, the Human Development Index (HDI) (Bos and Gupta 2019, 2; Caney 2016, 27–31; Lenferna 2018, 219; Roser and Seidel 2016, 156). Finally, efficiency matters too. The carbon intensity of a fossil fuel refers to the amount of carbon that is emitted per unit of energy its combustion delivers. This is determined by the carbon intensity of the fossil fuel itself, the impact of the extraction process and geographical factors like the distance between the place of extraction and a country's supply and demand centres (Caney 2016, 24; McGlade and Ekins 2015, 189). If we realized the 2°C target in the most efficient way, approximately one-third of the world's oil reserves, half of the world's gas reserves and more than 80% of the world's coal reserves would stay unproduced. The Middle East would carry half of the unburnable oil and gas, the USA and the former Soviet Union states would own half of the unburnable coal (McGlade and Ekins, 2015, 3). Most reserves, thus, should be stranded in these regions and in highly developed countries.

2. Legitimate Expectations: A Harm-Based Account

If a state or company's reserves become stranded, and the stranding is justified given the discussed normative considerations, should they simply accept this and leave their fossil fuels and the benefits they could have yielded underground? One way in which these unrealised benefits could still have some normative relevance is through the expectations one had about them. The focus on expectations is evident if one considers the concept of stranded assets: assets that suffer devaluations that are unanticipated (unexpected) or premature (earlier than expected). Having stranded assets, thus, is nothing more than having unfulfilled expectations. If one wants to investigate the relevance of stranded fossil fuel reserves, therefore, one has to investigate the normative relevance of these reserve owners' expectations.

An expectation that cannot be ignored, that is morally relevant or that counts normatively is called legitimate. It should be fulfilled, or some kind of compensation or an apology is required (Meyer and Sanklecha 2014, 371–372). Imagine, for instance, that two housemates, A and B, have enjoyed for a long time having dinner together on Fridays. They take turns in preparing dinner, and since this Friday it is A's turn, A expects that B will turn up if she prepares dinner. If B does not turn up, this seems morally wrong (Meyer and Sanklecha 2014, 370; Green 2020a, 402). However, not all expectations seem to be normatively relevant. Considering a case originally presented by Simmons (1996, 258) and recently discussed by Green (2020a, 402), imagine that Kant's daily walks create expectations in the Konigsberg housewives that they will be able to set their clocks by his passing. One day, however, Kant decides to stay home to read Rousseau. It seems that Kant did not do anything wrong and that the housewives cannot make any normative claim.

There are a variety of views on what makes an expectation legitimate. My *harm-based account* of LE starts from the bedrock principle of morality that one should not wrongfully harm others. I argue that by causing false expectations, one harms the expecting agent and that this is wrongful if one is morally responsible for having caused the harm. An expectation is legitimate or normatively relevant, consequently, if another agent is morally responsible for having caused the expectation and the subsequent harm. One should either fulfill the expectation in order to avoid having wrongfully harmed the other agent or one should compensate the harm.

In this way, my account focuses on the normative relevance of actual expectations. Most other LE accounts, in contrast, argue that the expecting agent gained a right or entitlement to the content of the expectation because the law (cf. Brown 2018, 54–57), the justice of the expectation (cf. Matravers 2017; Meyer and Sanklecha 2014, 377–387), a society with a just basic structure (cf. Rawls 1971, 10) or a legitimate authority (cf. Meyer and Sanklecha 2014, 375–377) has created the entitlement. In practice, the expector may also expect that he will enjoy these benefits, but having the actual expectation is not necessary for having the entitlement. Green (2020a, 418) calls this an expectation-independent model of LE. Instead of using the term legitimate *expectations*, I argue, it would be better to speak of a specific kind of (legitimate) entitlements that follow from previous acts or interactions if certain conditions are fulfilled. The kind of interactions that generate these entitlements, arguably, are promises or agreements. This chapter does not focus on these promissory entitlements to save assets from being stranded. After all, if we accept that contracts and promises should be just before they give rise to entitlements, this way of arguing could not justify fossil fuel owners benefitting more than their fair share, nor would it work for any other losses that follow from just transitions.

To indicate the normative relevance of actual expectations and how this leads to my harm-based account, I first explain the importance of expectations in people's lives. Expectations are important for how well people's lives go as they enable agents to make plans (Brown 2017, 440; Buchanan 1975, 419–422; Green 2020a, 398; Hodgson 2012, 315; Meyer and Sanklecha 2014, 375). Making plans is deciding on future actions in advance, for instance, whether one will buy a house. Making plans, in turn, is valuable because it enables one to realise them or, in other words, achieve goals: by planning to buy a house, one could save money for it. Realising plans or achieving goals, finally, is valuable as it can further the fulfillment of one's needs/interests, like having a home. Realising these goals can also be a way of exercising autonomous agency (Rawls 1971, 358–360). The goals of states are determined by the citizens they represent and could be, for instance, having a good health care system or having sufficient energy supply. A company's goals could be providing quality customer service or being financially healthy and are determined by its owners.

How do normative claims arise from this? Moore (2017) contends that the human interest in having stable background conditions and future-oriented projects gives rise to the entitlement that one's expectations become fulfilled (Moore 2017, 235–236). In the dinner example, this would mean that B should show up because A has an interest in enjoying his company. However, Moore only explains why B has a positive duty to fulfill the expectation – why it would be laudable if B showed up. In other words, she only grounds a duty of charity/supererogation. The normative relevance of A's expectation, however, seems stronger: B's duty to show up is a matter of justice, A has a *right* to expect that B will show up. My harm-based account of LE conceives the duty to fulfill one's expectations as a stringent negative duty of justice by inferring it from the principle that one should not wrongfully harm others, in which harming refers to the 'thwarting, setting back, or defeating of an interest' (Feinberg 1987, 32). Given the value of having correct expectations for one's ability to make and execute plans, providing false expectations harms the expecting agent. An expectation is legitimate or normatively relevant, therefore, if fulfillment is required to avoid that the expectation-creator wrongfully harmed the expector by having created a false expectation. An important implication is that LE claims require that there is an agent that caused the expectation.

Having a LE does not always imply, however, that the expectation should be fulfilled. If the content of the expectation is unjust, it is impermissible for the expectation-creator to fulfill the expectation, even if this implies that he wrongfully harmed the expector. Having created an expectation in someone, after all, does not entitle one to commit injustices. In these cases, the expectation-creator should compensate the costs he incurred on the expector by causing the false expectation. Another implication of my harm-based account, therefore, is that in cases of just transitions, LE claims can only justify compensation for the costs of having acted upon false expectations, which have been called preliminary losses (Colla 2017, 298–299) or reliance losses (Robertson 1998, 361). The expected but unrealised benefits themselves, the so-called primary losses (Colla 2017, 298–299) or expectation losses (Robertson 1998, 361) cannot be claimed.

Neither does having a LE imply, thus, that the expector's (disrupted) plans were not morally objectionable. One could object that, if this is the case, the expectation loses its normative relevance and the need for compensation is nullified because the expector does not deserve to be compensated. I disagree with this. The expector's bad intentions, I contend, do not alter the fact that others should not wrongfully harm him and that he is owed compensation if it does happen. The duty not to cause false expectations, in other words, also holds towards morally imperfect agents. If the unjust expectation were fulfilled, the expector would have to compensate the costs he incurred on his victims, but due to the changed circumstances, this is not the case. The expector, then, has *moral luck*: even though his actions were determined by factors beyond his control and not by moral considerations, they can still be assessed as morally decent (Nelkin 2021). He still might be blameworthy, however, for his past actions. Allocating compensatory duties for having committed past injustices is not incompatible with claiming that the same agent could also make normative claims himself.

However, merely harming the expector by causing a false expectation is not enough for giving rise to normative claims: the expectation and the subsequent harm must also be caused wrongfully. An act of harming is wrongful and therefore requires compensation if one is morally responsible for causing the harm (Denaro 2012, 150; Feinberg 1987, 34–35; Feinberg 1990, xxvii–xxix; Thomson 1986, 383). Brown (2017, 2018) has already focused on the responsibility for creating an expectation. He does not, however, clearly articulate how the creation of an expectation is normatively relevant, nor does he explain when one is responsible. Instead of providing criteria for responsibility, he presents four illustrative, inexhaustive ways or 'modes' in which an agent can be responsible. In doing so, he aims to 'step outside the narrow constraints of the doctrine of legitimate expectations as it occurs in both English administrative law and European Community law, for example' (Brown 2018, 64), but his less narrow account, in my view, attributes responsibility and LE claims too easily. Inadvertently, negligently or intentionally causing an expectation or causing an expectation in bad faith can lead to responsibility but does not necessarily do so. These modes explain how one can be causally responsible, but to be morally responsible, additional criteria must be fulfilled. By adding these criteria, it can be explained how intuitively appealing criteria like reasonableness or justice are relevant. Brown's theory, in contrast, has counterintuitive implications, as Green (2020b, 465) indicates: 'If an agent acts irrationally, unreasonably, or viciously in forming an expectation, or if the content of their expectation is unreasonable, immoral, or unjust, this would not affect the determination of legitimacy'. However, instead of rejecting the focus on responsibility for creating expectations, as Green does, the concept of responsibility should be refined. In this way, fossil fuel owners' expectation-related harms justify compensation claims less easily.

According to most philosophers, an agent can be considered morally responsible for an action if the person has control over the action (if the action was avoidable) and if the person could have foreseen its consequences (Rudy-Hiller 2018). Mostly, the creation of an expectation and the subsequent harm is avoidable, as one usually has control over the behavioural and communicative acts that lead to expectations in other agents. The foreseeability condition is less often fulfilled. How people form expectations is determined by many factors that are difficult to foresee: one's other beliefs, cognitive abilities, critical attitude, etc. The expecting agent might also be responsible if she could have avoided the creation of her false expectation by being more careful in forming her beliefs. If A, for instance, forms the expectation that B will invite her to a four-star restaurant because he asked whether she was hungry, B could not have foreseen this and A could have avoided this, making it her responsibility if she gets harmed by having this false belief. In this way, the epistemic validity or reasonableness of the expectation matters: if an expectation is unreasonable (if there are no good reasons for it), its creation is more likely to be unforeseeable by the expectation-creator and avoidable by the expecting agent. Before an expectation can be normatively legitimate, thus, it should fulfill a certain baseline of epistemic validity: epistemic validity limits the responsibility of the expectation-creator. In this way, the focus on responsibility explains our intuition that reasonableness matters, a criterion that is also important in other LE accounts (cf. Gosseries and Hungerbühler 2006, 111; Green 2019; Meyer and Sanklecha 2014, 370; Moore 2017).

While Brown (2017, 2018) is too generous in assessing the legitimacy of expectations, Green (2020a) is too restrictive. Instead of focusing on whether one is responsible for having caused an expectation, Green's practice-dependent LE account only considers an expectation normatively relevant if it follows from a shared norm that governs a social practice in which the relevant actors partake. The social practice of taking turns in preparing dinner, for instance, is governed by the shared norm that the other person will show up. The expectation that B will show up when it is A's turn to cook, then, is legitimate because it follows from this norm. The duty to fulfill the expectation of adhering to the mutually recognised norm, Green contends, is 'similar in nature to the special rights that arise from a promise or contract' (2020a, 404). According to this practice-dependent account, expectations can only be legitimate on the interpersonal level, Green goes further, as expectations about large-scale societal transitions (or stability) do not follow from shared norms that govern social practices in which the regulator and regulated agent mutually participate.

The promissory obligations that follow from participating in social practices ground important rights, and Green provides a valuable framework for identifying the conditions for this. However, these promissory claims should be distinguished from claims that focus solely on the normative relevance of the expectations themselves. Expectations that do not arise from participating in social practices might not ground promissory obligations but can still be normatively relevant in other ways. For my claim that expectation-related harm is normatively relevant if another agent is morally responsible for causing the expectation, it is not clear why the expectation should follow from a social practice or shared norm. If I assert to a friend that it will not rain because I do not want our trip to be postponed and it rains so heavily that his clothes are soaked and damaged, my friend's expectation is legitimate in the sense that I am responsible for the harm he suffers, even though there was no social practice governed by a rule that it should not rain. That the expectation should be reasonable and therefore might require a fine-grained assessment of the context, moreover, does not imply in any way that the involved agents must participate in a shared social practice, as Green asserts. Rejecting this practice-dependent condition extends the applicability of the LE concept to expectations like those of fossil fuel owners about their future productions.

Green's (2020a, 416–419) second concern about the usefulness of LE for responding to transitional losses has been referred to as the moral costs problem. Real-world decision-makers, Green contends, do not possess the epistemically privileged position normative theorists assume and will face significant costs in identifying the relevant expectations. To do so, they need to infer agents' expectations from their testimony and conduct. While financier moral costs refer to the economic costs of states to conduct these investigations, agent moral costs refer to the costs associated with the intrusive practice of people having their expectations investigated. However, these costs can be reduced significantly by placing the burden of proof on the expecting agent, who is in an epistemically better position than the state. A governmental agency, then, only has to evaluate the LE claim based on the evidence provided. In this way, agent moral costs are mitigated as the expecting agents can decide for themselves which information they want make available or not. Financier moral costs are also reduced as the relevant information is directly accessible to the expectors. Additionally, only the agents that might have LE will make a claim, not everyone that is negatively affected by the transition needs to be investigated, as Green (2020a, 417) asserts. For some claims, still, it might not be worth doing the efforts to enforce them, but this also holds for other kinds of claims and is not peculiar to LE. Moreover, this is only problematic for trivial claims, not for claims of fossil fuel owners, which concern huge amounts of money.

Green (2020a, 417) also claims that the moral costs problem is particularly great when it comes to legal transitions (or large-scale transitions in general, regardless of how they are triggered) because these affect large numbers of agents with heterogeneous expectations. However, this does not only lead to more investment costs, there is also more at stake in terms of justice claims that can be realised. The more agents that make similar claims, moreover, the more cost-effective it is to enforce them (possibly by the assistance of representative structures). The costs, thus, of identifying the relevant expectations and the harms that follow from this are not such that it is better to throw the concept overboard, also not (certainly not) when large-scale transitions are at stake.

3. Expectations about Fossil Fuel Productions: Legitimate for Companies, Illegitimate for States

Now, I investigate whether fossil fuel reserve owners (companies and states) could use the LE argument to justify transitional aid. A first implication of my harm-based account, as explained, is that there should be an agent that caused the false expectation about future production benefits. The expectation that the value of fossil fuel reserves will remain the same can be divided into expectations about regulatory, economic and physical stability, as fossil fuel reserves can be stranded by regulatory, economic or physical events. In case of regulatory stranding, asset devaluation is caused by the introduction of governmental regulations like carbon pricing or subsidy removals or by litigation and changing statutory interpretations. Economic stranding occurs when assets become stranded due to a change in relative costs or prices, due to technological changes or evolving social norms or behaviour. Physical stranding, finally, is the result of environmental changes (Caldecott 2017, 5).

So far, the literature has focused solely on LE about regulatory stability, because only the fulfillment of these expectations is under human control: 'Moral, political, and legal philosophers of LE … are not interested in predictive expectations writ large—encompassing the weather, or the laws of gravity, for instance—but, rather, in predictive expectations *about the behaviour of other agents'* (Green 2020a, 398). Meyer and Sanklecha (2014, 370) also stress that the fulfillment of an expectation must be under human control before it can be considered legitimate. However, what should be under human control according to my harm-based account of LE is not the fulfillment but the creation of an expectation. Consider my friend's expectation that it will not rain. There was no human control over the fulfillment of his expectation, but as its creation was under human control, it is still normatively relevant.

Nevertheless, I also focus on the causation of false expectations about regulatory stability. One reason for this is that the energy transition is most likely to occur due to regulations: 'more and more literature recognizes the potential stranding of assets and resources due to future climate change mitigation regulations' (Bos and Gupta 2019, 4). Another reason is that in the case of causing false beliefs of regulatory stability, it is more likely that the expectation is caused by another agent. An obvious candidate is the regulatory body that decides about the policies that affect a fossil fuel owner's future production. In the case of companies, there is a strong, centralised regulatory body that decides on the regulations to which they will be subject: states. Previous governmental policies and (lack of) decisions could make companies expect that the rules are going to stay the same (whether states are responsible for it still needs to be proven). States, as explained, also own fossil fuels themselves. For them, there is no similar strong and centralised institution on the international or global level that can enforce regulations and thereby cause expectations. If a state's reserves become stranded, it is more likely to be a result of environmental organisations that sue governments by reinterpreting existing laws or non-regulatory stranding, rather than the introduction of new regulations (van Asselt 2021). States can be blamed for having caused LE, but they cannot use the argument themselves.

The second implication of my harm-based account is that fossil fuel owners cannot claim the expected production benefits, but only compensation for the costs of having relied on this expectation, since causing an expectation cannot entitle someone to commit injustices. These costs consist of investing in extractive infrastructure and in explorations to find new reserves. For companies, these costs are high. They have been pumping billions of dollars into fossil fuel projects, buying offshore platforms, building new pipelines and extending lifelines to coal power plants (Tabuchi 2021). Additionally, they have been investing substantial sums in fossil fuel explorations, which is especially costly as they have to pay the countries in which they do the explorations. According to some estimates, investor-owned companies invest over $700 billion per year in explorations and productions. States, in contrast, own a lot of reserves but invest relatively little in explorations. For them, the energy transition does not require so much that they stop investing, but that they do not exploit their already proven reserves (Heede and Oreskes 2016, 18–19). Unfortunately for fossil fuel-rich countries, however, LE claims are not relevant for their unrealised benefits but only for investment costs.

The compensation owed for fossil fuel companies' useless investments might be relatively small compared to the compensation they should pay themselves for their past injustices. Companies that have produced more fossil fuels than permissible should compensate the agents who can now produce less in order to reach the climate target and/or the victims of climate change. Moreover, fossil fuel companies might not only exceed permissible production levels: they are also responsible for high consumption levels. The fossil fuel industry, after all, does not only cause emissions by producing fossil fuels that get burnt later, manufacturing and transportation also release a huge amount of emissions. As Grasso (this volume) explains, just 15 oil companies are responsible for 30% of the global industrial greenhouse gas emissions between 1965–2018, which creates duties of reparation and decarbonisation. While the Norwegian government might have to compensate Norwegian oil companies' useless investments after unexpectedly withdrawing subsidies to carry out oil exploration, thus, these oil companies themselves, arguably, also bear significant compensatory duties as a consequence of their past productions.

That companies have been provided with false expectations, however, is not enough to claim compensation. The expectation-creator should also be responsible for having caused the expectation. This can only hold if the expectation is reasonable. Importantly, the belief should not be reasonable given the expector's actual set of evidence but given the evidence that he *should* have. Only then can the creation of an expectation be considered foreseeable by the expectation-creator and unavoidable by the expector, allowing the former to be held responsible. An agent, thus, cannot make his expectations legitimate by ignoring relevant announcements or by failing to consider relevant evidence about future changes in regulations. The evidence one should have, then, depends on the availability of information and the means to acquire information. The capacities of states, multinationals and large companies are larger than those of medium- and small-sized companies. In this way, the epistemic validity condition is another reason why LE claims are more difficult to make for states, especially for developed states that have a greater capacity to assess carbon risks compared to developing states (Bradley, Lahn and Pye 2018).

Green (2020a, 406) suggests there are never good reasons for expecting that regulations will stay the same: 'It is in the nature of a legislature that it make, and hence change, laws … and this renders untenable the idea … that any particular law will never change'. The idea that a particular law will never change, indeed, might be untenable, but that it will stay the same in the near future, for a certain amount of time, can definitely be reasonable. Regulatory stability, I contend, is likely when existing regulations, given certain knowledge about relevant circumstances and facts, correspond with the conception of justice of the regulatory system. Colla (2017, 287–288) refers to this conception of justice as internal normativity. It should be distinguished from the expector's or society's conception of justice (external normativity) and from objective justice – the objective truth about what is just (insofar such a truth exists). When relevant things happen, however, this affects the reasonableness of the expectation of regulatory stability (Colla 2017, 290–291; Lemos 2007, 19). Expectations about the stability of permissions to produce fossil fuels might be illegitimate because of the changed knowledge about the effects of greenhouse gas emissions on the climate. This has created, in the words of Meyer and Sanklecha (2011, 454), 'a time of radically changed circumstances in which all predictions are suspect'. However, that they are suspect does not mean they are illegitimate. Changing circumstances are not enough to make the expectation of regulatory stability illegitimate. It should also be shown that, because of this new knowledge, the existing (lack of) regulations no longer fit with the justice conception of the regulatory body.

While my harm-based LE account focuses on internal normativity, many existing LE accounts consider objective justice a condition for legitimacy (cf. Gosseries and Hungerbühler 2006, 111–115; Moore 2017). This position faces several problems. First, there is the metaphysical task of justifying that there is an objective truth about what justice entails and the difficulty of sorting out how that truth can be known. Secondly, the condition that transitional losers' expectations must be just before they can be normatively relevant would make the concept of LE quite useless. If their expectations are just, after all, they should not have transitional losses and if they did have transitional losses, it is not necessary to invoke their LE, as they can refer to general justice principles. The biggest difficulty, however, lies in explaining why objective justice would matter for the normative relevance of an expectation in the first place. Imagine that Donald Trump caused the expectation in a developing country that the United States (US) will not mitigate climate change and that, therefore, the country prepares for the devastating consequences of global warming by investing huge amounts in adaptation measures. If eventually the US does mitigate under the new president and avoids climate change together with the rest of the world, the developing country's investments were useless, and compensation seems justified. That the content of the expectation was unjust (that the US will not mitigate climate change) does not seem to be relevant, it is not immoral to take this unjust expectation into account when making plans for the future.

Trump and the US should not be released from their duty to compensate for the caused harm.

Only the justice conception of the regulatory body matters, thus, as it is relevant for assessing the reasonableness of the expectation of regulatory stability. If the regulator provided evidence for the belief that it does not care about intergenerational justice or avoiding climate change, a fossil fuel owners' expectation of regulatory stability is legitimate and justifies compensation if it is frustrated. On the global level, there is no strong, centralised regulator that can be held responsible for causing such beliefs, in contrast to the domestic level. States, I contend, have been providing clear signals that no strong regulations will be implemented that require fossil fuel owners to keep their reserves underground. One indication is how a state has responded to similar cases in the past. If one lives, for instance, in an extremely traditionalist society in which rules hardly ever change or in which there is always strong opposition to change, one could expect the same rigidity in the future (Gosseries and Hungerbühler 2006, 111). Usually, laws or policies relating to environmental issues are highly political and subject to change. However, because of its intergenerational and global aspect, climate change is a unique case. As Jamieson (2010, 83) notes, 'Our current value system presupposes that harms and their causes are individual, that they can readily be identified, and that they are local in space and time'. Therefore, one can better look at how a state has been responding to climate change itself.

Unfortunately, since the dangerous effects of climate change have become known, states have done disappointingly little. The longer they wait to act, the more legitimate it becomes for companies to expect that no strict regulations will be implemented. Moreover, states have continued granting licenses to companies to conduct explorations, and they have continued supporting the activities of fossil fuel industries by paying subsidies (Kartha et al. 2016). Perhaps the Paris Agreement, in which states agreed that global warming should not exceed 2°C, might be relevant. However, as this did not lead to binding national targets (Arellano and Roberts 2017), it cannot be seen as a clear sign that mitigation measures will be imposed either. In this way, states are foreseeably causing expectations in companies that they will be able to keep benefitting from fossil fuel productions. If the energy transition eventually succeeds, they have to pay compensation for the harm that follows from these unfulfilled expectations.

Conclusion

The energy transition is one of the most important challenges humanity has ever faced. However, like many other transitions, it brings adverse consequences for some agents, which gives rise to the question of whether transitional losers are owed any kind of transitional aid. This chapter focused on states and companies that will see their fossil fuel reserves becoming worthless if the transition succeeds. Inter- and intragenerational justice require that two-thirds of proven recoverable reserves stay underground, mostly in countries that are wealthy and/or have a lot of carbon-intensive fossil fuels. As stranded assets concern expectations by definition, the concept of LE can be a powerful argument to save fossil fuel owners' reserves from being stranded or to justify transitional aid.

The chapter presented an account of LE that is in equilibrium both with the general principle that one should not wrongfully harm others and with intuitions that, for instance, reasonableness and justice (at least the regulator's conception of it) matter. According to this account, LE claims follow from the duty not to wrongfully harm others by creating false expectations. In particular, I argued that if an agent is responsible for the creation of expectations in another agent, she should fulfill these expectations to avoid having wrongfully harmed the other agent or compensate the harms that these false expectations brought about if fulfillment is not permissible, which is the case when just transitions are at stake. It follows that states – which own the lion's share of the world's fossil fuel reserves – cannot use the LE argument because there is no strong, centralised regulatory body at the global level that can be blamed for having caused false expectations, because their expectations led to relatively few investments and because they should have known about the stranding. While states cannot complain, however, they can be blamed, as they are powerful structures that have a lot of control over the legislation they impose. Despite the consensus that climate change requires strong mitigation measures, by not taking action, states caused expectations in companies that existing regulations will continue to exist. If the energy transition takes place, companies should be compensated for the costs of having these false expectations, which consists of investments in exploring fossil fuel reserves and preparing to extract them.

The concept of LE is also relevant if the energy transition does not succeed. The adverse consequences of climate change would already ground compensation duties, but false expectations about the avoidance of these harms could provide additional claims. After all, if the victims would have known that climate change would not be mitigated, they could have invested in adaptation. Green investors could also complain that they have been provided with false expectations that their investments would be profitable. One could have LE about conserving the existing rules, thus, but also about the implementation of new mitigation measures – which is actually also a kind of conservation: conservation of the climate. Referring to one's LE, in fact, does not need to be conservative in any way, as one's expectations are not necessarily based on the existing state of affairs. Future research, therefore, could elaborate on both the theoretical development and the wider application of this concept and make us better at realising transitions fairly.

*I am grateful to my colleagues at the venues I presented this chapter: the DK Final Workshop CliMatters at the University of Graz and the Midi de l'Ethique of the Hoover Chair at the University of Louvain-la-Neuve. Specials thanks go to Lukas Meyer, Axel Gosseries and Kian Mintz-Woo. I also want to thank the editors and two anonymous reviewers. Finally, I acknowledge the Austrian Science Fund (FWF), which funded this work under Research Grant W1256 (Doctoral Programme Climate Change: Uncertainties, Thresholds and Coping Strategies).

References

Arellano, Angelica, and Timmons Roberts. 2017. "Is the Paris Climate Deal Legally Binding or Not?". *Climate Home News*. 2 November 2017. https://www.climatechangenews.com/2017/11/02/paris-climate-deal-legally-binding-not/

Asselt, Harro van. 2021. "Governing Fossil Fuel Production in the Age of Climate Disruption: Towards an International Law of 'Leaving It in the Ground'". *Earth System Governance* 9 (September): 100118. https://doi.org/10.1016/j.esg.2021.100118

Beitz, Charles R. 1979. *Political Theory and International Relations: Revised Edition*. Princeton University Press.

Bos, Kyra, and Joyeeta Gupta. 2019. "Stranded Assets and Stranded Resources: Implications for Climate Change Mitigation and Global Sustainable Development". *Energy Research & Social Science* 56 (October): 101215. https://doi.org/10.1016/j.erss.2019.05.025

Bovens, Luc. 2011. "A Lockean Defense of Grandfathering Emission Rights". In *The Ethics of Global Climate Change*, edited by Denis G. Arnold, 124–44. Cambridge University Press.

BP. 2018. "BP Statistical Review of World Energy 2018". https://www.bp.com/en/global/corporate/news-and-insights/press-releases/bp-statistical-review-of-world-energy-2018.html

Bradley, Siân, Glada Lahn, and Steve Pye. 2018. "Carbon Risk and Resilience". *Chatham House – International Affairs Think Tank*, 12 July 2018. https://www.chathamhouse.org/2018/07/carbon-risk-and-resilience

Brown, Alexander. 2011. "Justifying compensation for frustrated legitimate expectations". *Law and Philosophy* 30 (6): 699–728.

Brown, Alexander. 2012. "Rawls, Buchanan, and the Legal Doctrine of Legitimate Expectations". *Social Theory and Practice* 38 (4): 617–44.

Brown, Alexander. 2017. "A Theory of Legitimate Expectations". *Journal of Political Philosophy* 25 (4): 435–60. https://doi.org/10.1111/jopp.12135

Brown, Alexander. 2018. *A Theory of Legitimate Expectations for Public Administration. A Theory of Legitimate Expectations for Public Administration.* Oxford University Press.

Buchanan, Allen. 1975. "Distributive Justice and Legitimate Expectations". *Philosophical Studies* 28 (6): 419–25. https://doi.org/10.1007/BF00372903

Caldecott, Ben, Nicholas Howarth, and Patrick McSharry. 2013. "Stranded Assets in Agriculture: Protecting Value from Environment-Related Risks". https://ora.ox.ac.uk/objects/uuid:4496ac03-5132-4a64-aa54-7695bfc7be9d

Caldecott, Ben. 2017. "Introduction to special issue: stranded assets and the environment". *Journal of Sustainable Finance & Investment* 7 (1): 1–13. https://doi.org/10.1080/20430795.2016.1266748

Caney, Simon. 2009. "Justice and the distribution of greenhouse gas emissions". *Journal of Global Ethics* 5 (2): 125–46. https://doi.org/10.1080/17449620903110300

Caney, Simon. 2016. "Climate change, equity, and stranded assets". *Oxfam America Research Backgrounder Series*. http://www.oxfamamerica.org/explore/research-publications/climate-change-equity-and-stranded-assets

Carbon Tracker Initiative. 2011. "Unburnable Carbon: Are the World's Financial Markets Carrying a Carbon Bubble?" https://carbontracker.org/reports/carbon-bubble/

Carbon Tracker Initiative. 2013. "Wasted Capital and Stranded Assets". https://carbontracker.org/reports/unburnable-carbon-wasted-capital-and-stranded-assets/

Carbon Tracker Initiative. 2017. "Stranded Assets". https://www.carbontracker.org/terms/stranded-assets/

Carnes, Thomas S. 2020. "Unauthorized Immigrants". *Social Theory and Practice* 46 (4): 681–707. https://doi.org/10.5840/soctheorpract20201026102

Colgan, Jeff D., Jessica F. Green, and Thomas N. Hale. 2021. "Asset Revaluation and the Existential Politics of Climate Change". *International Organization* 75 (2): 586–610. https://doi.org/10.1017/S0020818320000296

Colla, Anne-France. 2017. "Elements for a General Theory of Legitimate Expectations - Dimensions". *Moral Philosophy and Politics*, 4 (2): 283–305. https://doi.org/10.1515/mopp-2017-0040

Damon, Maria, Daniel H. Cole, Elinor Ostrom, and Thomas Sterner. 2019. "Grandfathering: Environmental Uses and Impacts". *Review of Environmental Economics and Policy* 13 (1): 23–42. https://doi.org/10.1093/reep/rey017

Denaro, Pietro. 2012. "Moral Harm and Moral Responsibility: A Defence of Ascriptivism". *SSRN Scholarly Paper* ID 2064418. Rochester, NY: Social Science Research Network. https://doi.org/10.1111/j.1467-9337.2012.00508.x

Edmond, Charlotte. 2020. "These Are the Top Risks Facing the World in 2020". World Economic Forum, 2020. https://www.weforum.org/agenda/2020/01/top-global-risks-report-climate-change-cyberattacks-economic-political/

Fay, Marianne. 2015. *Decarbonizing Development*. http://archive.org/details/decarbonizing

Feinberg, Joel. 1987. *The Moral Limits of the Criminal Law Volume 1: Harm to Others*. New York: Oxford University Press.

Feinberg, Joel. 1990. *The Moral Limits of the Criminal Law Volume 4: Harmless Wrongdoing*. New York: Oxford University Press.

Generation Foundation. 2013. "Stranded Carbon Assets: How Carbon Risks Should Be Incorporated in Investment Analysis". https://www.genfound.org/media/1374/pdf-generation-foundation-stranded-carbon-assets-v1.pdf

Gosseries, Axel. 2004. "Historical Emissions and Free-Riding". *Ethical Perspectives* 11 (1): 36–60.

Gosseries, Axel, and Mathias Hungerbühler. 2006. "Rule change and intergenerational justice". In *Handbook of Intergenerational Justice*, edited by Joerg Chet Tremmel. Cheltenham, UK – Northampton, MA, USA: Edward Elgar

Green, Fergus. 2019. "Who Should Get What When Governments Change the Rules? A Normative Theory of Legal Transitions". PhD, The London School of Economics and Political Science (LSE). http://etheses.lse.ac.uk/3980/

Green, Fergus. 2020a. "Legal Transitions without Legitimate Expectations". *Journal of Political Philosophy* 28 (4): 379–420. https://doi.org/10.1111/jopp.12231

Green, Fergus. 2020b. "Book Review: Alexander Brown, A Theory of Legitimate Expectations for Public Administration (Oxford: Oxford University Press, 2017), pp. 226". *Law and Philosophy* 39 (4): 463–70. https://doi.org/10.1007/s10982-020-09385-4

Green, Fergus, and Ajay Gambhir. 2020. "Transitional assistance policies for just, equitable and smooth low-carbon transitions: who, what and how?". *Climate Policy* 20 (8): 902–21. https://doi.org/10.1080/14693062.2019.1657379

Heede, Richard. 2014. "Tracing Anthropogenic Carbon Dioxide and Methane Emissions to Fossil Fuel and Cement Producers, 1854–2010". *Climatic Change* 122 (1): 229–41. https://doi.org/10.1007/s10584-013-0986-y

Heede, Richard, and Naomi Oreskes. 2016. "Potential emissions of CO2 and methane from proved reserves of fossil fuels: An alternative analysis". *Global Environmental Change* 36 (January): 12–20. https://doi.org/10.1016/j.gloenvcha.2015.10.005

Hodgson, Louis-Philippe. 2012. "Why the Basic Structure?". *Canadian Journal of Philosophy* 42 (3–4): 303–34. https://doi.org/10.1080/00455091.2012.1071 6779

Hölscher, Katharina, Julia M. Wittmayer, and Derk Loorbach. 2018. "Transition versus Transformation: What's the Difference?". *Environmental Innovation and Societal Transitions* 27 (June): 1–3. https://doi.org/10.1016/j. eist.2017.10.007

IPCC. 2021. "Summary for Policymakers". In *Climate Change 2021: The Physical Science Basis. Contribution of Working Group I to the Sixth Assessment Report of the Intergovernmental Panel on Climate Change*, edited by Masson-Delmotte, V., P. Zhai, A. Pirani, S.L. Connors, C. Péan, S. Berger, N. Caud, Y. Chen, L. Goldfarb, M.I. Gomis, M. Huang, K. Leitzell, E. Lonnoy, J.B.R. Matthews, T.K. Maycock, T. Waterfield, O. Yelekçi, R. Yu, and B. Zhou. Cambridge University Press. In Press.

Jamieson, Dale. 2010. "Ethics, Public Policy, and Global Warming". In *Ethics, Public Policy, and Global Warming*. Oxford University Press. https://doi. org/10.1093/oso/9780195399622.003.0011

Kartha, Sivan, Michael Lazarus, and Kevin Tempest. 2016. "Fossil Fuel Production in a 2°C World: The Equity Implications of a Diminishing Carbon Budget", September. https://www.sei.org/publications/equity-carbon-budget/

Kenton, Will. 2019. "State-Owned Enterprise (SOE)". *Investopedia*. 4 May 2019. https://www.investopedia.com/terms/s/soe.asp

Knight, Carl. 2013. "What is grandfathering?". *Environmental Politics* 22 (3): 410–27. https://doi.org/10.1080/09644016.2012.740937

Knight, Carl. 2014. "Moderate Emissions Grandfathering". *Environmental Values* 23 (5): 571–92.

Lemos, Noah. 2007. *An Introduction to the Theory of Knowledge*. Cambridge University Press.

Lenferna, Georges Alexandre. 2018. "Can we equitably manage the end of the fossil fuel era?" *Energy Research & Social Science*, Energy and the Future, 35 (January): 217–23. https://doi.org/10.1016/j.erss.2017.11.007

Mao, Frances. 2021. "Climate Change: Why Australia Refuses to Give up Coal". *BBC News*, 22 October 2021, sec. Australia. https://www.bbc.com/news/world-australia-57925798

Matravers, Matt. 2017. "Legitimate Expectations in Theory, Practice, and Punishment". *Moral Philosophy and Politics* 4 (2): 307–23. https://doi.org/10.1515/mopp-2017-0017

McGlade, Christophe, and Paul Ekins. 2015. "The Geographical Distribution of Fossil Fuels Unused When Limiting Global Warming to 2 °C". *Nature* 517 (7533): 187–90. https://doi.org/10.1038/nature14016

Meinshausen, Malte, Nicolai Meinshausen, William Hare, Sarah C. B. Raper, Katja Frieler, Reto Knutti, David J. Frame, and Myles R. Allen. 2009. "Greenhouse-Gas Emission Targets for Limiting Global Warming to 2 °C". *Nature* 458 (7242): 1158–62. https://doi.org/10.1038/nature08017

Meyer, Lukas H. 2013. "Why Historical Emissions Should Count". *Chicago Journal of International Law* 13 (2). https://chicagounbound.uchicago.edu/cjil/vol13/iss2/15

Meyer, Lukas H., and Pranay Sanklecha. 2011. "Individual Expectations and Climate Justice". *Analyse & Kritik* 33 (2): 449–71.

Meyer, Lukas H., and Pranay Sanklecha. 2014. "How Legitimate Expectations Matter in Climate Justice". *Politics, Philosophy & Economics* 13 (4): 369–93. https://doi.org/10.1177/1470594X14541522

Moore, Margaret. 2017. "Legitimate Expectations and Land". *Moral Philosophy and Politics* 4 (2): 229–255.

Nelkin, Dana K. 2021. "Moral Luck". In *The Stanford Encyclopedia of Philosophy*, edited by Edward N. Zalta, Summer 2021. Metaphysics Research Lab, Stanford University. https://plato.stanford.edu/archives/sum2021/entries/moral-luck/

Pye, Steve, Siân Bradley, Nick Hughes, James Price, Daniel Welsby, and Paul Ekins. 2020. "An equitable redistribution of unburnable carbon". *Nature Communications* 11 (3968): 1-9. https://doi.org/10.1038/s41467-020-17679-3

Rawls, John. 1971. *A Theory of Justice*. Cambridge: Harvard University Press.

Robertson, Andrew. 1998. "Reliance and Expectation in Estoppel Remedies". Legal Studies 18 (3): 360–68. https://doi.org/10.1111/j.1748-121X.1998.tb00022.x

Roser, Dominic, and Christian Seidel. 2016. *Climate Justice: An Introduction*. Taylor & Francis.

Rudy-Hiller, Fernando. 2018. "The Epistemic Condition for Moral Responsibility". In *The Stanford Encyclopedia of Philosophy*, edited by Edward N. Zalta, Fall 2018. Metaphysics Research Lab, Stanford University. https://plato.stanford.edu/archives/fall2018/entries/moral-responsibility-epistemic

Schuessler, Rudolf. 2017. "A Luck-Based Moral Defense of Grandfathering". In *Climate Justice and Historical Emissions*, edited by Lukas H. Meyer and Pranay Sanklecha, 141–64. Cambridge: Cambridge University Press.

Simmons, A. John. 1996. "Associative Political Obligations". *Ethics* 106 (2): 247–73.

Singer, Peter. 2010. "One Atmosphere". In *Climate Ethics: Essential Readings*, edited by Stephen Gardiner, 1 edition. Oxford ; New York: Oxford University Press.

Society of Petroleum Engineers (SPE). 2008. "Petroleum Resources Management System 2007". https://www.spe.org/industry/docs/Petroleum_Resources_Management_System_2007.pdf

Tabuchi, Hiroko. 2021. "Private Equity Funds, Sensing Profit in Tumult, Are Propping Up Oil". *The New York Times*, 13 October 2021, sec. Climate. https://www.nytimes.com/2021/10/13/climate/private-equity-funds-oil-gas-fossil-fuels.html

The Independent. 2019. "Norway Refuses to Drill for Billions of Barrels of Oil in Arctic, Leaving 'Whole Industry Surprised and Disappointed'", 9 April 2019. https://www.independent.co.uk/climate-change/news/norway-oil-drilling-arctic-ban-labor-party-unions-a8861171.html

Thomson, Judith Jarvis. 1986. "Feinberg on Harm, Offense, and the Criminal Law: A Review Essay". *Philosophy & Public Affairs* 15 (4): 381–95.

Waligore, Timothy. 2017. "Legitimate Expectations, Historical Injustice, and Perverse Incentives for Settlers". *Moral Philosophy and Politics* 4 (2): 207–28. https://doi.org/10.1515/mopp-2016-0032

Weinman, Michael D. 2018. "Arendt and the Legitimate Expectation for Hospitality and Membership Today". *Moral Philosophy and Politics* 5 (1): 127–49. https://doi.org/10.1515/mopp-2016-0043.

14

From Food to Climate Justice: How Motivational Barriers Impact Distributive Justice Strategies for Change

SAMANTHA NOLL

Climate change is one of the most important and complex problems of the modern age. The sheer scale of the harm produced, coupled with the fact that the changes are human-induced, necessitates a duty to prevent climate-induced impacts. There is a growing literature exploring how costs and benefits should be shared at national, state and generational levels. This chapter adds to this literature by exploring how normatively guided plans could be hindered by barriers beyond distributive justice frameworks and their subsequent applications. Even if we recognise a *prima face* duty to mitigate climate change impacts, there are motivational barriers that could block people from making the requisite changes – motivational barriers that ultimately curtail the effectiveness of climate mitigation strategies. After outlining each of these barriers, this chapter then argues that insights from local food movements could provide novel strategies to potentially address each of the barriers above, as they have leveraged normative arguments to motivate individual and collective action. Lessons from food-focused activism, coupled with normatively guided strategies for sharing costs and benefits, has the potential to help communities work towards effectively addressing climate change.

Introduction

Climate change is one of the most important and complex problems of the modern age. The extraordinary scope of this crisis was recently emphasised in 2018 and 2021 reports by the Intergovernmental Panel on Climate Change (IPCC). These reports urge countries around the globe to make drastic changes to avoid dire environmental and social consequences. In light of these impacts, several scholars have issued a normative call to action, as the scale of the harm to humans and the environment necessitates a duty or responsibility to prevent this harm (Bell 2013; Blau 2017; Quirico 2018). For example, according to Kyllonen (2018, 737), 'the well acknowledged "no-harm principle" directly necessitates a correlative pro tanto duty to refrain from causing the harm and, when that is not possible, to repair or compensate for the harm inflicted'. However, due to the scale and complicated nature of the problem, there is intense debate concerning the distribution of responsibility to address climate change.

Even when recognizing this *pro tanto* duty, 'individualists' claim that each person has a responsibility to change their behaviour, such as by reducing their carbon footprint and thus limiting their contribution to climate change (Almassi 2012; Cripps 2013; Fahlquist 2009). In contrast, 'collectivists' argue that responsibility should be recognised at the collective level (Hiller 2011; Sinnott-Armstrong 2010; Vance 2017; Vanderheiden 2011), as the problem was largely caused by collective agents. For the latter position, the amount of changes needed to make a difference can only be performed by groups. However, as Kyllonen (2018) argues, a sense of personal responsibility is needed to act as a motivation for individuals to engage in collective action. Thus, it is only through the recognition of personal duty that we can address climate impacts and the clear distinction between individualist and collectivist approaches breaks down. This analysis is timely, as it illuminates the key role that personal motivations (be those ethical, social, political or economic) play on the ground, as solutions are actualised.

The philosophical debate on the proper application of the no-harm principle is important and on-going. However, the purpose of this chapter is to explore a lacuna that could negatively impact the effectiveness of individualist and collectivist applications of the no-harm principle: motivational barriers. If individualists are correct that each person as a duty to change their behaviour in such a way as to mitigate their contributions to climate change, then one could argue that barriers to discharging this duty are important factors that need to be considered. In addition, if Kyllonen (2018) is correct that motivations play an important role in collective action, then one could make a similar argument concerning this position. Work on barriers to climate action is not new in the scholarly literature. In addition to extensive research in psychology (Shalev 2015; Baumeister and Bargh, 2014), environmental philosophers have explored the role that participation plays in nurturing the attitudes necessary for recognising a duty to bring about environmental change (Light 2006; Taylor 1981).

This chapter begins from the position that the individualist and collectivist applications of the no-harm principle could provide the normative justification necessary to motivate individuals to act. However, it adds to this work by arguing that normative arguments need to be wedded to interdisciplinary work on other motivational barriers (beyond the lack of a normative argument) – barriers that could limit a person's ability to effectively discharge this duty. Specifically, it argues that even if we recognise a *prima face* duty to address climate impacts, there are obstacles that could ultimately curtail the effectiveness of such arguments. These include, but are not limited to, goal barriers, ambiguity barriers, threat barriers and structural justice barriers. This chapter then draws from theoretical work in environmental philosophy in general and local food movements (LFMs) in particular to outline potential strategies for addressing each of the barriers above. The aim of this chapter is to provide the foundation necessary for further work on normative duty and motivational barriers. I hope this analysis leads to the development of a robust philosophical framework (combining duty-based arguments with interdisciplinary work) that could help to effectively address what has been called the most wicked problem of the modern age: climate change.

1. Climate Change, Wicked Problems and Motivational Barriers

While factual evidence on climate change impacts is increasing (Tollefson 2021), attempts to encourage people to change their behaviour to address this crisis have had limited success (Shalev 2015; Whitmarsh and O'Neill, 2010; Moser and Ekstrom 2010). According to Moser and Ekstrom (2010, 22026), 'adaptation to climate change has risen sharply as a topic of scientific inquiry, in local to international policy and planning, in the media, and in public awareness… Yet climatic events in Europe, the United States, and Australia in recent years have also led to critical questioning of richer nations' ability to adapt to climate change'. According to Shalev (2015), one of the barriers to climate change adaption is linked to human motivation, which is understood as the process that moves individual people to act. Human motivation can be roughly divided into two categories: motivations that do not involve conscious awareness and those that require conscious thought, as these help individuals plan for future contingencies (Bargh 1997; Baumeister and Bargh 2014). In the second category, motivational barriers can be roughly broken down into three subsets: goal barriers, threat barriers and ambiguity barriers (Shalev 2015). Concerning the first type, clearly identifying a desired endpoint and using this as a reference to direct behaviour is imperative for pursuing solutions (Kruglanski et al. 2002; Shah et al. 2002). In other words, a clear sense of direction is necessary for providing the personal impetus to realise particular goals (Shalev 2015; Higgins 1989).

However, climate change is a 'wicked problem', or a complex and systemic issue that has no simple solution (Luwig 2001; Whyte and Thompson 2011). Such challenges involve interactions between biological processes and a diverse array of human conduct. Thus, they are not simply complicated but are uniquely challenging, as the way we formulate the problem is nebulous (is climate change an economic, environmental or social problem?) and solutions are varied. As Whyte and Thompson (2011, 441-442) argue:

> To describe climate change as an economic problem means that one has already limited oneself to particular economic solutions to addressing it. Because proposed solutions are so closely tied to problem formulations, disagreements among stakeholders who foresee themselves as being impacted differently by the solutions can take the form of ontological debates. Unlike problems where there is little disagreement about its basic formulation, wicked problems are characterized by deep ambiguity in the ontological assumptions and metaphysical categories used in their articulation.

As such, wicked problems are exceptionally complex, multifaceted and difficult to address. It follows from this categorisation that there is no one clear solution or goal to guide individual behaviour. Even when we recognise a duty to address climate-induced impacts, there is intense debate concerning how this responsibility should be distributed, and this influences the prioritisation of specific goals (Kyllonen 2018). Due to this complexity, local, national and international discourses often focus on weighing various goals and solutions (Shalev 2015). According to Shalev (2015), there is a plethora of research coming out of social psychology that connects the prolonged evaluation and assessment of goals to inaction and decision-making paralysis (Kruglanski et al. 2010; Shalev and Sulkowski 2009). If this is the case, then climate change decision-making may be hampered by a motivational goal barrier. Additionally, threat barriers also come into play in this context.

Similar to goal barriers, threat also inhibits a person's ability to change their behaviour in new ways and try new experiences. Due to this fact, 'individuals who feel they are under threat, therefore, tend to neglect their long-term, future planning goals in favor of the short-term goal of self-defense' (Shalev 2015, 131). Threat then acts as a motivation to reaffirm typical behaviour rather than changing it (Steele 1988; Cohen and Sherman 2014). As a motivational factor, paradoxically, this can reduce an individual's ability to modify behaviour or judgements when faced with new evidence (Shalev 2015). Likewise, ambiguity also increases resistance to change (Jost et al.,

2003), as do situations where a rapid response is required (Kruglanski and Webster 1996; Kruglanski 2004). When taken as a whole, the 'wicked' nature of climate change induces a triad of motivational barriers that actively work against collective action. Goal barriers, threat barriers and ambiguity barriers each play a role in limiting the ability of individuals and the communities they make up to address climate impacts.

What this translates to in a philosophical context is that even if we recognise a *prima face* duty to address climate impacts, there are also motivational barriers to encouraging people to make the requisite changes – motivational barriers that could ultimately curtail the effectiveness of such arguments. This is especially important if Kyllonen (2018) is correct that collectivist arguments need individualist motivations (specifically, a normative sense of duty) to provide sufficient reason for collective action. If this is the case, then motivational barriers that impact an individual's ability to alter their behaviour in ways necessary to meet the challenge of climate change could have a direct impact on whether they can meet their moral duty. In addition to motivational barriers, 'structural injustice barriers' should be included, as they also impact climate change mitigation strategies. As Harlan et al. (2015, 1) argue, climate change is marked by a myriad of justice concerns. First, the causes of climate change are driven by social inequalities, as marginalised nations and communities typically use vastly less fossil fuels. Second, the poor experience climate impacts more dramatically than the rich at the local, national and global levels. Third, climate change mitigation policies will also 'have starkly unequal impacts within and across societies' (Harlan et al. 2015, 128). Such inequalities could impact an individual's ability to take actions necessary to address a changing climate. These include, but are not limited to, the 'unequal distribution of impacts, unequal responsibility for climate change, and unequal costs for mitigation and adaptation' (Sowers 2007, 140). Additionally, the history of unequal global development could also negatively impact the ability of individuals to act, depending on their context, as this history often translates into a greater vulnerability to climate impacts and/or the limitation of available options to curtail carbon emissions (Adger et al. 2003). This has led some scholars, such as Parks and Roberts (2010), to argue that micro-discussions of rational choice need to be integrated into larger discussions concerning structural insights and barriers. Barnett and Adger (2007) have also argued that, in certain contexts, climate adaption strategies should incorporate the wider goals of regenerating ecological, social and human capital needed to address climate impacts. An important take-away from such critiques is that a) wider justice issues need to be brought into climate change discussions and b) burden-sharing should be explored in contexts where development is curtailed by climate change (Parks and Roberts 2010). Thus, these wider structural justice issues could also play a major role in undermining individual action and thus form a fourth barrier: the structural justice barrier.

2. What Are Local Food Movements and Why Do They Matter?

When taking these barriers into account, one could easily adopt a pessimistic attitude concerning whether individuals and the collective agents they make up can institute the strategies necessary for climate change mitigation. Fortunately, however, philosophical work on LFMs could provide key insights concerning how to address each of the barriers above. This may be surprising to some readers. However, both LMFs and other environmental movements have a long history of bringing about ethically motivated change while grappling with wicked problems. Scholarly work on LFMs often highlights a) how ethically and ontologically centered concerns are actualised to motivate individuals and b) how commitments to address larger justice concerns can guide individual decision-making (Delind 2011; Alkon 2012; Noll and Murdock 2020). The ways that LFMs couple larger systemic issues with local action is evident in the theoretical structure guiding these initiatives. For example, supporters of LFMs typically argue that they provide an alternative to industrialised food systems by reconnecting communities to agricultural processes, including the production, processing and distribution of foodstuffs (Noll and Werkheiser 2017; Levkoe 2011). As industrial food systems produce food on a large scale, they essentially reduce the wide spectrum of food choices that individuals make to 'shallow' choices concerning brands in a supermarket (Delind 2011). In the cultural sphere, these shallow choices translate into the loss of food knowledge and practices (Alkon 2012). Due to these factors, LFMs are often driven to address larger food system concerns by focusing on local production, reviving heritage breeds and seeds, supporting and rekindling local food customs and helping to increase community control over food systems. As such, LFMs have effectively coupled individual action (buying organic, supporting local farmers, etc.) with addressing complicated food system problems. Depending on the type of LFM, these system-focused problems could be deeply concerned with justice issues (Alkon 2012; Noll and Murdock 2020; Schanbacher 2010). Thus, I argue that the literature on LFMs could provide fruitful lessons for those working to motivate individuals to mitigate climate change. These insights coupled with duty-focused arguments, such as applications of the no-harm principle, could provide the philosophical foundations necessary for making new climate change mitigation strategies a reality.

It is imperative that normative arguments and strategies to negate motivational barriers be integrated if we hope to mitigate climate change. With this goal in mind, the next section of this chapter provides a brief overview of different LFMs before highlighting key strategies they use to address motivational barriers. It should be noted that this analysis is not meant to provide a fully fleshed out solution to this complex dilemma. Rather, it is meant to be the start of a much larger conversation on this topic.

The number of LFMs has been increasing steadily since the middle of the twentieth century. These initiatives primarily aim to realise a wide range of food-related goals, such as connecting local food producers and consumers, developing more resilient food systems, improving local economies and bringing about positive impacts to specific communities (Delind 2011; Alkon 2012; Schanbacher 2010). More generally, they are united by a desire to provide alternatives to industrialised food systems and place environmental and structural injustices at the forefront of discussions concerning food (Levkoe 2011). LFMs have more than 40 years of experience tackling wicked problems in the realm of agriculture (Whyte and Thompson 2011). As Pirog et al. (2014, 1) state, 'the local food movement... has evolved over the past 25 years, including a more recent convergence with movements supporting food access and health, food justice, environment, food sovereignty, and racial equity'. Due to the wide range of goals associated with local food movements, Werkheiser and Noll (2014) place them into three sub-movements, each defined by their unique goals and ontological conceptualisations of 'people' and 'food' that guide group action. In short, LFMs actualise normative commitments, ontological frameworks and justice mandates as they push us to rethink our very relationship with food, society and ourselves.

According to Werkheiser and Noll (2014), there are three types of LFMs, each with their own distinct philosophical commitments. The individual-focused sub-movement is the largest subset of these initiatives. Members of this sub-movement predominantly focus on personal choice or lifestyle politics, conceptualising the food we eat as one choice among many with far-ranging impacts. In contrast, the systems-focused sub-movement conceptualises change as happening at the larger structural or systems level. What separates this sub-movement from the individual-focused sub-movement is that it shifts the conversation from the level of individual choice to push for larger changes that could alter food-based subsidies and policies, such as those in the United States (US) Farm Bill. Supporters argue that, by framing the issue beyond individual choice, systems-focused initiatives can change 'many of the environmental, economic, and political problems of the current system' and have on-the-ground impacts, such as increasing food security (Werkheiser and Noll 2014, 203).

Finally, the community-focused sub-movements can be understood as the intersection of the LFM and food sovereignty initiatives. Rather than seeing food as an interchangeable commodity bought and sold on the market, this sub-movement conceptualises 'food' and 'people' as intertwined. Food is not simply a product. It is an important catalyst for creating and reproducing personal identity, community and culture (Desmarais et al. 2010; Schanbacher 2010). As such, food-related issues can and have become important bases for activism around the globe. These movements conceive of people as members of communities with distinct cultures, including local food customs. And 'if communities and food are co-constituted, then the particular culture and the particular food become much more important, as does the symbolic nature of food' (Werkheiser and Noll 2014, 207). For members of this sub-movement, activism is equally co-constituted with communities and shared customs Desmarais et al. 2010). As such, strategies for bringing about change are often framed as social justice issues. According to Schanbacher (2010, ix), 'The food sovereignty model considers human relationships in terms of mutual dependence, cultural diversity, and respect for the environment... Ultimately, if food sovereignty's demands are not met, the current global food system constitutes a massive violation of human rights'. The community-focused sub-movement places particular importance on actualising justice frameworks when working towards food-related goals.

3. Local Food and Motivational Barriers

The no-harm principle can be understood to necessitate a *pro tanto* duty to refrain from harm and/or repair harm that has been inflicted (Kyllonon 2018). This requires that each person change their behaviour to help address the wicked problem of climate change. The next section outlines how LFMs could provide strategies for mitigating the specific barriers outlined above. For example, the individual-focused sub-movement has been quite successful in addressing the goal motivational barrier. While wicked problems associated with food production are vast and complicated (Whyte and Thompson 2011), these LFMs have distilled myriad possible solutions down into a single message. If you 'buy local', then you help bring about positive change 'one meal, and one family at a time' (Delind 2011, 277). While simplistic, this provides individuals with a clear goal or way to discharge their ethical duties, such as increasing food sustainability, protecting the environment, improving animal welfare, creating a more equitable food system, etc. While a prolonged evaluation and assessment of goals is an important part of adequately addressing wicked problems, such as climate change, providing an easily identifiable action point could help marshal individualist and collectivist responses.

These could include the following: the recognition of a moral obligation to reduce individual emissions, the desire to lessen personal support for carbon-intensive industries by 'voting with your dollar', a commitment to not participate in environmentally harmful group activities, etc. (Vance 2017). Each are individual actions, but they could help mitigate the goal motivational barrier, as citizens committed to making personal changes are more likely to support collective action. For example, Thompson (2015) argues that 'buying local' can act as a heuristic that helps individual consumers better understand the wider impacts of food choice, thus gradually placing them on a path of supporting wider collectivist and justice-focused initiatives. A similar argument could be made concerning climate change mitigation that balances individual responsibility and collective action.

In fact, the literature on LFMs illustrates how active participation in local initiatives has the potential to increase motivation to work towards system-focused change. For example, LFMs are, by their very definition, focused on connecting people to local food-ways and systems. This translates to community-based initiatives that connect citizens to some segment of their local production system. This could take the form of creating farmers markets in food insecure neighbourhoods, community-supported agricultural projects, 'meet your farmer' events, field-to-fork programmes that highlight local products, etc. What LFMs offer, then, in addition to the promise of addressing food-related issues, are opportunities to have meaningful interactions with local components of the food system and the environments where food is grown. Both scientists (Church 2018; Colding and Barthel 2013; Krasny et al. 2014) and environmental philosophers (Light and Higgs 1996; Light 2009) have argued that these types of experiences contribute to strengthening moral relationships and building robust connections with the natural world. According to Church (2018, 879), engagement with the environment 'has the potential to contribute to building human connections with nature, to facilitate increased understanding of natural systems, and to influence individual environmental values and behavior'. Similarly, according to Light (2009), public participation helps increase social capital, whereby members of a community are more likely (for multiple reasons) to lay claim to their contexts and engage in collective action projects to maintain these spaces. As climate change threatens ecosystem services (Grimm et al. 2013) and biodiversity (Bellard et al. 2012; Botkin et al. 2007) worldwide, direct contact with what is being lost may help reduce the influence of the ambiguity barrier.

In addition, this analysis illustrates how LFMs could provide insight into how to begin to mitigate the threat barrier. In particular, they could also help to provide a blueprint for developing the 'radical hope' that Williston (2012) argues is necessary for bringing about change in increasingly dire times. Williston (2012, 165) defines radical hope as striving 'to retain our ability to flourish as moral agents'. However, the climate crisis creates a context where the ability of humans to flourish is being called into question. Thus, the object of radical hope is the desire to successfully avoid 'total catastrophe', and this desire grounds our motivations for addressing climate impacts. Part of maintaining radical hope is recognising 'the vital interests of members of the moral community' (Williston 2012,165). As such, one could argue that participating in collective action projects could be one way to cultivate this positive attitude. These types of projects are opportunities to build moral relationships with others and often includes expanding moral communities to encompass the environment and other species (depending on the type of project). They also involve recognising the vital interests of these newly recognized others. As such, these experiences may help to 'find a way for us to flourish in the teeth of the climate crisis' (Williston 2012, 183). Rather than be paralysed by fear, this could nudge individuals to build community ties and retain their ability to flourish in the face of potentially catastrophic climate-related impacts, thus lessening the impact of the threat motivational barrier.

LFMs also have a long history of balancing larger justice-related goals with small-scale projects. Specifically, community-focused LFMs have devoted years both to actively working to achieve food related goals and also addressing larger social justice concerns (Desmarais et al. 2010; Werkheiser and Noll 2014). We need only turn to Declaration of Nyéléni's (2015) definition of food sovereignty (an important type of community-focused LFM) to illustrate the plethora of structural injustices currently on the table:

> Food sovereignty is the right of peoples to healthy and culturally appropriate food produced through ecologically sound and sustainable methods, and their right to define their own food and agriculture systems. It puts those who produce, distribute and consume food at the heart of food systems and policies rather than the demands of markets and corporations. It defends the interests and inclusion of the next generation... It ensures that the rights to use and manage our lands, territories, waters, seeds, livestock and biodiversity are in the hands of those of us who produce food. Food sovereignty implies new social relations free of oppression and inequality between men and women, peoples, racial groups, social classes and generations.

The definition includes goals that are clearly in the agricultural context, such as increasing sustainability, improving agricultural processes and better managing land and water use. These types of projects are typical of most LFM sub-movements (Desmarais et al. 2010; Noll and Werkheiser 2017; Schanbacher 2010). However, one of the unique components of community-focused initiatives is that they often employ an expanded justice framework when working towards change. According to Murdock and Noll (2020, 4), 'food sovereignty focuses... on the larger structures and procedures that problematically create injustice and one of those injustices is food insecurity. In terms of solutions, then, food sovereignty has to be sensitive and aware of the different models of justice and the different ways in which harms can be perpetuated'. This sensitivity and awareness, coupled with community-based action, could provide useful insights for climate change mitigation.

There are myriad inequalities associated with climate change that could impact an individual's ability to act. Structural inequalities include the 'unequal distribution of impacts, unequal responsibility for climate change, and unequal costs for mitigation and adaptation' (Sowers 2007, 140). Additionally, the history of unequal global development negatively impacts the ability of communities to address harms effectively, as this history often translates into a greater vulnerability to climate impacts and/or a limitation of available options to curtail carbon emissions (Adger et al. 2003). If Barnett and Neil (2007) are correct that climate adaption strategies need to recognise wider socially relevant goals, then food sovereignty movements may be able to provide blueprints for how to meaningfully bring about change in such contentious contexts. Indeed, one could argue that the recognition of structural injustices by local movements is a precursor to meaningful deliberation concerning how to address them. At the very least, LFMs provide a model for how wider justice issues can be brought into climate change initiatives, even as communities work towards addressing short-term problems on the ground. As these wider structural justice issues could play a role in potentially curtailing individual action, such a blueprint may be imperative for helping communities across the globe begin to mitigate climate impacts.

Conclusion

One criticism of the above analysis that needs to be discussed is the argument that it moves beyond philosophy, as it draws from empirical work on motivations, environmental movements and local food. If this the case, then motivational barriers and strategies for potentially addressing these barriers should be considered empirical questions and thus outside of the purview of philosophical investigation. One reply to this critique is that environmental philosophers have been discussing the connections between collective action and the desire to protect nature for over 30 years. As such, this paper is not separate from philosophy but is building on this robust theoretical literature. In addition, as climate change is a 'wicked problem', it is complex, multifaceted and difficult to address (Whyte and Thompson 2011). These types of problems require solutions that are interdisciplinary by design. This chapter is also interdisciplinary, illustrating how philosophy can contribute to solutions and how this work could be useful in other disciplines.

This chapter began from the position that the individualist and collectivist applications of the no-harm principle provides the normative justification necessary to motivate individuals to act (Kyllonen 2018). However, these normative arguments need to be wedded to interdisciplinary work on motivational barriers. Even if we recognise a *prima face* duty to mitigate climate change impacts, certain motivational barriers could block people from making the requisite changes. Drawing from work in psychology, this chapter outlined four key barriers that could negatively impact a person's ability to discharge their duty to mitigate climate change. These include, but are not limited to goal barriers, ambiguity barriers, threat barriers and structural justice barriers. Local food and other environmental movements have a long history of bringing about ethically motivated change while grappling with wicked problems. Scholarly work on LFMs often highlights a) how ethically and ontologically centered concerns are actualised to motivate individuals and b) how commitments to address larger justice concerns can guide individual decision-making (Delind 2011; Alkon 2012; Noll and Murdock 2020). Concerning goal barriers, LFMs have developed solutions, such as providing clear directives, while attempting to address complex, systemic issues. Concerning the ambiguity barrier, community-based initiatives provide direct contact with what is being lost. In addition, this analysis illustrated how LFMs could provide insight into how to begin to mitigate the threat barrier. In particular, they could help provide a blueprint for developing the 'radical hope' that is needed to remain motivated in dire times. Finally, there are several inequalities associated with climate change that could impact an individual's ability to act. I hope this analysis leads to the development of a robust philosophical framework (combining duty-based arguments with motivational barriers) that could help to address effectively what has been called the most wicked problem of the modern age: climate change.

References

Adger, N., Saleemul H., Brown, K., Conway, D., and Hulme, M. 2003. "Adaptation to Climate Change in the Developing World." *Progress in Development Studies* 3 (3): 179–95.

Alkon, A. H. 2012. *Black, White, and Green: Farmers Markets, Race, and the Green Economy*. University of Georgia Press.

Almassi, B. 2012. "Climate Change, Epistemic Trust, and Expert Trustworthiness." *Ethics and the Environment* 17: 29–49. https://doi.org/10.2979/ethicsenviro.17.2.29

Bargh, J. A. 1997. "The Automaticity of Everyday Life,." In *The Automaticity of Everyday Life: Advances in Social Cognition*, edited by R.S. Wyer, 1–61. Mahway, NJ: Erlbaum.

Barnett, J., and W. Neil Adger. 2007. "Climate Change, Human Security and Violent Conflict." *Political Geography*, Climate Change and Conflict, 26 (6): 639–55. https://doi.org/10.1016/j.polgeo.2007.03.003

Baumeister, R. F., and J. A. Bargh. 2014. "Conscious and Unconscious: Toward an Integrative Understanding of Human Mental Life and Action." In *Dual-Process Theories of the Social Mind*, edited by J Sherman, B Gawronski, and Y Trope, 35–49. New York: Guilford Press.

Bell, D. 2013. "Climate Change and Human Rights." *WIREs Climate Change* 4 (3): 159–70. https://doi.org/10.1002/wcc.218

Bellard, C, Bertelsmeier, C., Leadley, P., Thuiller, W., and Courchamp, F. 2012. "Impacts of Climate Change on the Future of Biodiversity." *Ecology Letters* 15 (4): 365–77. https://doi.org/10.1111/j.1461-0248.2011.01736.x

Blau, J. 2017. *The Paris Agreement: Climate Change, Solidarity, and Human Rights*. London: Palgrave Macmillan UK.

Botkin, D. B., Saxe, H., Araújo, M. B., Richard Betts, R. H., Bradshaw, W., Cedhagen, T., Chesson, P., et al. 2007. "Forecasting the Effects of Global Warming on Biodiversity." *BioScience* 57 (3): 227–36. https://doi.org/10.1641/B570306

Church, S. 2018. "From Street Trees to Natural Areas: Retrofitting Cities for Human Connectedness to Nature." *Journal of Environmental Planning and Management* 61 (5–6): 878–903.

Cohen, G. L., and D. K. Sherman. 2014. "The Psychology of Change: Self-Affirmation and Social Psychological Intervention." *Annual Review of Psychology* 65: 333–71. https://doi.org/10.1146/annurev-psych-010213-115137

Colding, J., and S. Barthel. 2013. "The Potential of 'Urban Green Commons' in the Resilience Building of Cities." *Ecological Economics*, Sustainable Urbanisation: A resilient future, 86 (February): 156–66. https://doi.org/10.1016/j.ecolecon.2012.10.016

Cripps, E. *Climate Change and the Moral Agent: Independent Duties in an Interdependent World.* Oxford University Press.

Dahlberg, K. 1993. "Regenerative Food Systems: Broadening the Scope and Agenda of Sustainability." In *Food for The Future: Conditions and Contradictions of Sustainability*, 75–103. Hoboken: Wiley.

DeLind, L. B. 2011. "Are Local Food and the Local Food Movement Taking Us Where We Want to Go? Or Are We Hitching Our Wagons to the Wrong Stars?" *Agriculture and Human Values* 28 (2): 273–83.

Desmarais, A. A.,Wiebe, N., and H. Wittman. 2010. *Food Sovereignty: Reconnecting Food, Nature & Community*. Fernwood.

Fahlquist, Jessica Nihlén. 2009. "Moral Responsibility for Environmental Problems—Individual or Institutional?" *Journal of Agricultural and Environmental Ethics* 22 (2): 109–24. https://doi.org/10.1007/s10806-008-9134-5

Grimm, N. B, Chapin, F. S., Bierwagen, B., Gonzalez, P., Groffman, P.M., Luo, Y., and F. Melton. 2013. "The Impacts of Climate Change on Ecosystem Structure and Function." *Frontiers in Ecology and the Environment* 11 (9): 474–82. https://doi.org/10.1890/120282

Harlan, S., Pellow, D., Roberts, J.T., Bell, S.E., Holtt, W.G., Nagel, J. 2015. "Climate Justice and Inequality." In Climate Change and Society: Sociological Perspectives (eds. Riley E. Dunlap and Robert J. Brulle), 1-55. Oxford: Oxford University Press.

Higgins, E. T. 1989. "Continuities and Discontinuities in Self-Regulatory and Self-Evaluative Processes: A Developmental Theory Relating Self and Affect." *Journal of Personality* 57 (2): 407–44. https://doi.org/10.1111/j.1467-6494.1989.tb00488.x

Hiller, V. 2011. "Climate Change and Individual Responsibility." *The Monist* 94 (3): 349-368.

Jost, J. T., Glaser, J., Arie., Kruglanski, W. and F. J. Sulloway. 2003. "Political Conservatism as Motivated Social Cognition." *Psychological Bulletin* 129 (3): 339–75. https://doi.org/10.1037/0033-2909.129.3.339

Krasny, M. E., Russ, A., Tidball, K. G., and T. Elmqvist. 2014. "Civic Ecology Practices: Participatory Approaches to Generating and Measuring Ecosystem Services in Cities." *Ecosystem Services* 7 (March): 177–86. https://doi.org/10.1016/j.ecoser.2013.11.002

Kruglanski, A. W., Orehek, E., Higgins, E. T., Pierro, A. and I. Shalev. 2010. "Modes of Self-Regulation: Assessment and Locomotion as Independent Determinants in Goal-Pursuit." In *Handbook of Personality and Self-Regulation*, edited by R Hoyle, 375–402. London: Blackwell Publishing.

Kruglanski, A. W., Thompson, E. P., Higgins, E. T., Atash, M. N., Pierro, A., Shah, J. Y., and S. Spiegel. 2000. "To 'Do the Right Thing' or to 'Just Do It': Locomotion and Assessment as Distinct Self-Regulatory Imperatives." *Journal of Personality and Social Psychology* 79 (5): 793–815. https://doi. org/10.1037//0022-3514.79.5.793

Kruglanski, A. W., and D. M. Webster. 1996. "Motivated Closing of the Mind: 'Seizing' and 'Freezing.'" *Psychological Review* 103 (2): 263–83. https://doi. org/10.1037/0033-295x.103.2.263

Kruglanski, A. W. 2004. *The Psychology of Closed Mindedness*. The Psychology of Closed Mindedness. New York, NY: Psychology Press.

Kyllönen, S. 2018. "Climate Change, No-Harm Principle, and Moral Responsibility of Individual Emitters." *Journal of Applied Philosophy* 35 (4): 737–58. https://doi.org/10.1111/japp.12253

Levkoe, C. 2011. "Towards a Transformative Food Politics." *Local Environment* 16 (7): 687–705.

Light, A. 2009. "Ecological Restoration and the Culture of Nature." In *Readings in the Philosophy of Technology*, edited by David M. Kaplan, 398–411. New York, NY: Rowman & Littlefield.

Light, A, and E. S. Higgs. 1996. "The Politics of Ecological Restoration." *Environmental Ethics* 18 (3): 227–47.

Ludwig, D. 2001. "The Era of Management Is Over." *Ecosystems* 4 (8): 758–64.

Moser, Susanne, and Julia Ekstrom. 2010. "A Framework to Diagnose Barriers to Climate Change Adaptation." *Proceedings of the National Academy of Sciences of the United States of America* 107: 22026–31. https:// doi.org/10.1073/pnas.1007887107

Noll, S., and E. Murdock. 2020. "Whose Justice Is It Anyway? Mitigating the Tensions Between Food Security and Food Sovereignty." *Journal of Agricultural and Environmental Ethics* 33: 1–14. https://doi.org/10.1007/s10806-019-09809-9

Noll, S., and I. Werkheiser. 2017. "Local Food Movements: Differing Conceptions of Food, People, and Change." In *Oxford Handbook of Food Ethics.* Oxford: Oxford University Press.

Parks, B., and T. Roberts. 2010. "Climate Change, Social Theory and Justice." *Theory, Culture & Society* 27 (2–3): 134–66.

Pirog, R., Miller, C., Way, L., Hazekamp, C., and E. Kim. 2014. *The Local Food Movement; Setting the Stage for Good Food.* Lansing, Michigan: MSU Center for Regional Food Systems.

Quirico, O. 2018. "Climate Change and State Responsibility for Human Rights Violations: Causation and Imputation." *Netherlands International Law Review* 65 (2): 185–215. https://doi.org/10.1007/s40802-018-0110-0

Schanbacher, W. D. 2010. *The Politics of Food: The Global Conflict Between Food Security and Food Sovereignty.* Santa Barbara: ABC-CLIO.

Shah, J. Y., Friedman, R., and A. W. Kruglanski. 2002. "Forgetting All Else: On the Antecedents and Consequences of Goal Shielding." *Journal of Personality and Social Psychology* 83 (6): 1261–80. https://doi.org/10.1037/0022-3514.83.6.1261

Shalev, I. 2015a. "The Climate Change Problem: Promoting Motivation for Change When the Map Is Not the Territory." *Frontiers in Psychology* 6: 131. https://doi.org/10.3389/fpsyg.2015.00131

Shalev, I. 2015b. "The Climate Change Problem: Promoting Motivation for Change When the Map Is Not the Territory." *Frontiers in Psychology* 6 (February): 131. https://doi.org/10.3389/fpsyg.2015.00131

Shalev, I., and M. L. Sulkowski. 2009. "Relations between Distinct Aspects of Self-Regulation to Symptoms of Impulsivity and Compulsivity." *Personality and Individual Differences* 47 (2): 84–88. https://doi.org/10.1016/j.paid.2009.02.002

Shukla, P.R., E Skea, Buendia Calvo, H Masson-Delmotte, D.C. Portner, P. Roberts, R. Zhai, et al. 2018. "IPCC, 2019: Climate Change and Land: An IPCC Special Report on Climate Change, Desertification, Land Degradation, Sustainable Land Management, Food Security, and Greenhouse Gas Fluxes in Terrestrial Ecosystems." Governmental Website. IPCC Climate Report. 2018. https://www.ipcc.ch/

Sinnott-Armstrong, W. 2010. "It's Not My Fault; Global Warming and Individual Moral Obligations." In *Climate Ethics: Essential Readings*, edited by Stephen Gardiner, Simon Caney, Dale Jamieson, and Henry Shue, 332–46. New York: Oxford University Press.

Sowers, J. 2007. "The Many Injustices of Climate Change." *Global Environmental Change* 7 (4): 140–46.

Steele, C. M. 1988. "The Psychology of Self-Affirmation: Sustaining the Integrity of the Self." In *Advances in Experimental Social Psychology, Vol. 21: Social Psychological Studies of the Self: Perspectives and Programs*, 261–302. San Diego, CA, US: Academic Press.

Taylor, P. W. 1981. "The Ethics of Respect for Nature." *Environmental Ethics* 3 (3): 197–218. https://doi.org/10.5840/enviroethics19813321

Tollefson, J.. 2021. "IPCC Climate Report: Earth Is Warmer than It's Been in 125,000 Years." *Nature* 596 (7871): 171–72. https://doi.org/10.1038/d41586-021-02179-1

Thompson, P. 2015. *From Field to Fork: Food Ethics for Everyone*. Oxford: Oxford University Press.

Vance, C. 2017. "Climate Change, Individual Emissions, and Forseeing Harm." Journal of Moral Philosophy 14(5): 562-584.

Vanderheiden, S. 2011. "Globalizing Responsibility for Climate Change." *Ethics & International Affairs* 25 (1): 65–84. https://doi.org/10.1017/S089267941000002X

Whitmarsh, L., and S. O'Neill. S2010. "Green Identity, Green Living? The Role of pro-Environmental Self-Identity in Determining Consistency across Diverse pro-Environmental Behaviours." *Journal of Environmental Psychology*, Identity, Place, and Environmental Behaviour, 30 (3): 305–14. https://doi.org/10.1016/j.jenvp.2010.01.003

Whyte, K., and P. Thompson. 2011. "Ideas for How to Take Wicked Problems Seriously." *Journal of Agricultural and Environmental Ethics* 25 (4): 441–45. https://doi.org/10.1007/s10806-011-9348-9

Zhai, P., Pirani, A., Connors, S.L., Pean, C., Berger, S., Caud, N., Chen, Y., Goldfarb, L., Gomis, M.I., Huang, M., Leitzell, K., Lonnoy, E., Matthews, J.B.R., Maycock, T.K., Waterfield, T., Yelekci, O., Yu, R., and B Zhou. "IPCC, 2021: Climate Change." Governmental Website. IPCC Climate Report. 2021. https://www.ipcc.ch/report/ar6/wg1/

15

Gender and Climate Change

VERA TRIPODI

In this chapter, I focus on the negative consequences of climate change for women's rights. This chapter is divided into two parts. In the first part, I point out that poorer women in developing countries are one of the social groups most exposed to the effects of climate change. In the second part, I show why climate change affects women's lives and their rights.

Introduction

Global climate change is a matter in which gender issues intersect with those related to care and social justice. As the European Parliament Resolution of 20 April 2012 reports (Dankelman 2010), the impact of climate change on human beings has gender-differentiated effects. There is evidence that extreme natural phenomena have different consequences on people based on their gender identity. Thus, the impact of climate change on women is not the same as on men. Gender discrimination, much like the intersection between race discrimination and other factors, such as poverty or geography, hinders considerably a person's access to the means and resources they need to face and minimise the effects of climate change on a global level. These effects include natural disasters such as drought, floods and hurricanes. Poverty, along with social, economic and political barriers, make women increasingly unable to cope with the negative impacts of climate change. Thus, environmental crises, which often lead to humanitarian crises, pose a serious question of climate justice. Nevertheless, many governments still ignore the impact of gender inequality and women's socio-economic disadvantages. Hence, the underlying issue of gender inequality remains a critical challenge for ecofeminists to address. In this chapter, I show that women are one of the social groups most exposed to 'climate injustice'. The aim of this chapter is twofold. First, I show that poorer women in developing countries are highly susceptible to climate change. The responsibilities these women assume or are forced to assume during natural disasters render them more exposed to the effects of climate change. Second, I illustrate why climate change poses such a threat to women's lives and how these changes significantly affect their social, political and economic rights. Climate change affects human rights, and the negative consequences of climate change can be seen as violating fundamental rights of women in particular.

1. Women and the Negative Effects of Climate Injustice

Climate change increases the number and intensity of natural disasters such as tropical cyclones, prolonged dry spells, intense rainfall, heatwaves, tornadoes, thunderstorms, severe dust storms, rising sea levels, deforestation and water and environmental pollution. These ecological disasters have dramatic consequences for human populations. The most serious of these consequences include the risk of disease due to an increasing number of pathogenic microorganisms, malnutrition due to the destruction or failure of crops and the forced migration of both people and animals. Statistically, 'climate refugees' and the victims of natural disasters are more likely to be women and children than men or adults (Alston and Whittenbury 2013). Women – especially heads of households in rural, less-developed countries – are among those whom 'climate injustice' affects most severely. This term relates to the inability of these women to deal with the harmful effects of climate change effectively (Masika 2002; Tuana and Cuomo 2014). Factors such as a person's social status, gender, poverty and access to and control over economic resources affect how much climate change impacts their lives. As many studies show (Neumayer and Plümper 2007; Castañeda and Gammage 2011; Swarup 2011; Atkinson and Bruce 2015), since women in developing countries have reduced economic, political and legal power, they are less able to manage the negative effects of climate disasters. In rural areas, women are particularly vulnerable to climate change because they are highly dependent on local resources for their livelihoods. In these countries, women are predominantly in charge of securing clean drinking water, food and fuel for cooking and heating. Gender inequality increases after events, such as extreme drought, in places where it is the responsibility of women and girls to collect water every day, especially when water resources become scarcer. As Stern (2009, 70) maintains, in drought-affected countries, 'many poor people, but particularly women and girls, will have to spend more time and energy fetching water from further away'.

Several factors explain the differences between women and men in terms of their exposure and vulnerability to risks from climate change. First, gender differences affect a person's ability to access goods and services. Certain markets and institutions limit women's access to credit and labour opportunities. As Cuomo (2011) explains, the global gender gap in earnings and productivity is considerable: women earn only 30–80% of what men earn each year on average. According to a World Bank survey carried out in 141 countries, 103 countries impose legal restrictions on women that hamper their economic opportunities (Mishra and Mohapatra 2017, 151). The available data from developing countries indicates that more women work in agriculture and basic crop production than men. More than their male counterparts, these women often reinvest their income into the family, too. Even though women represent half of the agricultural workforce, they only own between 10–20% of the land.

Secondly, unlike men, women face significant challenges in influencing any level of policy or decision-making within their communities. Indeed, women are often underrepresented or completely absent from the institutions that make or negotiate decisions regarding how their society might tackle climate change. Unlike men, women in certain rural areas also lack important financial resources. This economic deprivation not only causes income inequality for women but also renders them less capable of contributing towards the policies and programmes that could impact their lives.

Third, socio-cultural norms often limit women's knowledge of the information and skills they need to avoid danger, including the ability to swim or climb trees when escaping rising water levels. Likewise, these same norms often impose dress codes on women that limit their mobility in times of danger, as well as their ability to assist small children who cannot swim or run. Climate problems and environmental disasters also significantly impact women's education, health and migration. Environmental catastrophes reduce education rates considerably: children often drop out of school to help their mothers work in fields, and the strain on resources forces women to work harder to obtain drinking water, food and energy (Stern 2009, 70). Even if these women experience the greatest impacts of environmental disasters, they often lack the information and knowledge needed to act effectively in emergencies. Furthermore, the percentage of illiterate women in the world is much higher than that of men. Since women experience illiteracy at a higher rate on average, they are more likely to lack the information and means of communication to act accordingly in the event of a natural and catastrophic event.

Moreover, social influences make women disproportionately vulnerable to disasters and the negative effects of climate change. During and after extreme weather events, girls face an increased risk of violence and exploitation, including sexual and physical abuse and trafficking (Alston et al. 2014). These risks intensify when women are out collecting food, water and firewood or when disasters force them to stay in temporary shelters. In climate-affected areas, married women in farming families report that their obligations significantly increase compared to climate-unaffected area. In fact, drought reduces agricultural income and forces many women to seek job opportunities outside the farm (Afridi, Mahajan and Sangwan 2021, 2022). Nevertheless, although some of these women have to travel far away or relocate temporarily for work, most women face care responsibilities within the home, which form a significant barrier to their opportunities for progress. In contrast, their male partners respond to the loss of financial security and livelihoods with increased rates of depression and domestic violence (Alston 2008). Climate change escalates the likelihood of physical distress for women and girls along with other effects, such as decreased opportunities to attend school and increased risk of assault (Stern 2009; United Nations News Centre 2009). Ultimately, then, men and women experience different levels of vulnerability to the consequences of climate change.

Fourth, the lack of gender distinctions in data related to livelihoods, disaster preparedness, environmental protection, health and well-being often leads to this data misrepresenting women's roles and contributions. The result is often a gender-blind climate change policy that does not consider differences in gender roles and norms, such as the fact that women and men have distinct needs, constraints and priorities. In particular, in the cases of forced or voluntary migration induced by climate change, women and men need different provisions regarding health and personal security and protection. In poor countries, droughts and water shortages resulting from climate change force women to work more than men to provide water, food and energy, and young girls frequently abandon school to help their mothers in these tasks. Therefore, such policies might have the unintended effect of increasing women's vulnerability (I am not saying or denying here that climate change policies, which are often gender-blind and do not recognise women's contributions to strategies for changing environmental conditions, do not indirectly affect and damage men, too).

Therefore, it is not surprising to learn that how climate change disproportionately affects women – but also people of colour and the poor – is a matter of feminist concern (Masika 2002; Hemmati and Röhr 2007; Dankelman 2010). Natural disasters affect women in relation to the role they play as producers and procurers of food, caregivers and economic subjects. Climate change, both natural and human-induced, makes poorer women in developing countries even more vulnerable. The responsibilities these women adopt either voluntarily or through necessity during disasters make them defenceless to the effects of such changes. Not only do women consistently face shortages of water and food as a result of climate change, but they also face risks to their health and lives (UNFCCC 2007). Moreover, the high mortality and infectious diseases rates of women in affected areas are precisely due to the fact that they are mainly caregivers.

Contemporary environmental problems influence the lives and status of women (Gaard and Gruen 2005). Thus, the negative effects of climate change, which include water and environmental pollution, are a feminist issue. In particular, eco-feminists have examined how patriarchal, misogynistic, racist and anthropocentric cultures have partaken in eco-destructive forms of development and progress (Griffin 1978; Merchant 1980; Haraway 1990; Warren 1990; Cuomo 1992; Gaard and Gruen 1993; Mies and Shiva 1993; Plumwood 1994; Cuomo 1998). According to these critics, feminists should take the issue of vulnerability and gender inequality seriously in relation to the goals of climate justice. Furthermore, eco-feminists emphasise that their perspectives offer alternatives to eco-destructive cultures. Some eco-feminists even use the slogan 'nature is a feminist issue' to define their environmental philosophy (Warren 2000). The next section examines how climate change disproportionately impacts the rights of people in vulnerable groups and situations, such as women who are experiencing poverty.

2. Climate Justice and Human Rights

As some authors maintain, climate change is a social injustice issue that threatens one's ability to enjoy the full range of their human rights. According to the United Nations Universal Declaration of Human Rights, 'everyone has the right to life, liberty and security of person'. Article 1 of the Declaration states that 'all human beings are born equal in dignity and rights'. Additionally, Article 19 asserts that 'everyone has the right to freedom of opinion and expression; this right includes freedom to hold opinions without interference and to seek, receive and impart information and ideas through any media and regardless of frontiers'. Consequently, the Declaration maintains that all states should, at the very least, take effective measures against predictable or preventable loss of life. Moreover, these states should respect, protect, promote and fulfil the right to life along with equal access to information and education and liberty to participate in the cultural and economic life of their community.

However, one might wonder how the weather effects of climate change and global warming, such as hurricanes, cyclones, typhoons, drought, floods, rain and snow affect human rights. Moreover, it might be argued that climate is a matter of weather and that the phrase 'weather effects of climate change' is redundant. I might reply that there are effects of climate change that do not manifest themselves as weather phenomena, such as, for example migration, violence and exploitation as negative consequences induced by some effects of climate change. As Lane and McNaught (2009) show, women and children are particularly vulnerable to abuse during times of drought. Also, during these events and in emergency situations, many young women are forced to labour for cash in local towns or have to take care of their siblings (Swarup 2011). In doing so, they do not attend school or renounce education opportunities, and their exposure to potential abuse and exploitation increases. In addition to weather phenomena, climate change impacts important rights, such as the rights to life, health, food, water and sanitation, adequate housing and, in some communities, the very right to self-determination. Further detrimental impacts include the rights to development, education and meaningful and informed participation in social, cultural and economic spheres. The rights of those most affected by climate change are those of future generations. Human rights and climate justice are therefore interdependent issues.

Everyone should have the right to participate actively, freely and meaningfully in planning and decision-making processes that impact global climate and weather. The participation should also include the right to receive a prior assessment of the consequences of the proposed actions on climate change and human rights. It should similarly include the right to a fair hearing and the right of underrepresented groups to participate in the protection of their land, natural resources, property rights and cultural heritage.

However, many governments deny social groups such as indigenous people and women the right to participate equally and meaningfully in climate change discussions. In particular, climate change intensifies gender inequality and discrimination and adversely affects women's rights. Many countries deny women the right to be informed and to participate in decision-making processes about changes to their physical and natural environments – the very land on which they rely for their health and survival. In many social contexts, governments deny women the right to receive timely, clear, understandable and accessible information without financial burden. Furthermore, these women often lack the right to have, express and disseminate their opinions about climate change issues. In addition, authorities in poor and disaster-affected areas often deny women the right to an education, a climate education and the ability to learn from multiple perspectives.

As highlighted above, women play an important role in managing natural resources along with other productive and reproductive activities within the household and wider community. However, authorities do not always recognise this role, meaning that women are often unable to contribute to livelihood strategies adapted for changing environmental conditions. Many authors argue that society should recognise women's knowledge and skills and then utilise them in strategies for mitigating the catastrophic effects of climate change (Adger et al. 2013; Leisheret al. 2016; Tanjeela and Rutherford 2018; Wamsler and Brink 2018; Garikipati and Kambhampati 2021). There are good reasons to believe that allowing women to participate meaningfully in these policies would improve the effectiveness and sustainability of climate change projects and programmes. Not only would recognising the role of women have the advantage of addressing existing social inequalities, but it would also satisfy international agreements on the need to address women's equality and empowerment. Therefore, eco-feminists should promote a better understanding of how climate change impacts the human rights of women, namely of how climate change entails the violation of some women's rights or prevents women from fully enjoying them – I should specify that, by women, I mean anyone who identifies as a woman and/or who is perceived or treated as a woman.

Following de Beauvoir (1972), I adopt the practice of using gender as social notion and sex as biological one and the view according to which gender and sex are not necessarily linked. More precisely, a violation of fundamental human 'capabilities' that – following Nussbaum – are essential to a good life - as it is noted, Nussbaum (1995, 85) distinguishes a list of 10 central capabilities, and she maintains that 'a life that lacks any one of these capabilities, no matter what else it has, will fall short of being a good human life'. To achieve this goal, we should further examine the variegated data that focuses specifically on gender and its intersection with other factors, such as age, disability and ethnicity. Gender equality is a fundamental human right. As many authors argue (Denton 2002; Terry 2009; Tuana 2012, 2022), it is imperative that governments promote equality when devising their climate-development policies. In all efforts to reverse or prevent climate change, world governments must centre gender equality along with the full and equal participation of women. I am not saying that men are not indirectly damaged from authorities that do not recognise the importance of women's contributions to livelihood strategies adapted for changing environmental conditions. Rather, I am saying that climate change policies that do not recognise or do not take into account women's contributions from the very beginning amplify inequalities and have a negative impact on gender equality and women's rights.

3. Climate Change and Women Rights

The philosophical issue of women's rights has long concerned feminist scholars. Since the end of the seventeenth century, feminist thinkers have discussed these rights especially in terms of the legal status of women. Two texts are of seminal interest here: Olympe de Gouges's (1791) *The Declaration of the Rights of Woman and the Female Citizen* and Mary

Wollstonecraft's (1792) *A Vindication of the Rights of Woman: With Strictures on Political and Moral Subjects*. John Stuart Mill (1869) also examines the question of women's legal status in *The Subjection of Women*, a text that later became a point of reference for liberal feminist thought. In the latter text, Mill brilliantly and passionately defends the moral, civil and political rights of women. All of the above texts are decidedly current due to their polemical tone and how they openly denounce the subjugation of women. However, only the Vienna Declaration truly attempts to articulate women's rights and formulate an agenda to ensure that women can fully enjoy them.

The representatives of the states present at the United Nations World Conference on Human Rights approved the Vienna Declaration on 25 June 1993. The Declaration proclaims that 'the human rights of women and of the girl-child are an inalienable, integral and indivisible part of universal human rights' (Article 18). Furthermore, the Declaration maintains that each signatory must defend 'the full and equal participation of women in political, civil, economic, social and cultural life, at the national, regional and international levels, and the eradication of all forms of discrimination on grounds of sex are priority objectives of the international community'. However, we must now determine how women's rights constitute human rights and why some of them relate only to the female gender. Feminist philosophers have argued that the standard lists of human rights included in treaties and documents is not a sufficient remedy for gendered oppression and dehumanisation (Cudd 2005; Meyers 2016). Men and women face different risks. The lists of human rights do not sufficiently take into account these differences and do not make a prominent place some issues like domestic violence, reproductive choice and trafficking of women and girls for sex work. The lists of human rights have to be expanded 'to include the degradation and violation of women' (Okin 1998; Bunch 2006, 58; Lockwood 2006).

On 10 December 1948, the United Nations General Assembly guaranteed human equality with the Universal Declaration of Human Rights. However, although the 1948 Declaration marked an important first step towards international recognition of human dignity, it proved to be ineffective in protecting the dignity of women. In fact, many interpretations of the Declaration regard the violation of women's rights within the family and at home as a private matter that does not concern states or international organisations. In addition to affirming the principle of non-discrimination between the sexes, the Action Program of the Vienna Declaration specifies, for the first time in the history of international law, that gender-based violence due to culture or religion, harassment or sexual exploitation is detrimental to the dignity and value of the human person. Moreover, the Action Program defines mass violence against women such as ethnic rape, forced pregnancy and sex trafficking as crimes against humanity. According to the 1993 Vienna

Declaration, defending women's rights is therefore an integral part of protecting human rights in general. The Vienna Declaration thus addresses shortcomings in the original 1948 document. In line with the Vienna Declaration, the Program of Action at the Fifth World Conference of Women in Beijing (1995) recognised women's sexual and reproductive freedoms as integral to human rights in general - women's reproductive rights include, for example, the right to abortion, control over reproduction, birth control, freedom from coerced sterilisation, the right to receive education about sexually transmitted infections, menstrual health and protection from practices of genital mutilation (Freedman-Isaacs 1993). I am not denying that also men have the right to sexual and reproductive freedom (Ziegler 2020).

For this reason, it is legitimate to ask which rights count as human rights and which relate specifically to the lives of women. Current debates about which to include among human rights, along with recent controversies over their risk of devaluation due to constant demands and pressures, expanding the list that contains these rights is a difficult task. By definition, human rights are the inalienable rights of person. They include, among others, the right to life, to freedom and security, to freedom of opinion and expression, to a fair trial, to a dignified existence, to religious freedom and to gather in organisations or associations. However, we should distinguish these human rights from civil and political rights. While human rights are universal in their distinction from citizenship or national laws, civil and political rights apply only to the organisations and associations of citizens. Thus, each state's national laws guarantee and govern civil rights. All individuals possess human rights for the simple fact that they belong to the human species. But civil rights depend upon an individual's citizenship within an organized society. Subsequently, when we speak of the rights of women, we generally mean that they are a set of particular human rights. However, one may object that to speak of human rights as particular to a certain group is a contradiction in terms. In fact, human rights are universal and belong to everyone, both women and men. In this way, one might question the legitimacy of reserving a set of human rights to a specific part of the human species.

However, two issues motivate the need to claim women's rights as specific human rights. For political and historical reasons, we cannot separate the need to affirm equality between the sexes from a recognition of their differences. Since the 1990s, the international community has affirmed the need to reread people's fundamental freedoms from a gendered perspective. To ensure the protection of women's rights, it is not enough to condemn sexual discrimination and recognise gender equality between men and women. Rather, we need to radical reconceptualise people's rights in light of sex and gender differences. In this context, though, one need not understand these differences in an essentialist or identitarian sense. On the one hand, the essentialisation of differences between the sexes and, on the other, the claim that some communities have made for collective cultural liberties has resulted in serious violations of women's rights. A pertinent example of such violations includes the acts of genital mutilation or child marriage that occur within some religious communities. Thanks to the significant work of international organisations since the 1990s, we have begun to reconceptualise universal human rights. Namely, we assert that feminists need not compromise the dignity of women with respect to the cultural or religious beliefs that govern their role in a given society. In this way, we have affirmed a new dimension of personal rights.

One of the aims of the above trend is to place women's rights on a foundation that reconciles cultural and religious specificities with respect for the universal dignity of all women. The rights in question here also include those of ethnic groups and other social minorities. By identifying among human rights those that belong specifically to women, feminists have also questioned the historical trend of using a male model to formulate and elaborate upon individual liberties. It is only through a review of the relationship between equality and gender difference that a defence of universal rights is possible.

As many international institutions and organisations assert, all future policies for preventing natural disasters must guarantee that women have fair participation in planning and implementing decision-making projects (Cuomo 2011; Posner and Weisbach 2010; UNDP 2009; Vanderheiden 2008). In fact, respect for human rights and equal opportunities can greatly enhance policies that aim to develop a sustainable economy and protect the natural environment. Therefore, as some authors maintain, there is a need to consider the issue of gender when devising policies for climate change, development and the fight against poverty. In other words, we need a policy that can address the unique skills that women have developed over the years, such as managing drinking water for domestic and agricultural use. Thus, since climate change concerns, and must be addressed from, areas as diverse as the political-economic to the social and the cultural, the strategies for combatting this global problem must also be attentive to and respectful of gender.

Climate problems intensify gender inequalities and discrimination. In poorer countries, where the effects of climate change are dramatically more evident, the outcomes also negatively affect women's rights. Alongside financing or energy-saving strategies, as well as agricultural policies that respect biodiversity, activists must also raise awareness of a more equitable division of family care and household management. They must also work to protect women and their health by highlighting women's individual needs and rights, such as those of education. In efforts to combat climate change, eco-feminists can achieve their goal of 'climate justice' only by fighting against poverty and gender inequalities while also campaigning for a reformulation of care work. Furthermore, since the above types of marginalisation exclude particular groups and social categories from enjoying their fundamental rights, feminists assert that human rights require the characteristics of a distinction. Nevertheless, we can still include the protection of women in the broader defence of universal human rights. In this way, one should not understand the task of safeguarding women's rights as something that benefits only women. Rather, as many critics assert, eco-feminism is a political-social fight that is necessary for the progress of all humanity. Feminists have historically outlined the demand for women's rights primarily through the fight for equal access to education and legal equality between spouses. However, this legal claim did not in fact put an end to gender inequality. Therefore, as many theorists argue, the defence of women's rights today requires a complete philosophical redefinition that includes the concept of care (Okin 1998; Nussbaum 2000; Nussbaum 2007). In fact, the issue of care raises important questions regarding social justice, gender inequality and climate change. Since most cultures allocate caring responsibilities primarily to women, it is predominantly women who care for children and families in areas that suffer the harmful effects of climate emergencies, crises and breakdowns. Obviously, this does not mean that women have human rights as caregivers. Rather, women have to be granted specific human rights as women. The fact that, in many countries, they are mainly caregivers due to gendered roles and patriarchal culture is merely de facto.

References

Adger W. N., Barnett J., Brown K., Marshall N., O'Brien K. 2013. "Cultural dimensions of climate change impacts and adaptation". *Nature Climate Change* 3: 112–117.

Afridi F., Mahajan K., Sangwan N. 2022. "The Gendered Effects of Droughts: Production Shocks and Labor Response in Agriculture". *Labour Economics* 78: 102227.

Afridi F., Mahajan K., Sangwan N. 2021. "The Gendered Effects of Climate Change: Production Shocks and Labor Response in Agriculture". IZA Discussion Paper, no. 14568. https://www.iza.org/publications/dp/14568/the-gendered-effects-of-climate-change-production-shocks-and-labor-response-in-agriculture

Alston M. 2008. "The big dry: The link between rural masculinities and poor health outcomes for farming men," *Journal of Sociology* 44 (2): 133–47.

Alston M., Whittenbury K. 2013. *Research, Action and Policy: Addressing the Gendered Impacts of Climate Change*, Springer.

Alston M., Whittenbury K., Haynes A., Godden N. 2014. "Are climate challenges reinforcing child and forced marriage and dowry as adaptation strategies in the context of Bangladesh?". *Women's Studies International Forum* 47: 137–144.

Atkinson H. G., Bruce J. 2015. "Adolescent Girls, Human Rights and the Expanding Climate Emergency," *Annals of Global Health* 81 (3): 323–330.

Bunch C. 2006. "Women's Rights as Human Rights," in B. Lockwood (ed.), *Women's Rights: A Human Rights Quarterly Reader*, Baltimore: Johns Hopkins University Press.

Caney S. 2010. "Climate Change, Human Rights and Moral Thresholds," in Humphreys, S. (ed.), *Human Rights and Climate Change*, Cambridge: Cambridge University Press.

Castañeda I., Gammage S. 2011. "Gender, global crises, and climate change," in Jain, D., Elson D. (eds), *Harvesting Feminist Knowledge for Public Policy,* New Delhi, India: Sage Publications.

Cuomo C. J. 2011. "Climate Change, Vulnerability, and Responsibility". *Hypatia. A Journal of Feminist Philosophy* 26 (4): 690–714.

Cuomo C. J. 1992. "Unraveling the problems in ecofeminism". *Environmental Ethics* 15 (4): 351–63.

Cuomo C. J. 1998. *Feminism and Ecological Communities: An Ethic of flourishing*. New York: Routledge.

Dankelman I. 2010. *Gender and Climate Change: An Introduction*. London: Earthscan.

De Beauvoir S. 1972. *The Second Sex*, Harmondsworth: Penguin.

Denton F. 2002. "Climate change vulnerability, impacts, and adaptation: Why does gender matter?". *Gender and Development* 10 (2): 9–20.

Freedman Lynn P., Isaacs S. L. 1993. "Human Rights and Reproductive Choice." *Studies in Family Planning* 24 (1): 18-30.

Garikipati S. and Kambhampati U. 2021. "Leading the Fight Against the Pandemic: Does Gender Really Matter?". *Feminist Economics* 27 (1–2): 401–418.

Gaard G., Gruen L. (1993). "Ecofeminism: Toward global justice and planetary health". *Society and Nature* 2 (1): 1–35.

Gouges Olympe de. 1791 / 2011. *Les droits de la femme. A La Reine.* Translated as *The Rights of Woman. To the Queen*, in *Between the Queen and the Cabby: Olympe de Gouges's Rights of Woman* (J H. Cole, Trans). Montreal: McGill-Queen's University Press, 28–41.

Griffin S. 1978. *Woman and Nature: The Roaring Inside Her.* New York: Harper & Row.

Haraway D. J. 1990. *Simians, Cyborgs, and Women: The Reinvention of Nature.* Routledge, New York.

Hemmati M., Röhr U. 2007. "A huge challenge and a narrow discourse: Ain't no space for gender in climate change policy?". *Women and Environment International* 74–75: 5–9.

Lane R., McNaught R. 2009. "Building gendered approaches to adaptation in the Pacific". *Gender and Development* 17 (1): 67–80.

Leisher C., Temsah G., Booker F. et al. 2016. "Does the gender composition of forest and fishery management groups affect resource governance and conservation outcomes? A systematic map," *Environmental Evidence* 5 (6).

Lockwood, B. (ed.). 2006. *Women's Rights: A Human Rights Quarterly Reader*, Baltimore: Johns Hopkins University Press.

Masika R. 2002. *Gender, Development, and Climate Change.* Rugby: Oxfam Publishing.

Merchant C. 1980. *The Death of Nature: Women, Ecology and the Scientific Revolution*. San Francisco: Harper and Row.

Meyers D T. 2016. *Victims' Stories and the Advancement of Human Rights*, Oxford: Oxford University Press.

Mies M., Shiva V. 1993. *Ecofeminism*. London: Zed Books.

Mishra A. K., Mohapatra U. 2017. "Is Climate Change Gender Neutral?", *International Journal of Humanities and Social Sciences (IJHSS)* 6 (5): 149–154.

Mill, J. S. 1984. *The Subjection of Women*. In J. Robson (Ed.), *Essays on Equality, Law, and Education: Collected Works of John Stuart Mill, Volume XXI* (259–348). Toronto: Toronto University Press.

Neumayer E., Plümper T. 2007. "The gendered nature of natural disasters: the impact of catastrophic events on the gender gap in life expectancy, 1981–2002". *Annals of the Association of American Geographers* 97: 551–566.

Nussbaum M., Glover J. (eds). 1995. *Women, Culture and Development. A Study of Human Capabilities*, Clarendon Press: Oxford.

Nussbaum M. 2000. *Women and Human Development: The Capabilities Approach*, Cambridge, MA: Harvard University Press.

Nussbaum M. 2007. *Frontiers of Justice*, Cambridge, MA: Harvard University Press.

Okin, S. 1998. "Feminism, Women's Human Rights, and Cultural Differences". *Hypatia*, 13: 32–52.

Plumwood V. 1994. *Feminism and the Mastery of Nature*. Routledge: London.

Posner E., Weisbach D. 2010. *Climate change justice*, Princeton: Princeton University Press.

Stern N. 2009. *The Global Deal: Climate Change and the Creation of a New Era of Progress and Prosperity*. New York: Public Affairs.

Swarup A *et al.* 2011. *Weathering the Storm: Adolescent Girls and Climate Change,* Plan International, London.

Sweetman C. (ed.) 2009. "Climate Changes and Climate Justice", *Gender and Development* 17 (1): 1–178.

Tanjeela M., Rutherford S. 2018. "The influence of gender relations on women's involvement and experience in climate change adaptation programs in Bangladesh," *Sage Open* 8 (4): 1–9.

Terry G. 2009. "No climate justice without gender justice: An overview of the issues". *Gender & Development* 17 (1): 5–18.

Tuana N. 2012. "Climate change and human rights," in Thomas Cushman (eds), *Handbook of Human Rights*. London and New York: Taylor and Francis, 410–418.

Tuana N. 2022. *Racial Climates, Ecological Indifference: An Ecointersectional Approach*. Oxford University Press.

Tuana N., Cuomo C. J (Eds). 2014. "Climate Change". *Hypatia: A Journal of Feminist Philosophy"* 29 (3), 533–719.

United Nations News Centre 2009. *With Better Stoves, UN Aims to Cut Risk of Murder, Rape for Women Seeking Firewood.* https://news.un.org/en/story/2009/12/324532

United Nations Development Programme (UNDP). 2009. *Resource Guide on Gender and Climate Change.* http://www.un.org/womenwatch/downloads/Resource_Guide_English_FINAL.pdf

Vanderheiden S. 2008. *Atmospheric justice: A political theory of climate change*. Oxford: Oxford University Press.

Wamsler C., Brink E. 2018. "Mindsets for sustainability: exploring the link between mindfulness and sustainable climate adaptation", *Ecological Economics*, 151: 55–61.

Wollstonecraft M. 1972. *A Vindication of the Rights of Woman with Strictures on Political and Moral Subjects* (M. Brody Kramnick, Ed.). Harmondsworth: Penguin.

Ziegler M. 2020. "Men's Reproductive Rights: A Legal History", *Pepperdine Law Review*, vol. 47 (3), 665–730.

16

The Rights of the Ecosystem and Future Generations as Tools for Implementing Environmental Law

GIANLUCA RONCA

The threats of ecological destruction and the annihilation of the very possibility of conservation of the way of being-in-the-world, which is caused by climate change, poses a theoretical challenge to both political philosophy and applied ethics in the investigation on the normative assumptions of intergenerational justice. The first part of this chapter examines the difference between a right of the ecosystem and a right to the environment under a framework of the third generation of rights. Do non-living beings enjoy the right of inviolability? What kind of collective right is involved? How does the right to a healthy environment relate to the system of rights promoted at the international level? Is it a human right? The second part of this chapter outlines the difference between the right to the future and the right of future generations. Starting from this double taxonomy, I defend a cogent interpretation of collective responsibility, which is politically justified and inferred from the new doctrine born around the international responsibility of states.

Introduction

Nowadays, the threat of global warming is not only recognised by the vast majority of the scientific publications and by mainstream national political agendas, but it has also become central to public debate at all levels. This chapter assumes that anthropogenic climate change exists and is an urgent problem. This chapter aims to question philosophically the relevance of the environment for human rights. More precisely, I ask whether recent legal developments related to climate change might prompt us as philosophers and jurists to justify the rights of non-human entities, on the one hand, and the rights of future generations on the other. As can be seen, here climate change is linked explicitly to the concept of environment (I specify this term below). The aim is to answer a series of questions that straddle the line between political and legal philosophy. I examine these two disciplines together because I believe that in the literature they often get confused. In order to sustain a critique of anthropogenic ecological exploitation, it is essential to distinguish between the rights to the future, rights of future generations and rights of the ecosystem. The latter two set of rights constitute the object of this work: if national constitutions tend, in general, towards a protection of the latter, international charters underpin the existence of the former.

1. Why Does the Environment Matter?

1.1. Environment and Values

A preliminary question is how to conceive of the 'environment' when we talk of 'the right to the environment'. And it is not idle to wonder whether there exists a right of entities of nature *strictu sensu* that differs from the right to the enjoyment of the environment by human beings.

The question of value is crucial in also determining the type of rights with which we are dealing. As Brennan and Yeuk-Sze (2021, 3) state:

> In the literature on environmental ethics the distinction between *instrumental value* and intrinsic value (in the sense of 'non-instrumental value') has been of considerable importance. The former is the value of things as *means* to further some other ends, whereas the latter is the value of things as *ends in themselves* regardless of whether they are also useful as means to other ends. For instance, certain fruits have instrumental value for bats who feed on them, since feeding on the fruits is a means to survival for the bats. However, it is not widely agreed that fruits have value as ends in themselves... [A] certain wild plant may have instrumental value because it provides the ingredients for some medicine or as an aesthetic object for human observers. But if the plant also has some value in itself independently of its prospects for furthering some other ends such as human health, or the pleasure from aesthetic experience, then the plant also has intrinsic value.

Insofar as an entity is endowed with intrinsic value, it is reasonable to consider certain effort to protect, caring for and preserve it. If we take the example cited above, we would perhaps instinctively be inclined to consider plants useful for our well-being. Human beings are one of the classic examples of entities endowed with intrinsic value not only in the philosophical literature, but also in international declarations of human rights. The position of theoretical privilege can take the guise of a strong or weak anthropocentrism depending on the exclusivity or not of the intrinsic value assigned to human beings (Brennan and Yeuk-Sze 2021). A typical intermediate solution proposed by environmental ethicists is the adherence to what has been labelled enlightened or prudential anthropocentrism, which assumes an immediate practical implication and is justified as necessary and sufficient for preserving the environment (see Norton 1991; de Shalit 1994; Light and Katz 1996).

Beyond the different internal positions, it is useful to schematise the theses at the basis of enlightened or prudential anthropocentrism so as to compare them with the positions advocated in this work:

1. Duties to the environment are not *prima facie* duties, but rather derive from duties to other human beings.
2. Such duties provide necessary and sufficient reasons to construct more pragmatic policies and actions and provide moral grounds for environmental protection.

Within this overall approach, it would not be necessary to inflate the value of non-human environmental entities to justify our actions towards them.

1.2 The Environment and European Law

The Treaty of Rome (1957) did not contain any explicit mention of community competences in the environmental sector. Growing awareness of environmental problems, beginning the early 1970s, has led to the recognition of the need to develop a continental policy in this field. As a result of this change in awareness, the European Council adopted the 'First Action Programme on the Environment' in 1973, which was soon followed by others. However, although through this activity the European Economic Community (EEC) had taken on an increasingly important role in environmental protection, the legal bases for EEC-wide competence in environmental matters remained uncertain. Consequently, the only viable way to anchor the requirements of environmental protection in the EEC's legal order seemed to be to resort to an extensive interpretation of some provisions of the Treaty of Rome, such as Articles 100 and 235.

The first article of the Treaty of Amsterdam (1997) explicitly includes the 'principle of sustainable development' in the seventh recital of the Preamble of the Treaty of the European Union, in the framework of the promotion of 'economic and social progress' of the peoples of the member states and more explicitly as an objective of the union in order to achieve 'balanced and sustainable development' (Treaty of Amsterdam, art. 2 ex art. B, 8). The needs of the environment become a necessary element in the evaluation of policies and actions likely to damage the environment. Article 191 of the Treaty on the Functioning of the European Union specifies the objectives of EEC's environmental policy by defining the basic principles on which the EEC's environmental action should be based, listing the factors that should be taken into account in environmental policy development.

The Single European Act (1987), the Regulation of 7 May 1990 No. 1210/90/EEC and related documents until the Regulation (EC) No. 401/2009 state that the preservation, protection and improvement of the quality of the environment – including the conservation of natural habitats and of wild fauna and flora – constitute a core mission of general interest pursued by the EEC. These objectives include and refer to the protection of human health and the prudent and rational utilisation of natural resources.

Fundamental principles of the Treaty state the following: i) preventive action is needed; ii) damage caused to the environment must be rectified at the source; iii) the polluter pays principle must be applied; iv) requirements of environmental protection are a component of other EEC policies, which the EEC must take into account in developing its own environmental policy, establishing the respective powers of the EEC and its member-states in the field of environmental protection; and v) environmental policy should be set at the core of international relations . The centrality of environmental protection within the fundamental principles of the European Union has also been recognised by the well-established jurisprudence of the European Court of Justice (see, e.g., van Zeben and Rowell 2020; Lee 2014; Gunningham 2009; Cichowski 1998).

1.3 Towards Constitutional Rights for Natural Entities?

As we noted previously, the use of notions such as 'resources' in connection with natural objects and systems has been accused of treating wild and free things as tools at the disposal of human beings. The objection is that such language could promote a 'property bias': natural goods are in human hands, and this generates human ownership (Plumwood 1993; Sagoff 2004). If natural things are endowed with intrinsic value and recognised by constitutional laws and, thus, create international obligations, is it possible to imagine a non-anthropocentric discourse on environmental law? Is it possible to claim an autonomous status of rights for those who are not yet born and will inhabit the planet earth?

Following the acknowledgment of there being 'moral, ethical, cultural, aesthetic, and purely scientific reasons for conserving wild beings' (WCED 1987, Overview, paragraph 53, 20) over time, strong sustainability has come to be focused not only on the needs of human and other living things, but also on their rights (Redclift 2004, 218). Thomas Berry (1999), in particular, has endeavoured to outline a new conception of human being by translating recent scientific insights – primarily those of ecology – into narrative form with the goal of expanding the reach of human ethics and elaborating an ecocentric cosmology. Cormac Cullinan (2002; 2011) has further developed this approach specifically in relation to the idea of law. He emphasises how 'any given legal order is constrained by its tacit frame of reference, by the deep structure of the prevailing social values' (Cullinan 2002, 45). This is evident as regards the theme of property in this field of so-called 'Earth Jurisprudence', or EJ (Bell 2003; Burdon 2010; Burdon 2011). Western law 'thinks' in terms of 'property rights and property relations: land and nature are automatically conceived as consisting of parcels and objects to be owned'(De Lucia 2013, 173).

The starting point of EJ reflects a reversal of legal reasoning, overturning the main anthropocentric assumption 'that nature is here to be exploited for human ends, but [needs] to be protected when the destruction of nature threatens human survival or some other human interest' (Filgueira and Mason 2009, 4). By contrast, nature is inviolable, so departures from this principle are to be considered exceptional '[i]n that sense [EJ] is a radical departure from the norms of modern legal thought' (Filgueira and Mason 2009, 4; see also De Lucia 2013). The codification of rights granted to natural things goes precisely in the direction indicated by this approach.

If we place ourselves in the dilemma of whether there is a right of natural entities or of the ecosystem as such, we should establish the extent to which these rights can be defined as inviolable. Obviously, we would not be willing to accept that every organism or natural entity on the planet has inviolable rights. Recently, there has emerged a recognition of the principles that do not seem to reduce the possession of rights to Homo sapiens, except in explicit and necessary references to the culture and local traditions of a community. The first attempts to protect the rights of non-human and non-animal entities saw (some) states as promoters of an articulated process of promulgation and constitutional revision. We are not yet at the supranational level, but we must note how some traditional and local recognitions are transmuted to the constitutional level.

Article 5 of the Bolivian *Ley de derechos de la Madre Tierra* (2010) defines nature as a 'collective subject of public interest'. Article 7 of the same law affirms the rights to life, diversity of life, water, clean air, balance and restoration. Article 5 affirms that additional rights may be provided.

Article 10 of Ecuador's 2008 constitution states: '[N]ature shall be the subject of those rights recognized by the Constitution'. Articles 71 and 72 lay the foundations for the rights of nature, distinguishing between those rights that concern existence and those that concern restoration (on 'Pachamama politics' see, e.g., Humphreys 2017). In New Zealand, the Whanganui River (Te Awa Tupua) is recognised as a living being through an agreement entered into on 5 August 2014 between representatives of the Whanganui River Iwi People and the New Zealand government. Through further agreements and acts, the same river is recognised as a legal person. In 2014, Te Urewera National Park became a legal entity with the rights, powers, duties and responsibilities of a legal person (Te Urewara Act 2014, s. 11(1)). In Australia, the Yarra River and its surrounding territories are defined as a 'living and integrated natural entity' (Yarra River Protection Bill 2017, 5). In Colombia, a 2016 ruling of the Constitutional Court held that the Atrato River and its its basin and tributaries is bearer of rights (*sujeto de derechos*). It remains up to the state and the ethnic communities to provide legal representation to the rights of the river and implement the environmental protections and the restoration in case of damage.

In 2017, the High Court of Uttarakhand at Nainital in India stated that the Ganga and Yamuna Rivers are legal and living persons. In 2019, the High Court Division of the Supreme Court of Bangladesh recognised all rivers in the country as living entities with legal personalities. With Mexico City's 2017 constitution, natural ecosystems and species were recognised, in Article 13, as a 'collective entity subject to rights'. In Brazil in 2017, the Bonito City Council amended Article 236 of the *Lei Orgânica No. 01/2017* to recognise nature's right to exist, prosper and evolve. Article 1 of the 2010 Pittsburgh City Code read, 'All residents, natural communities and ecosystems in Pittsburgh possess a fundamental and inalienable right to sustainably access, use, consume, and preserve water drawn from natural water cycles that provide water necessary to sustain life within the City'. In Tamaqua and Schuylkill counties in Pennsylvania, Section 7.6 of Ordinance No. 612 recognises natural communities and ecosystems as persons for the purpose of civil rights enforcement. In Lafayette, Colorado, Section 1(a) of Ordinance No. 02 ensures the right of residents to a healthy climate and ecosystem. In Uganda, Article 4 of the National Environment Act 2019 states, 'Nature has the right to exist, persist, maintain and regenerate its vital cycles, structure, functions and its processes in evolution' (see Perra 2020, 459–464 for all references).

I am aware that many domestic laws do not provide such protections, but we should not underestimate the ability of jurisdictions to influence each other and reshape national principles of law through an ongoing, two-way engagement with international law (see Shaffer and Bodansky 2012). Based on the legal evidence gathered from the above-mentioned cases, and drawing from EJ assumptions, I advance the following thesis:

> Thesis: Every ecosystem possesses inviolable rights in terms of preservation, existential continuity and conservation of biotic capacities.

> Definition: An ecosystem 'consists of the biological community that occurs in some locale, and the physical and chemical factors that make up its non-living or abiotic environment. (University of Michigan 2017, 3).

Defining the ecosystem in this way implies the ecosystem is a systemic concept in itself, in which *Homo sapiens* are fully inserted and, as a moral and political subject, she/he has a duty of care towards ecosystems (see Corrigan 2021; Sumudu and Andrea 2019; Boyd 2017). Recent development in Environmental law shows us it is possible to engage the international community in codifying ecosystems rights as human rights of the future. Here, I use the term 'future' in reference to both the right of an ecosystem to have a future and for humans of the future (for the last, see next section).

> Argument 1: Every ecosystem has a natural tendency to regenerate its biotic capacity, and any reduction in this capacity represents direct harm to the biosphere in all its components.

> Corollary 1: As members of the ecosystem, humans suffer damage to their present capacity and expectations for life and well-being.

> Corollary 2: The human species in its world inhabitation must have reasonable expectations of coexistence with other species and in a system of multiple relationships as balanced as possible and suitable for the preservation of life itself on earth.

> Conclusion: At the constitutional level, the protection of ecosystem rights is justified in order to preserve local biospheres.

2. The Role of the Future: Us and Them, Today and Tomorrow: Some Considerations on International Law of the Environment

So far, I have dealt with the normative and principle-based assumptions that allow us to justify the protection of local ecosystem rights. Now, I turn to international agreements to see whether and how the human rights of the unborn are treated. If the protection of local biospheres can already be implemented at the national level, the place where positions on the rights of future generations are expressed can only be that of international law.

Among the first constitutions to recognize the rights of future generations are the Japanese constitution ('these fundamental human rights guaranteed to the people by this Constitution shall be conferred upon the people of this and future generations as eternal and inviolable rights'); the 1922 Norwegian constitution ('every person has a right to an environment that is conducive to health and to a natural environment whose productivity and diversity are maintained. Natural resources should be managed on the basis of comprehensive long-term considerations whereby this right will be safeguarded for future generations as well'); and the Bolivian constitution ('[all citizens have a fundamental right] to enjoy a healthy environment, ecologically well balanced, and appropriate to her well-being, while keeping in mind the rights of future Generations' (Gosseries 2008, 448).

Otherwise, the Paris Agreement (2015) constitutes an international law. All contracting parties also constitute parties to at least one of the major international human rights treaties recognised by the United Nations. We must point out that some of the explicit language referring to future generations in the Paris Agreement was expunged prior to the approval of the final text of the agreement. This is the case with Article 2, for example, which referred to future generations in draft (see Lewis 2018, 7–8), but which ultimately read:

> This Agreement will be implemented to reflect equity and the principle of common but differentiated responsibilities and respective capabilities, in the light of different national circumstances (Paris Agreement 2015, art. 2).

A similar fate befell more stringent references to human rights in the same article. The following passage – here indicated with brackets – was removed from the final text of the agreement:

> This Agreement shall be implemented on the basis of equity and science, in [full] accordance with the principles of equity

> and common but differentiated responsibilities and respective
> capabilities [in light of national circumstances ... the principles
> and provisions of the Convention], while ensuring the integrity
> and resilience of natural ecosystems, [the integrity of Mother
> Earth, protection of health, a just transition of the workforce
> and creation of decent work and quality jobs in accordance
> with nationally defined development priorities] and the respect,
> protection, promotion and fulfillment of human rights for all,
> including the right to health and sustainable development,
> [including the right of people under occupation] and to ensure
> gender equality and the full and equal participation of women
> [and intergenerational equity] (Lewis 2018, 6).

This context reveals an explicit link between climate and rights entitlement, not only in terms of an enjoyment of the right to a healthy environment but also in the relationship of promoting fundamental human rights through environmental obligations. We saw here in action the propagative force of the rights system. Future generations are named as beneficiaries of current actions for environmental protection in United Nations documents preceding the Paris Agreement (WCED 1987; Rio Declaration 1992). These documents mention the 'abilities' of future generations to pursue their respective needs. To understand the normative scope of these passages, it is necessary to refer to the capabilities of future generations.

The aspect that remains for us to consider, now, is in what terms a rights-centered approach can involve future human beings:

> [R]ights serve as a label of significance. Upgrading an interest
> to the status of right is a sign of its special importance in the
> same way as constitutionalising a merely legal right is. It
> signals something about the significance of the duties to which
> it correlates. Second, correlating a duty to a right tells us
> something about the purpose of such a duty. It gives a
> direction to that duty. Using the language of rights is thus
> quantitatively (significance) and qualitatively (purpose)
> important (Gosseries 2008, 453).

Note that references to 'the future' need not be limited to the future of human beings only. In keeping with the non-anthropocentric focus of much environmental philosophy, a concern with sustainability and biodiversity can embrace a concern for opportunities available to non-human beings. The future we have to take care of is the future of the biosphere (and its multiple ecosystems) as such, as well as its human inhabitants. If we dwell on the ecological belonging of human beings, some contradictions in the arguments against taking on an environmental responsibility for future generations would emerge. Let us take for granted the well-justified view that we cannot reason in terms of separating our needs, desires and motivations from our surroundings. Thornier is the question that justifies the terms of political-legal responsibility we are defining. In particular, I focus on two criticisms of environmental law's emphasis on protection of future generations.

A. Intergenerational Overlap

As Gosseries (2008, 447) maintains:

> The intergenerational context exhibits a unique set of features that make it especially challenging. The temporal direction of causation generates problems of asymmetry of power as well as restrictions to the possibility of giving back to the past. The lack of coexistence among remote generations raises the question whether obligations of justice obtain at all between non-overlapping generations. Distance between some of the generations increases uncertainty as to the effects of our actions or the nature of future generations' preferences or their environment.

Preconception torts illustrate the fact that future rights can make perfect sense whenever the existence of the victim and the wrongdoer overlap after harm has taken place. 'Preconception torts' refer to situations where wrongful actions take place before conception, i.e., before victims exist (Gosseries 2008, 465; see also Bambrick 1987; Banashek 1990; Kennedy 1991). If an agent *X*, through his or her conduct, causes harm that compromises the health of a human being who has not yet been born, he or she is responsible for that conduct because he or she has infringed not only the rights of the parents, but also the expectations, desires and possibilities that the newborn could have developed in the absence of the harmful conduct. Such a right can be invoked after the unborn child has come to inhabit the world: 'We can say that at the moment the harmful action takes place, the baby's right is still a future one (hence, it does not exist); but when the injury takes place, there is contemporaneity of the harmdoer and the victim' (Gosseries 2008, 466).

A similar right therefore can be configured as present even in the absence of the bearer and claimed directly by the bearer in the future. This is the kind of condition I pose to justify the viability of the environmental rights of future generations: '[T]he future-rights-of-future-people one. Here, the right-holders are also the primary interest-holders themselves. And the idea of future rights can do a significant amount of the work that present rights do' (Gosseries 2008, 455).

B. Actionability of Environmental Rights

First, recent developments in international law and state responsibility suggest that the polluter pays principle has become increasingly limited and that there has been an attempt to direct more attention towards a global takeover of prevention, protection and deterrence against environmental risks (Davanlou et al. 2018, 199).

Prevention also plays a significantly greater role in the framework of international law and, specifically, in international environmental law. Commitments are clearly expressed in the latest relevant documents in this field: the United Nations Framework Convention on Climate Change (1992), the Kyoto Protocol (1997) and the Paris Agreement (2015). The commitment required is to be implemented through a multilateral system (Dupuy and Jorge 2018). The skepticism about this framework would seem to lie in the voluntaristic nature of international agreements, which remain little more than declarations of intent. Davanlou et al. (2018, 200) state: 'Of course, given the voluntary nature and lack of a guarantee of the implementation of these documents, we cannot expect any positive developments to take place. As stated, international responsibility is one of the areas of international law with many uncertainties'.

To this fear, we can answer with some encouraging data from the recent jurisprudence of the International Court of Justice (ICJ). The ICJ has somewhat modernised the conception of what is reparable and indeed compensable in Costa Rica v. Nicaragua (2018), where the court noted that 'damage to the environment, and the consequent impairment or loss of the ability of the environment to provide goods and services, is compensable under international law' (para. 42, 14). The court recognised, without implying any change in customary law, that the ordinary rules of reparation are flexible enough to encompass notions such as ecosystem services, as argued by Costa Rica.

The state is responsible for damage caused to the climate, not just for damage to current or future citizens, but also towards other states and the international community as a whole. If an obligation of protection and prevention is recognised, then any serious breach of this obligation implies sanctions and duties, the nature of which is debated.

The approach to climate change in international law has slowly changed from a horizontal to a community approach in which International Law Commission's Articles on the Responsibility of States for Internationally Wrongful Acts (ARSIWA 2009) attempt to provide a response:

> [T]he UN International Law Commission (ILC) completed the first reading of the draft articles on responsibility of international organizations in August 2009. The Commission submitted a full set of draft articles together with a commentary to governments and international organizations for comments and observations by 1 January 2011. One important part of the ILC's work relates to the responsibility of an international organization in connection with the act of a state. In particular, this touches upon sensitive questions of attribution and responsibility when a member state carries out its membership obligations (Hoffmeister 2010, 723).

Article 1 of the ARSIWA states that 'any internationally wrongful act of a State shall give rise to the international responsibility of that State'. There is an 'internationally wrongful act of a State when: (a) conduct consisting of an act or omission is attributable to the State, and (b) such conduct constitutes a violation of an international obligation' (ARSIWA, Article 2).

With respect to the triggering event, the attribution and diligence-related problems discussed in relation to the bilateral framing remain relevant, and they are further complicated by the need to determine what primary rules protecting the environment may amount to a 'peremptory norm' and what composite acts/omissions may amount to a 'serious breach' under Article 40 of the ARSIWA. There is evidence that the prevention principle entails *erga omnes* obligations (Responsibilities in the Area 2011, para. 180), but there is also evidence that it is not a peremptory norm (ILC, Study on the Fragmentation of International Law 2006, Conclusion 38 ILC; Responsibilities in the Area 2011, para. 125–135; Indus Waters Kishenganga case, Final Award, 2013, para. 111). But if we look carefully, we could agree that:

> a) With climate change, a state's failure to formulate and implement climate change laws and industry regulations constitutes an omission (ILC 1971, 264, 283); and,

> b) On the other hand, the causal chain of voluntary actions (or omissions) can also lead to the recognition of state responsibility for climate change. This is the so-called 'cumulative causation' argument supported, as I suggest, by

Article 15 of ARSIWA, where the duty could be breached through a series of acts and omissions. It is added that it is not essential in international law to prove the 'fault element':

[T]he law of state responsibility does not require proof of subjective knowledge about the wrongdoing state. The advantage of this argument is that, even if the harm caused is not reasonably foreseeable by any human agent of the wrongdoing state, it does not preclude a finding of state responsibility to mitigate climate change (Tsang 2021, 15).

Relying on these two interpretations of responsibility, I accept not only retrospective state responsibility, but also I defend

Thesis 2: Prospective responsibility to future generations' rights to live in a reasonably healthy ecosystem that should be placed under protection at the local level by the institutions that represent previous generations.

Conclusions

This chapter offered critical insights to answer the questions posed at the beginning. Specifically, I have outlined a right of the ecosystem and a right to the environment by evaluating in national and international legal frameworks whether there could be defended an autonomous entitlement of non-living beings to enjoy the right of inviolability and how the right to a healthy environment relates to the system of rights promoted at the international level.

I also outlined the difference between our right to the future and the right of future generations. I opted for a cogent interpretation of collective responsibility, justified in politics and international law, and the positively binding expression of the new doctrine born around the International Responsibility of States (ARSIWA). The topic is by no means exhausted, and legal developments will need to be watched carefully. However, I the time is right to advance more robust arguments for the existence of ecosystem rights *per se* and the rights of future generations regarding environmental issues, as well as robust arguments for asserting such rights in appropriate international legal courts.

References

Bambrick, G. 1987. "Developing maternal liability standards for prenatal injury". *St. John's Law Review* 61 (4): 592–614.

Banashek, K. 1990. "Maternal prenatal negligence does not give rise to a cause of action". *Washington University Law Quarterly* 68: 189–202.

Bell, M. 2003. "Thomas Berry and an Earth Jurisprudence: An Exploratory Essay." *The Trumpeter: Journal of Ecosophy* 19 (1): 69–96.

Berry, T. 1999. *The Great Work: Our Way into the Future.* New York: Bell Tower.

Boyd, David R. 2017. *The rights of nature: a legal revolution that could save the world.* Toronto: ECW Press.

Certain Activities carried out by Nicaragua in the Border Area, Costa Rica v Nicaragua, Compensation owed by Nicaragua to Costa Rica, ICJ GL No 150, [2018] ICJ Rep 15, ICGJ 520 (ICJ 2018), 2nd February 2018, United Nations [UN]; International Court of Justice [ICJ]

Corrigan, D.P., and M. Oksanen. 2021. *Rights of nature a re-examination.* London New York: Routledge.

Council of the European Union, Council Regulation (EEC) No 1210/90 of 7 May 1990 on the establishment of the European Environment Agency and the European Environment Information and Observation Network.

Cullinan, C. 2002. *Wild Law: A Manifesto for Earth Justice.* South Africa: Siber Ink.

Brennan, Andrew and Yeuk-Sze Lo- 2021. "Environmental Ethics". In *The Stanford Encyclopedia of Philosophy* (Winter 2021 Edition), edited by Edward N. Zalta (ed.), https://plato.stanford.edu/archives/win2021/entries/ethics-environmental/

Encyclopedia Britannica. "ecosystem", August 11, 2021. https://www.britannica.com/science/ecosystem

Burdon, P.D. 2010. "Wild Law: The Philosophy of Earth Jurisprudence." *Alternative Law Journal.* 35 (2).

Burdon, P.D. 2011. *Exploring Wild Law: The Philosophy of Earth Jurisprudence. Kent Town.* Wakefield Press.

Chong, Chit. 2006. "Restoring the Rights of Future Generations." *Int. J. Green Economics* 1 (1–2): 103–120.

Cichowski, Rachel A. 1998. "Integrating the environment: the European Court and the construction of supranational policy". *Journal of European Public Policy* 5 (3): 387–405.

Cullinan, C. "A History of Wild Law" in Burdon, P.D. 2011. *Exploring Wild Law: The Philosophy of Earth Jurisprudence. Kent Town:* Wakefield Press.

Davanlou, Mona, Seyed Abbas Poorhashemi, Ali Zare, and Mohsen Abdollahi. 2018. "Analysis of the International Responsibility System of Climate Change". *Current World Environment* 13 (2): 194–205.

De Lucia, V. 2013. "Towards an Ecological Philosophy of Law: a Comparative Discussion." *Journal of Human Rights and the Environment* 4 (2): 167–190.

de Shalit, A., 1994. *Why Does Posterity Matter?* London: Routledge.

Dupuy, Pierre-Marie, and Jorge E. Viñuales. 2018. *International environmental law.* Cambridge: Cambridge University Press.

European Union, *Treaty on European Union (Consolidated Version), Treaty of Amsterdam*, 2 October 1997. https://www.refworld.org/docid/3dec906d4.html

European Union, *Treaty Establishing the European Community (Consolidated Version), Rome Treaty,* 25 March 1957.

European Community, *Single European Act*, 29.6.87, Official Journal of the European Communities, No L 169/1

Filgueira, B. and Mason, I. 2009. "Wild Law: Is There Any Evidence of Principles Of Earth Jurisprudence In Existing Law And Legal Practice?". UKELA And Gaia Foundation Research Paper. https://www.ukela.org/common/Uploaded%20files/Wild%20Law%20Research%20Report%20published%20March%202009.pdf

Gosseries, Axel. 2008. "On future generations' future rights". *Journal of Political Philosophy* 16 (4): 446–474.

Gunningham, Neil. 2009. "Environment law, regulation and governance: Shifting architectures." *Journal of Environmental law* 21 (2): 179–212.

Hoffmeister, F. 2010. "Litigating against the European Union and Its Member States – Who Responds under the ILC's Draft Articles on International Responsibility of International Organizations?". *European Journal of International Law* 21 (3): 723–747.

Humphreys, D. 2017. "Rights of Pachamama: The emergence of an earth jurisprudence in the Americas". *Journal of International Relations and Development* 20 (3): 459–484.

Indus Waters Kishenganga Arbitration, Pakistan v India, Partial Award, ICGJ 476 (PCA 2013), 18th February 2013, Permanent Court of Arbitration [PCA].

International Law Commission. 1971. Yearbook of International Commission of Law Commission. Volume 2 Part 1. A/CN.4/SER.A/1971/Ad(U (Part 1).

International Law Commission 2006. ILC, *Report of the Study Group, Fragmentation of International Law: Difficulties Arising from the Diversification and Expansion of International Law: Conclusions* (A/CN.4/L.702) (18 July 2006).

International Law Commission. 2001. D*raft Articles on Responsibility of States for Internationally Wrongful Acts*, November 2001, Supplement No. 10 (A/56/10), chp.IV.E.1.

International Tribunal of the Law of the Sea. 2011. *Responsibilities and obligations of States with respect to activities in the Area, Advisory Opinion,* 1 February 2011, ITLOS Reports 2011, p. 10.

Kennedy, M. 1991. "Maternal liability for prenatal injury arising from substance abuse during pregnancy: the possibility of a cause of action in Pennsylvania". *The Pennsylvania Issue* 29 (3): 553–78.

Lee, M. 2014. *UE Environmental Law, Governance and Decision–Making*, Oxford: Hart Publishing, Second Edition.

Lewis, Bridget. 2018. "The rights of future generations within the post-Paris climate regime." *Transnational Environmental Law* 7(1): 69–87.

Light, A. and Katz, E. 1996. *Environmental Pragmatism*, London: Routledge.

Mayer, Benoit. 2014. "State Responsibility and Climate Change Governance: A Light through the Storm." *Chinese Journal of International Law* 13 (3): 539–575.

Norton, B.G. 1991. *Toward Unity Among Environmentalists*, New York: Oxford University Press.

Perra, L. 2020. "Protezione ambientale: abbandono dell'antropocentrismo giuridico e evoluzione del Diritto". *Revista Brasileira de Estudos Políticos* 121: 455–476.

Plumwood, V. 1993. *Feminism and the Mastery of Nature*. London: Routledge.

Redclift, M. 2005. "Sustainable Development (1987–2005): An Oxymoron Comes of Age". *Sustainable Development* 13: 212–27.

Regulation (EC) No 401/2009 of the European Parliament and of the Council of 23 April 2009 on the European Environment Agency and the European Environment Information and Observation Network (Codified version).

Sagoff, M. 2001 "Consumption". In *Companion to Environmental Philosophy*, edited by D. A. Jamieson. Oxford: Blackwell.

Shaffer, G., and Bodansky, D. 2012. "Transnationalism, Unilateralism and International Law". *Transnational Environmental Law* 1 (1): 31–41.

Sumudu, A., & Andrea, S. 2019. *Human Rights and the Environment: Key Issues* (1st ed.). London: Routledge.

Tsang, Vanessa S.W. 2021 "Establishing State Responsibility in Mitigating Climate Change under Customary International Law" (LL.M. Essays & Theses. 1. https://scholarship.law.columbia.edu/llm_essays_theses/1

United Nations- 1997. *Kyoto Protocol to the United Nations framework convention on climate change*, FCCC/CP/1997/L.7/Add.1. 10 December 1997.

United Nations. 2015. *Paris Agreement to the United Nations Framework Convention on Climate Change*, Dec. 12, 2015.

United Nations. 1987. *Report of the World Commission on Environment and Development: "Our common future" Our common future* [Brundtland report]. New York], Aug. 4, 1987.

United Nations- 1992. *Report of the United Nations Conference on environment and development*, A/CONF.151/26 (Vol. I) 31 ILM 874, General Assembly, Distr. GENERAL.12 August 1992.

United Nations, and Canada. 1992. *United Nations Framework Convention on Climate Change.* [New York]: United Nations, General Assembly.

University of Michigan. 2017. *The Ecosystem and how it relates to Sustainability*. https://www.globalchange.umich.edu/globalchange1/current/lectures/kling/ecosystem/ecosystem.html#:~:text=An%20ecosystem%20consists%20of%20the,%2C%20an%20estuary%2C%20a%20grassland

van Zeben, J., & Rowell, A. 2020. *A guide to EU environmental law.* (1 ed.) Berkeley: University of California Press. https://doi.org/10.1525/9780520968059

Note on Indexing

Our books do not have indexes due to the prohibitive cost of assembling them. If you are reading this book in paperback and want to find a particular word or phrase you can do so by downloading a free PDF version of this book from the E-International Relations website. View the e-book in any standard PDF reader and enter your search terms in the search box. You can then navigate through the search results and find what you are looking for. If you are using apps (or devices) to read our e-books, you should also find word search functionality in those.

You can find all of our books here: http://www.e-ir.info/publications

www.ingramcontent.com/pod-product-compliance
Lightning Source LLC
Chambersburg PA
CBHW060451030426
42337CB00015B/1543